Sustainable Agriculture and Climate Resilience

Sustainable Agriculture and Climate Resilience

Editors

Daniel El Chami
Maroun El Moujabber

MDPI

Basel • Beijing • Wuhan • Barcelona • Belgrade • Novi Sad • Cluj • Manchester

Editors

Daniel El Chami
Sustainability Research & Innovation
TIMAC AGRO Italia S.p.A.
Ripalta Arpina
Italy

Maroun El Moujabber
International Centre for Advanced Mediterranean Agronomic Studies (CIHEAM Bari)
Valenzano
Italy

Editorial Office
MDPI
St. Alban-Anlage 66
4052 Basel, Switzerland

This is a reprint of articles from the Special Issue published online in the open access journal *Sustainability* (ISSN 2071-1050) (available at: www.mdpi.com/journal/sustainability/special_issues/Agri_Climate_Resilience).

For citation purposes, cite each article independently as indicated on the article page online and as indicated below:

Lastname, A.A.; Lastname, B.B. Article Title. *Journal Name* **Year**, *Volume Number*, Page Range.

ISBN 978-3-7258-0422-1 (Hbk)
ISBN 978-3-7258-0421-4 (PDF)
doi.org/10.3390/books978-3-7258-0421-4

Cover image courtesy of Butusova Elena, Shutterstock Inc.

Contents

About the Editors

Daniel El Chami

Since 2016, Daniel El Chami has been leading the Department of Sustainability Research and Innovation, which he founded at TIMAC AGRO Italia S.p.A., a subsidiary of the French group Roullier, renowned for its forward-thinking and scientific look to the future of agriculture, and has established, for this purpose, the Global Innovation Centre (Centre Mondial de l'Innovation, CMI), the largest private research center in Europe dedicated to plant and animal nutrition.

Prior to assuming this role, Daniel El Chami amassed a transdisciplinary background across various countries. His experience encompassed academic research, teaching, expertise, and consultancy activities in numerous national and international research and development projects and case studies, primarily focusing on sustainable rural livelihoods. Throughout these endeavors, sustainable development has been the guiding principle uniting his diverse experiences.

Maroun El Moujabber

Since 2005, Maroun El Moujabber has been a senior officer at the International Centre for Advanced Mediterranean Agronomic Studies (CIHEAM Bari). Before this role, he was an associate professor in Lebanon and the National Project Director for several FAO projects focusing on protected agriculture.

During his tenure at CIHEAM Bari, he spearheaded a project addressing invasive pests in the Euro-Mediterranean region, showcasing extensive expertise in joint programming related to food security and natural resources management across Europe, the Mediterranean, and Africa. Notably, he oversaw a project dedicated to gender equality plans and currently coordinates the doctoral platform at CIHEAM Bari. Additionally, he contributes to the Global Service Facility, supporting the Directorate-General for Research and Innovation of the European Commission. Since 2022, he has held the position of Chair for the Scientific Advisory Committee of PRIMA.

Preface

Agriculture has been a driver and a victim of environmental change throughout history. Traditional farming methods often prioritize short-term gains over long-term sustainability, leading to soil degradation, water scarcity, and biodiversity loss.

In today's rapidly changing environment, where climate change poses significant challenges to food security, biodiversity, and the livelihoods of billions around the globe, the need for sustainable agricultural practices has never been more pressing. As we stand at the intersection of agricultural innovation and environmental stewardship, this textbook is a comprehensive guide to understanding and implementing sustainable practices that foster resilience in the face of climate variability.

In recent decades, there has been a paradigm shift towards more sustainable approaches; however, sustainable agriculture is still intensely debated. This textbook explores studies that underpin sustainable agriculture and climate resilience because it is a multifaceted subject requiring transdisciplinary collaboration and a holistic understanding of agrosystems. The purpose is to offer practical insights and evidence-based strategies for farmers, policymakers, students, and researchers alike.

As Guest Editors of this textbook, we are passionate about the potential of sustainable agriculture to address some of the most pressing challenges of our time. We hope that this book will inspire readers to embrace a more sustainable approach to agriculture and contribute to building a more resilient and equitable food system for future generations. Whether you are a student, a practitioner, or a policymaker, we invite you to join us on this journey towards a more sustainable and resilient agricultural future.

We would like to express our appreciation to our institutions, TIMAC AGRO Italia S.p.A. and the International Centre for Advanced Mediterranean Agronomic Studies (CIHEAM Bari), that have supported this project. Their investment in sustainable agriculture research and education has been instrumental in advancing our understanding of complex environmental challenges and promoting innovative solutions for a more resilient future.

We would like to extend our heartfelt gratitude to everyone whose contributions made this textbook possible. From researchers and practitioners to editors and designers, your expertise and support have been invaluable.

Thank you all for your dedication to sustainable agriculture and climate resilience.

Daniel El Chami and Maroun El Moujabber
Editors

sustainability

MDPI

Editorial

Sustainable Agriculture and Climate Resilience

Daniel El Chami [1] and Maroun El Moujabber [2,*]

1 TIMAC AGRO Italia S.p.A., S.P.13, Località Ca' Nova, I-26010 Ripalta Arpina, Italy; daniel.elchami@roullier.com
2 International Centre for Advanced Mediterranean Agronomic Studies (CIHEAM Bari), Via Ceglie 9, I-70010 Valenzano, Italy
* Correspondence: elmoujabber@iamb.it; Tel.: +39-080-4606-341; Fax: +39-080-4606-206

For decades, human-induced climate change has been scientifically predicted and observed to cause devastating global phenomena globally. It has also been assessed and reported under the Intergovernmental Panel on Climate Change (IPCC) umbrella. Since the beginning, adaptation, resilience, and mitigation have repeatedly been considered to be the main pillars of the response to climate change [1]. Even before anthropogenic climate change, adaptation was the central concept of evolution [2,3] since the publication of *The Origin of Species* by Charles Darwin because it includes all the physical, chemical, and biological mechanisms or behavioural changes by which organisms reduce their vulnerability to their environment [4]. Yet, to manage socioeconomic aspects and behaviours and sustainable development under climate change, resilience must be combined with adaptation [5] for organisms to self-organise and adapt quickly and efficiently [4].

Agriculture, which is fundamental for human survival, integrates biology, chemistry, and physics for food and feed supplies, and agrosystems largely depend on natural and climatic processes. This dependence increases their vulnerability to climate-related risks and uncertainties, be they biotic (i.e., pests, diseases, etc.), abiotic (i.e., temperature, humidity, radiation, etc.), or socio-economic conditions with different impact frequencies and intensities. Nevertheless, agriculture contributes a significant share of the GHG emissions that cause climate change, drastically threatening our existence.

The literature suggests that sustainable agriculture can preserve natural ecosystems and mitigate climate change. Yet, sustainable agriculture is, on one side, a transdisciplinary model that integrates adaptation practices and tools to make agrosystems more resilient [6,7]; however, on the another side, sustainable agriculture is still intensively debated in the scientific community, and there is no agreed definition. The concept is misused and misinterpreted [7].

Therefore, this Special Issue comprises a group of reviews and research papers analysing one or different aspects of agricultural resilience to climate change. The contributions include:

On one side, methodological reviews, such as systematic reviews and bibliometric analyses, are evidence-based and robust approaches with several applications in agrosystem and climate change science used to draw scientific conclusions for decision makers and identify the research gaps and opportunities. Thus, this book includes a systematic review assessing the tools for climate adaptation in coffee agrosystems worldwide (contribution 1) and a bibliometric network analysis determining the adaptation of global water systems to climate change (contribution 2).

Conversely, these research articles explore the interconnections and trade-offs between different agriculture and agrifood practices and their adaptation and resilience capability under climate change conditions.

In agrifood practices, Zahidi et al. (contribution 3) analysed the logic of proximity and its impacts on the resilience of the rose agrifood system in Kalâat M'gouna (Morocco). In addition, Alobid et al. (contribution 4) used a linear model to estimate the food shortage in Egypt and suggest the redistribution of crops in terms of production, food demand,

check for updates

Citation: El Chami, D.; El Moujabber, M. Sustainable Agriculture and Climate Resilience. *Sustainability* **2024**, *16*, 113. https://doi.org/10.3390/su16010113

Received: 13 December 2023
Accepted: 14 December 2023
Published: 21 December 2023

and land reallocation. Another group proposed adapting transformation methods in the agrifood industry (contribution 5). The agricultural practices include the assessment of different sustainable practices classified as nutrient (contributions 6–8), soil (contribution 9), water (contribution 11), and crop management (contributions 11–16). The results demonstrate the importance of these agricultural practices in improving the agrosystems' resilience and adaptation to climate change (Table 1).

Table 1. A summary of the published manuscripts.

Type	Manuscript	Method	Climate Science	Management Category	Main Results
Review	List of Contributions 1: Faraz et al. 2023	This article adopted a systematic approach to searching out information from the literature about different modelling approaches to assess the climate change impacts or/and adaptation on coffee crops worldwide. This review included all the scientific publications from the date of the first relevant article until the end of 2022 and screened 60 relevant articles.	Impacts and Adaptation	Knowledge Management	The selected manuscripts describe qualitative and quantitative modelling tools used to simulate the climate impact on crop suitability (55% of the results), crop productivity (25% of the studies), and pests and diseases (20% of the results). According to the analysed literature, MaxEnt is the leading machine learning model used to assess the climate suitability of coffee agrosystems. The most authentic and reliable model in pest distribution is the Insect Life Cycle Modelling Software (ILCYM) (version 4.0). The scientific evidence shows a lack of adaptation modelling, especially in shading and irrigation practices, which crop models can assess.
	List of Contributions 2: Sawassi and Khadra 2021	This research draws on a systematic bibliometric study of the data generated from the Web of Science research engine between 1990 and 2019, combined with statistical analysis, to explore the academic publication trends and identify the strategic gaps and opportunities in global scientific research dealing with the climate change impacts and uncertainties associated with the various dimensions of hydrologic variability; water system adaptation has risen to the top of global agenda.	Adaptation and Resilience	Knowledge Management	This analysis shows consistent national and international collaboration among authors, institutions, and countries. This statistical examination shows that the adaptation-informed literature on water systems remains fragmented and predominantly centred on framing water resource planning and management and water engineering and infrastructure. This analysis also revealed a relatively skewed understanding of various critical dimensions, such as governance, integrated water resources management, and stakeholder engagement, which are crucial for planning and implementing an efficient adaptation process. The observations reflect the need to build water-related adaptive approaches based on a thorough understanding of the potential climate uncertainties, rather than generically addressing all the uncertainties in one scenario analysis.
Research	List of Contributions 3: Zahidi et al. 2022	This methodology integrates the traditional and scientific knowledge by identifying the proximity and the valorisation modes of the territorial "rose" resource through analysing the actors in this system, while proposing improvements for sustainable development.	Adaptation and Resilience	Knowledge Management	The results of 57 semi-conductive interviews verify the domination of traditional proximities (geographical, organisational, relational, etc.) at the level of interactions between the actors. On the other hand, the relationships are sufficient based on the knowledge.
	List of Contributions 4: Alobid et al. 2021	The authors propose a linear model for crop redistribution in terms of production, food demand, and land reallocation to find the best solution to minimise the Food Gap (FG) under a set of constraints.	Adaptation and Resilience	Agrifood Management	The results found that the modelled FG value increased steadily from 2005 to 2017, and then it declined slightly. Furthermore, a significant water loss was noticed. There was a considerable difference, reaching around 25 billion m^3, between the water consumed by the studied crops and the total amount of renewable water. Moreover, the calculation of food demand with the estimated production and the redistribution of cropland reallocations were performed to achieve the best model fit between the crops in terms of minimising the FG in Egypt.
	List of Contributions 5: Benković-Lačić et al. 2023	The study assessed four drying methods of organic production and the laboratory analysis of polyphenolic compounds and antioxidants were conducted.	Adaptation and Resilience	Agrifood Management	The highest polyphenolic compound contents and antioxidant activity levels were measured in the flower samples dried in the sunlight. The most abundant compounds in all the samples were α-bisabolol oxide A (from 19.6 to 24.3%), bisabolol oxide B (from 19.3 to 23.2%), and β-farnesene E (from 15.9 to 25.5%).

Table 1. *Cont.*

Type	Manuscript	Method	Climate Science	Management Category	Main Results
Research	List of Contributions 6: Piscitelli et al. 2022	This study proposed a combination of phosphate rock with food processing by-products, such as olive pomace, barley spent grain, and citrus pomace, to increase phosphate rocks' solubility and the efficient use of P in organic agriculture.	Adaptation and Resilience	Nutrient Management	The mixtures' P release values ranged between 80% and 88%, whereas the phosphate rock lost 23% of its P over 30 days. Phosphate rock showed a constant water-soluble P fraction, whereas the mixtures exhibited a highly water-soluble P fraction that tended to decrease over time.
	List of Contributions 7: Leogrande et al. 2022	This study evaluated the effects of the foliar applications of a vegetable- and brown-algae-based extract (*Ascophyllum nodosum* (L.) Le Jol. on grapes (*Vitis vinifera* L. cv. Montepulciano) and olives (*Olea europaea* L. cv. Coratina) and its agronomic performance in two field experiments in the Apulia region.	Adaptation and Resilience	Nutrient Management	The results highlight that the crop responses differ in grape and olive orchards. The biostimulant application determined significant increases in the bunches' development (+9.5%) and weight (+10%) compared to those of the untreated control. In the olive orchard, the yield was not significantly influenced by biostimulant application, whereas we observed more quality improvements in the olive oil of the treated plants compared to those of the control.
	List of Contributions 8: El Handi et al. 2021	The study compares the leaf ionome of four important autochthonous Moroccan olive cultivars and eight Mediterranean varieties introduced in Morocco to develop hypotheses related to the resistance or susceptibility of the Moroccan olive trees to *Xylella fastidiosa* (Xf). Leaf ionomes, mainly Ca, Cu, Fe, Mg, Mn, Na, Zn, and P, were determined using inductively coupled plasma optical emission spectroscopy (ICP-OES). These varieties were also screened for their total phenolics and flavonoids content.	Adaptation and Resilience	Nutrient Management	The results showed that the varieties 'Leccino', 'Arbosana', and 'Arbequina' consistently contained higher Mn, Cu, and Zn and lower Ca and Na levels compared with those of the higher pathogen-sensitive 'Ogliarola salentina' and 'Cellina di Nardò'. Our findings suggest that 'Arbozana', 'Arbiquina', 'Menara', and 'Haouzia' may tolerate the infection by X. fastidiosa to varying degrees, which provides additional support for 'Leccino' having resistance to X. fastidiosa and suggests that both 'Ogliarola salentina' and 'Cellina di Nardò' are likely sensitive to X. fastidiosa infection.
Research	List of Contributions 9: Benković et al. 2023	The comparison of soil compaction with different tillage systems was performed, and the pressure of the front and rear tires of tractors during sowing was a subfactor of this study.	Adaptation and Resilience	Soil Management	The tillage systems applied resulted in different soil compaction outcomes. Thus, the deepest tillage had the lowest resistance, and the shallowest tillage had the highest resistance in all three experimental years. The penetrometer measurements showed the significant influence of the lowest tire pressure on reducing compaction and the highest pressure on increasing soil compaction.
	List of Contributions 10: Dragonetti and Khadra 2023	Hydrus-1D model calibration and scenario simulations were used to modulate the salt tolerance threshold.	Adaptation and Resilience	Water Management	With long-term irrigation, based on the irrigation frequency, the model demand (M) scenario achieved better root water and N uptake results than the farmer demand (F) scenario.
	List of Contributions 11: Japundžić-Palenkić et al. 2022	Greenhouse and field experiments on two pepper cultivars were used to test the effect of a Plasma-Activated Water (PAW) treatment of seeds on growth. The results were statistically studied using Fisher's test.	Adaptation and Resilience	Crop Management	The lowest measured parameter values were obtained in the open field without the PAW treatment. The peppers in a greenhouse grew faster and had a better quality after the PAW treatment.
	List of Contributions 12: Yasmeen et al. 2022	A pot experimental set-up was used to test the adaptation of two lentil cultivars to drought and melatonin treatment for adaptation. Measurements were performed and analysed statistically using ANOVA variance analysis.	Adaptation and Resilience	Crop Management	The results showed that increasing the levels of soil water deficit significantly decreased numerous crops' morphological and biochemical characteristics and increased the concentrations of malondialdehyde (MDA) and hydrogen peroxide (H_2O_2). Melatonin application helps contrast these impacts for better adaptation. Two different lentil varieties responded differently to drought; therefore, selecting crop variety is also an adaptation strategy.
	List of Contributions 13: Romanjek Fajdetić et al. 2022	Greenhouse and field experiments on two lettuce cultivars were conducted to test the effect of a Plasma-Activated Water (PAW) treatment of seeds on growth. The results were statistically studied using Fisher's test.	Adaptation and Resilience	Crop Management	The results found that the lettuces from the PAW-treated seeds had better results in the first measurement for both cultivars, with no statistically significant effects on the root system. It was also found that cultivation in a greenhouse resulted in a higher-quality plants than those in the open field.

Table 1. *Cont.*

Type	Manuscript	Method	Climate Science	Management Category	Main Results
Research	List of Contributions 14: Ostrowska et al. 2022	This study investigates specific responses (water content, gas exchange intensity, photosynthetic apparatus activity, chlorophyll content, plant height, and biological membrane integrity) of spring wheat (C3 photosynthesis) and maise (C4 photosynthesis) to drought and flooding stresses at the third leaf stage.	Adaptation and Resilience	Crop Management	A specific wheat response under drought and flooding conditions involved an ET_0/RC ratio increase, describing electron transport flux converted into a single reaction centre in PSII. Correlations between electrolyte leakage, the probability of electron transport beyond the plastoquinone Q_A, and the amount of energy used for the electron transport were also found. A Maise expressed an increase in stomatal conductance during flooding. A significant correlation between PN/Ci and relative water content was also exhibited.
	List of Contributions 15: Frem et al. 2022	This paper evaluates the effect of five safe biological treatments (olive soap, sodium bicarbonate, garlic extract, horsetail, and compost tea) in managing powdery mildew.	Adaptation and Resilience	Crop Management	The plants treated with sodium bicarbonate and garlic extract were the least affected by the powdery mildew regarding disease incidence and severity, while tea compost proved to be the least effective product.
	List of Contributions 16: El Handi et al. 2022	This research assessed the presence of *Xylella fastidiosa* (Xf), a pervasive emerging pathogen, in Morocco. The hosts were inspected and sampled randomly over different environments, including symptomatic and asymptomatic plants. The samples were screened using the DAS-ELISA commercial kit, while further analyses were carried out on the doubtful samples by PCR.	Impacts and Adaptation	Crop Management	The results of both the tests did not show any positive sample in the investigated areas. This finding is an update on the Xf situation in Morocco and confirms that this country is still free from this bacterium, at least in the monitored regions.

Finally, this collection of sustainable practices in agriculture fills a gap in the scientific literature that the authors have already addressed in previous works, where transdisciplinary approaches and integration of management practices are still missing. Yet we are still far from a holistic approach which takes into account the three dimensional model of sustainable development with the economic, social and environmental axes.

Funding: This research received no external funding.

Conflicts of Interest: The authors declare no conflict of interest.

List of Contributions:

1. Faraz, M.; Mereu, V.; Spano, D.; Trabucco, A.; Marras, S.; El Chami, D. A Systematic Review of Analytical and Modelling Tools to Assess Climate Change Impacts and Adaptation on Coffee Agrosystems. *Sustainability* **2023**, *15*, 14582. https://doi.org/10.3390/su151914582.
2. Sawassi, A.; Khadra, R. Bibliometric Network Analysis of "Water Systems' Adaptation to Climate Change Uncertainties": Concepts, Approaches, Gaps, and Opportunities. *Sustainability* **2021**, *13*, 6738. https://doi.org/10.3390/su13126738.
3. Zahidi, M.; Ayegou, J.; Ait Hou, M. Proximities and Logics of Sustainable Development of the Territorial Resource: The Case of the Localised Agro-Food System of Kalâat M'gouna in Morocco. *Sustainability* **2022**, *14*, 15842. https://doi.org/10.3390/su142315842.
4. Alobid, M.; Derardja, B.; Szűcs, I. Food Gap Optimization for Sustainability Concerns, the Case of Egypt. *Sustainability* **2021**, *13*, 2999. https://doi.org/10.3390/su13052999.
5. Benković-Lačić, T.; Orehovec, I.; Mirosavljević, K.; Benković, R.; Ćavar Zeljković, S.; Štefelová, N.; Tarkowski, P.; Salopek-Sondi, B. Effect of Drying Methods on Chemical Profile of Chamomile (*Matricaria chamomilla* L.) Flowers. *Sustainability* **2023**, *15*, 15373. https://doi.org/10.3390/su152115373.
6. Piscitelli, L.; Bennani, Z.; El Chami, D.; Mondelli, D. A Circular Economy Model to Improve Phosphate Rock Fertiliser Using Agro-Food By-Products. *Sustainability* **2022**, *14*, 16228. https://doi.org/10.3390/su142316228.
7. Legrande, R.; El Chami, F.G.; Di Carolo, M.; Piegari, G.; Elefante, M.; Perrelli, D.; Dongiovanni, C. Biostimulants for Resilient Agriculture: A Preliminary Assessment in Italy. *Sustainability* **2022**, *14*, 6816. https://doi.org/10.3390/su14116816.
8. El Handi, K.; Hafidi, M.; Habbadi, K.; El Moujabber, M.; Ouzine, M.; Benbouazza, A.; Sabri, M.; Achbani, E.H. Assessment of Ionomic, Phenolic and Flavonoid Compounds for a Sustainable Management of *Xylella fastidiosa* in Morocco. *Sustainability* **2021**, *13*, 7818. https://doi.org/10.3390/su13147818.
9. Benković, R.; Jug, D.; Šumanovac, L.; Jug, I.; Mirosavljević, K.; Domagoj Zimmer, D.; Teuta Benković-Lačić, T. Mechanical Soil Resistance Influenced by Different Tillage Systems and Tractor Tire Pressures. *Sustainability* **2023**, *15*, 10236. https://doi.org/10.3390/su151310236.
10. Dragonetti, G.; Khadra, R. Assessing Soil Dynamics and Improving Long-Standing Irrigation Management with Treated Wastewater: A Case Study on Citrus Trees in Palestine. *Sustainability* **2023**, *15*, 13518. https://doi.org/10.3390/su151813518.
11. Japundžić-Palenkić, B.; Benković, R.; Benković-Lačić, T.; Antunović, S.; Japundžić, M.; Romanjek Fajdetić, N.; Mirosavljević, K. Pepper Growing Modified by Plasma Activated Water and Growth Conditions. *Sustainability* **2022**, *14*, 15967. https://doi.org/10.3390/su142315967.
12. Yasmeen, S.; Wahab, A.; Saleem, M.H.; Ali, B.; Qureshi, K.A.; Jaremko, M. Melatonin as a Foliar Application and Adaptation in Lentil (*Lens culinaris* Medik.) Crops under Drought Stress. *Sustainability* **2022**, *14*, 16345. https://doi.org/10.3390/su142416345.
13. Romanjek Fajdetić, N.; Benković-Lačić, T.; Mirosavljević, K.; Antunović, S.; Benković, R.; Rakić, M.; Milošević, S.; Japundžić-Palenkić, B. Influence of Seed Treated by Plasma Activated Water on the Growth of *Lactuca sativa* L. *Sustainability* **2022**, *14*, 16237. https://doi.org/10.3390/su142316237.
14. Ostrowska, A.; Hura, T. Physiological Comparison of Wheat and Maize Seedlings Responses to Water Stresses. *Sustainability* **2022**, *14*, 7932. https://doi.org/10.3390/su14137932.
15. Frem, M.; Nigro, F.; Medawar, S.; El Moujabber, M. Biological Approaches Promise Innovative and Sustainable Management of Powdery Mildew in Lebanese Squash. *Sustainability* **2022**, *14*, 2811. https://doi.org/10.3390/su14052811.
16. El Handi, K.; Hafidi, M.; Sabri, M.; Frem, M.; El Moujabber, M.; Habbadi, K.; Haddad, N.; Benbouazza, A.; Abou Kubaa, R.; Achbani, E.H. Continuous Pest Surveillance and Monitoring

Constitute a Tool for Sustainable Agriculture: Case of *Xylella fastidiosa* in Morocco. *Sustainability* **2022**, *14*, 1485. https://doi.org/10.3390/su14031485.

References

1. IPCC. Climate Change: The 1990 and 1992 IPCC Assessment. In *IPCC First Assessment Report Overview and Policymaker Summaries and 1992*; IPCC Supplement; Intergovernmental Panel on Climate Change (IPCC): Geneva, Switzerland, 1992.
2. Thornhill, R. The concept of an evolved adaptation. In Proceedings of the Ciba Foundation Symposium, London, UK, 8–10 April 1997; Volume 208, pp. 4–13, Discussion 13–22. [CrossRef]
3. Williams, G.C. *Adaptation and Natural Selection: A Critique of Some Current Evolutionary Thought*; Princeton University Press: Princeton, NJ, USA, 1966.
4. IPCC. Climate change 2007: Appendix to synthesis report. In *Climate Change 2007: Synthesis Report*; Contribution of working groups I, II and III to the fourth assessment report of the Intergovernmental Panel on Climate Change (IPCC); Baede, A.P.M., van der Linden, P., Verbruggen, A., Eds.; IPCC: Geneva, Switzerland, 2007; pp. 76–89.
5. Denton, F.; Wilbanks, T.J.; Abeysinghe, A.C.; Burton, I.; Gao, Q.; Lemos, M.C.; Masui, T.; O'Brien, K.L.; Warner, K. Climate-resilient pathways: Adaptation, mitigation, and sustainable development. In *Climate Change 2014: Impacts, Adaptation, and Vulnerability*; Part A: Global and Sectoral Aspects. Contribution of Working Group II to the Fifth Assessment Report of the Intergovernmental Panel on Climate Change; Field, C.B., Barros, V.R., Dokken, D.J., Mach, K.J, Mastrandrea, M.D., Bilir, T.E., Chatterjee, M., Ebi, K.L., Estrada, Y.O., Genova, R.C., et al., Eds.; Cambridge University Press: Cambridge, UK; New York, NY, USA, 2014; pp. 1101–1131.
6. Howden, S.M.; Soussana, J.-F.; Tubiello, F.N.; Chhetri, N.; Dunlop, M.; Meinke, H. Adapting agriculture to climate change. *PNAS* **2007**, *104*, 19691–19696. [CrossRef] [PubMed]
7. El Chami, D. Towards sustainable organic farming systems. *Sustainability* **2020**, *12*, 9832. [CrossRef]

sustainability

Systematic Review

A Systematic Review of Analytical and Modelling Tools to Assess Climate Change Impacts and Adaptation on Coffee Agrosystems

Muhammad Faraz [1], Valentina Mereu [2], Donatella Spano [1,2], Antonio Trabucco [1,2], Serena Marras [1,2] and Daniel El Chami [3,*]

[1] Department of Agriculture Science, University of Sassari, Viale Italia 39/A, I-07100 Sassari, Italy
[2] Euro-Mediterranean Centre on Climate Change (CMCC) Foundation, Impacts on Agriculture, Forests and Ecosystem Services (IAFES) Division, Via De Nicola 9, I-07100 Sassari, Italy
[3] TIMAC AGRO Italia S.p.A., S.P.13, Località Ca' Nova, I-26010 Ripalta Arpina, Italy
* Correspondence: daniel.elchami@roullier.com; Tel.: +39-0373-669-111

Abstract: Several modelling tools reported the climate change impact on the coffee agrosystems. This article has adopted a systematic approach to searching out information from the literature about different modelling approaches to assess climate change impacts or/and adaptation on coffee crops worldwide. The review included all scientific publications from the date of the first relevant article until the end of 2022 and screened 60 relevant articles. Most results report research conducted in America, followed by Africa. The models assessed in the literature generally incorporate Intergovernmental Panel on Climate Change (IPCC) emission scenarios (80% of manuscripts), particularly Representative Concentration Pathways (RCP) and Special Report on Emission Scenarios (SRES), with the most common projection periods until 2050 (50% of documents). The selected manuscripts contain qualitative and quantitative modelling tools to simulate climate impact on crop suitability (55% of results), crop productivity (25% of studies), and pests and diseases (20% of the results). According to the analysed literature, MaxEnt is the leading machine learning model to assess the climate suitability of coffee agrosystems. The most authentic and reliable model in pest distribution is the Insect Life Cycle Modelling Software (ILCYM) (version 4.0). Scientific evidence shows a lack of adaptation modelling, especially in shading and irrigation practices, which crop models can assess. Therefore, it is recommended to fill this scientific gap by generating modelling tools to understand better coffee crop phenology and its adaptation under different climate scenarios to support adaptation strategies in coffee-producing countries, especially for the Robusta coffee species, where a lack of studies is reported (6% of the results), even though this species represents 40% of the total coffee production.

Keywords: coffee agrosystems; climate change (CC); impact; adaptation; modelling; IPCC scenarios

check for updates

Citation: Faraz, M.; Mereu, V.; Spano, D.; Trabucco, A.; Marras, S.; El Chami, D. A Systematic Review of Analytical and Modelling Tools to Assess Climate Change Impacts and Adaptation on Coffee Agrosystems. *Sustainability* **2023**, *15*, 14582. https://doi.org/10.3390/su151914582

Academic Editor: Hossein Azadi

Received: 10 August 2023
Revised: 13 September 2023
Accepted: 29 September 2023
Published: 8 October 2023

1. Introduction

The last Intergovernmental Panel on Climate Change (IPCC) Report [1] gathered extensive evidence that climate change had caused substantial damage and increasingly irreversible losses over terrestrial and marine ecosystems and natural resources. Agriculture is among the sectors most affected by climate change, mainly due to extreme events' increased frequency and intensity, with worsening expectations [2].

Climate change is estimated to increase agricultural production and food access pressures, especially in vulnerable regions, thus undermining food security and human nutrition [1]. It has altered hydrological cycles; extreme events such as droughts, floods, storms, heat waves, and other abnormalities on Earth are becoming more common [3,4]. The uncertainty in precipitation patterns, more intense rainfall, and the increase in soil erosion are regarded as direct climate change impacts on the agrosystems [5,6], which generate abiotic stress on biodiversity. Indeed, flooding and surface runoff are vehicles of soil nutrients, pesticides, and other harmful chemicals into freshwater, depleting soil fertility and polluting groundwater resources [7]. Water scarcity and temperature rise affect plants' biochemical and physiological processes [8]. Increasing temperatures have also caused a substantial decline in crop production and are considered a high risk for crops in the future [9,10], especially at mid and low latitudes. The impact of climate change on the productivity of several staple crops is foreseen to be critical in low-latitude tropical regions [11].

Coffee is one of the most important crops in low-latitude regions where climate changes are expected to impact agricultural systems heavily [2]. Thus, arable land in tropical and subtropical regions may lose a considerable amount of such areas by 2050; for example, South America may lose 1–21%, Africa 1–18%, Europe 11–17%, and India 2–4% [12]. Another study illustrated a critical level of water deficit (0.82 kPa) during the flowering stage of Arabica coffee, after which the yield significantly declined, and predicted that about 90% of countries will breach this benchmark if warming rises to 2.9 °C by 2095 [13].

Coffee is cultivated worldwide by about 20–25 million smallholder farmers on 11 million ha of arable land spread across 60 tropical regions [14]. The international trade in coffee commodities is ranked second after petroleum products. Developing countries contribute considerably to exports to more industrialised countries. The United States imports approximately 23% of total traded coffee beans, while the European Union imports about 43% [15]. The estimated coffee consumption is more than 400 billion cups per year, and almost 100 million people are engaged in this industry and derive their income directly or indirectly from coffee commodities [15,16].

The main commercial coffee species are Arabica (*Coffea arabica* L.) and Robusta (*Coffea canephora* L.), accounting for 99% of the total coffee production, where the individual share of both species is 60% and 40%, respectively [17]. Meanwhile, the worldwide prediction for the productivity of Arabica coffee is 35.5% higher than that of Robusta coffee. Brazil, Vietnam, Indonesia, and Colombia are the leading countries globally, contributing to 68% of the international market [15]. The production of the major coffee species and cumulative production by both species within the last five years, produced by the leading countries worldwide, is shown in Figure 1.

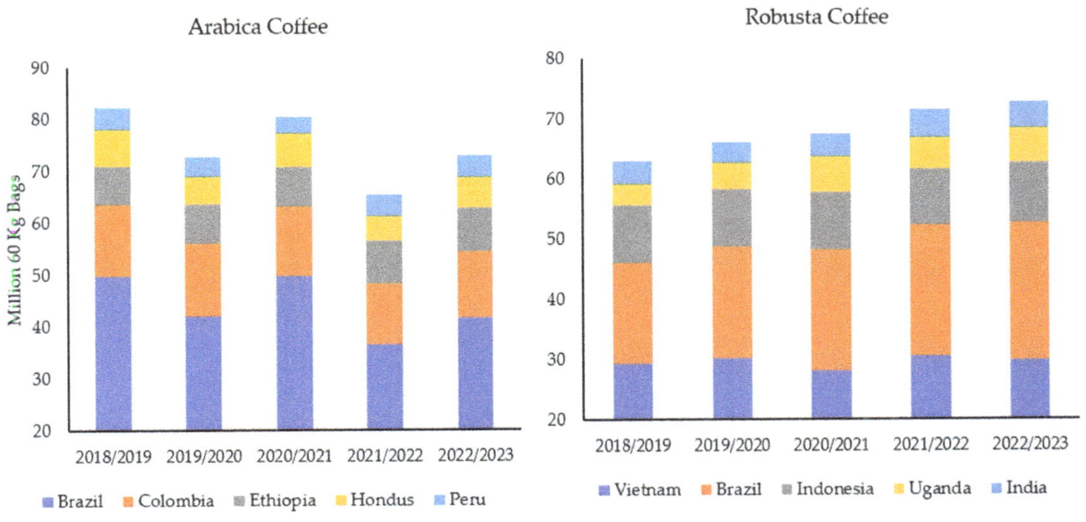

Figure 1. Coffee production in the top five countries worldwide for Arabica and Robusta coffee (data extracted from [18] for 2018–2022).

Climate changes affect coffee crop production due to more frequent insect and pest diseases induced by climate variability [19]. Moreover, high temperatures and reduced precipitation considerably affect the flowering and fruiting of coffee plants and the quality of the beans [20,21]. Coffee crops begin to bloom after the first spring rainfall, but under drought conditions, fewer flowers will sprout, consequently reducing fruit development. On the other hand, under heavy rain conditions, yield loss occurs as flowers and fruits fall off the tree [22,23]. The vegetative and reproductive phases are specifically dependent upon temperature. Temperature rise accelerates the berry ripening, reducing the bean filling duration. Low temperatures lead to defoliation and decreased photosynthesis, causing fluctuation in leaf mass [24]. In addition, low seasonal rainfall is causing branch death, reducing fruit setting resources and damaging coffee beans [25].

Arabica adapts better at high altitudes with an optimal temperature range of 18–22 °C; in contrast, Robusta thrives at lower altitudes with optimal temperatures between 22 and 28 °C. However, neither species can produce abundant yields under adverse conditions nor maintain beverage quality [26].

Several studies have modelled the impact of climate change on coffee production, depicting an upward shift of the crop along with yield losses at lower latitudes [27,28]. Climate uncertainty will affect the coffee production of 9.5 billion kg year^{-1} obtained in 2018, with a 50% reduction by 2050 in suitable environments, thus putting a heavy toll on the coffee trade as a threefold rise in demand is also expected [29]. Specifically, global warming will significantly affect coffee crop production worldwide, with a reduction in 2050 of up to 60% in southern Brazil [30], 90% in Nicaragua [19], and 30–60% in Kenya [31]. Both Robusta and Arabica will be negatively affected by increasing temperature: a 1 °C increase in minimum/maximum temperature (16.2/24 °C) could result in ≈14% or 350–460 kg ha^{-1} Robusta yield reduction [32], even though the Arabica favourable environment could be relocated to 300 m up the altitude gradient in Nicaragua [19]. In addition, high temperatures would make coffee farming susceptible to fungal attacks, such as coffee rust, at lower altitudes and borer damage at high elevations [33,34].

The available literature on coffee presents extensive insights and recommendations for using models and other analytical tools to study climate change impacts and adaptations in coffee production in different regions, such as [35] in Central America, [32] in Vietnam, [36] in Brazil, [37] in Colombia, [38] in Uganda, [39] in Ethiopia, and many more. While the impacts of climate change on coffee have been systematically studied [40], modelling tools still have not received enough attention in terms of systematic review and classification.

The current review is designed for a comprehensive view of the models and tools available to investigate the implications of climate change conditions on coffee growth and yield. The study will also help identify the potential gaps and future trends for research studies to improve modelling tools to guide farming towards sustainable and resilient management of coffee cultivation under climate change conditions. With this aim, a systematic review approach, already consolidated in the climate and agricultural sciences [41–43], is applied to explore the different modelling tools used to investigate climate change impacts and adaptation on the two major coffee species, Robusta and Arabica. We gave special attention to highlighting the eventual capacity of the available tools to assess the effectiveness of adaptation options.

2. Materials and Methods

The Collaboration for Environmental Evidence (CEE) described a systematic review guideline in which PECO or PICO elements demonstrate the research question in various components [44]. Based on a proper methodology, the research question was formulated, and proposed the following:

"What are the analytical tools for coffee crop modelling under climatic uncertainties?"

Based on this question, we developed the PICO elements and the search keywords in Table 1. Once created, we tested the keywords on different search engines, such as Web of Science, Scopus, and Science Direct, on 27 July 2021 (Table 2). To reduce incompatibilities between various search engines, we avoided the excessive use of operators (e.g., wildcard, Boolean, braces, etc.). We extracted the complete database on 13 February 2023.

Table 1. The breakdown of the research question into PICO components and related keywords.

PICO	Description	Keywords
Population	Coffee production, focusing on agrosystems and bean production but excluding the processing phases following post-harvest. The study includes impacts, adaptation, and resilience to all climate variables (temperature, rainfall, CO_2).	Coffee, crop, tree, production, agrosystems, farm.
Intervention	The intervention is the tools used to assess impacts, adaptation, and resilience to climate change—variability in temperature and precipitation. The review will consider no time scale.It will include all scenarios investigated in the literature.	Climate change, impact, adaptation, resilience, GHG emission, climate Variable.
Comparator	Qualitative vs. quantitative models; mathematical vs. biophysical models; spatial modelling.	
Outcome	Modelling techniques. Analytical tools. Programming.	Models, modelling, tools, programming.

Table 2. Development, trial, refinement, and screening of search terms. The keywords in bold represent the selected ones since they show a reasonable hit in all databases.

Search Term	Science Direct	WoS (All Fields)	Scopus (Title-Abs-Key)	Comments
"climate change" AND coffee	5573	536	538	The search term might include adaptation and resilience of coffee to climate change. It will also include other aspects related to impacts and mitigation or policy documents
"climate change" AND coffee AND (impact OR resilience OR adaptation)	4948	327	299	A good search term. A reasonable number of hits, which include all the words needed to answer the research question.
"climate change" AND coffee AND model AND (impact OR resilience OR adaptation)	3946	113	89	Somehow restrictive search term.
"climate change" AND coffee AND (model OR programme OR tool)	5212	217	187	A good search term. A reasonable number of hits which include all the words needed to answer the research question.
"climate change" AND coffee AND (model OR programme OR tool) AND (impact OR resilience OR adaptation)	4707	149	118	A good search term. A reasonable number of hits which include all the words needed to answer the research question.
climate AND coffee AND (model OR programme OR tool)	**15,161**	**387**	**381**	**A good search term. A reasonable number of hits which include all the words needed to answer the research question.**
climate AND coffee AND (model OR programme OR tool) AND (impact OR resilience OR adaptation)	11,453	187	147	A good search term. A reasonable number of hits which include all the words needed to answer the research question.

Besides database sources, the systematic review used search engines and organisation websites in which a maximum of 50 'hits' were recorded from each website (Table 3).

Table 3. List of academic database sources and websites used.

Database Sources	Search Websites	Organisation Websites
Web of Science (WoS) Scopus Science Direct	google.com (accessed on 13 February 2023). googlescholar.com (accessed on 13 February 2023).	World Bank FAO Consultative Group on International Agricultural Research (CGIAR) International Fund for Agricultural Development (IFAD) Natural Resources Institute Climate Institute Coffee & Climate International Trade Centre Fairtrade Coffee Research Institute International Coffee Organisation

For the literature screening, we adopted the following inclusion criteria: (i) subject relevant (anywhere in the world, small landholder farmer or commercial system); (ii) type of intervention (climate scenario available in the literature, tools to assess impact resilience to climate change); (iii) comparator (Spatial modelling); (iv) method (Qualitative vs. quantitative modelling); (v) outcome (studies that consider production modelling).

The effect modifier restricted access to limited primary data, and less variability in modelling and potential impacts (GHG emission scenarios, crop varieties, different production systems and techniques, different agro-ecological conditions, etc.) was unavoidable. Therefore, the review team agreed to adopt narrative analysis and, where possible, quantitative evidence instead of meta-analysis. Interpreting broad subjects with a narrative approach is more suitable, producing a disparate range of outcomes. The narrative analysis approach can acquire the attention of stakeholders and decision-makers by providing them with research gaps in targeted research areas [40–43]. The review team carefully reduced any source of biases in evaluating climate change mitigation and adaptation impacts on the coffee cropping systems.

The literature review did not include a timeframe and was extended until 31 December 2022, based on different search keywords tested on 27 July 2021. Available literature published in English was considered, without specific field restrictions. Keyword search outcomes were recorded and exported to "Mendeley" (a bibliographic software package, 2.100.0). The inclusion criteria were applied by selecting relevant title papers, then abstract evaluation, and, finally, reviewing full texts (Figure 2). Obtained data were tabulated using a common spreadsheet format (i.e., MS Excel). During data extraction, transparency was ensured to avoid heterogeneity in data documentation, and all the review steps were recorded using the PRISMA checklist.

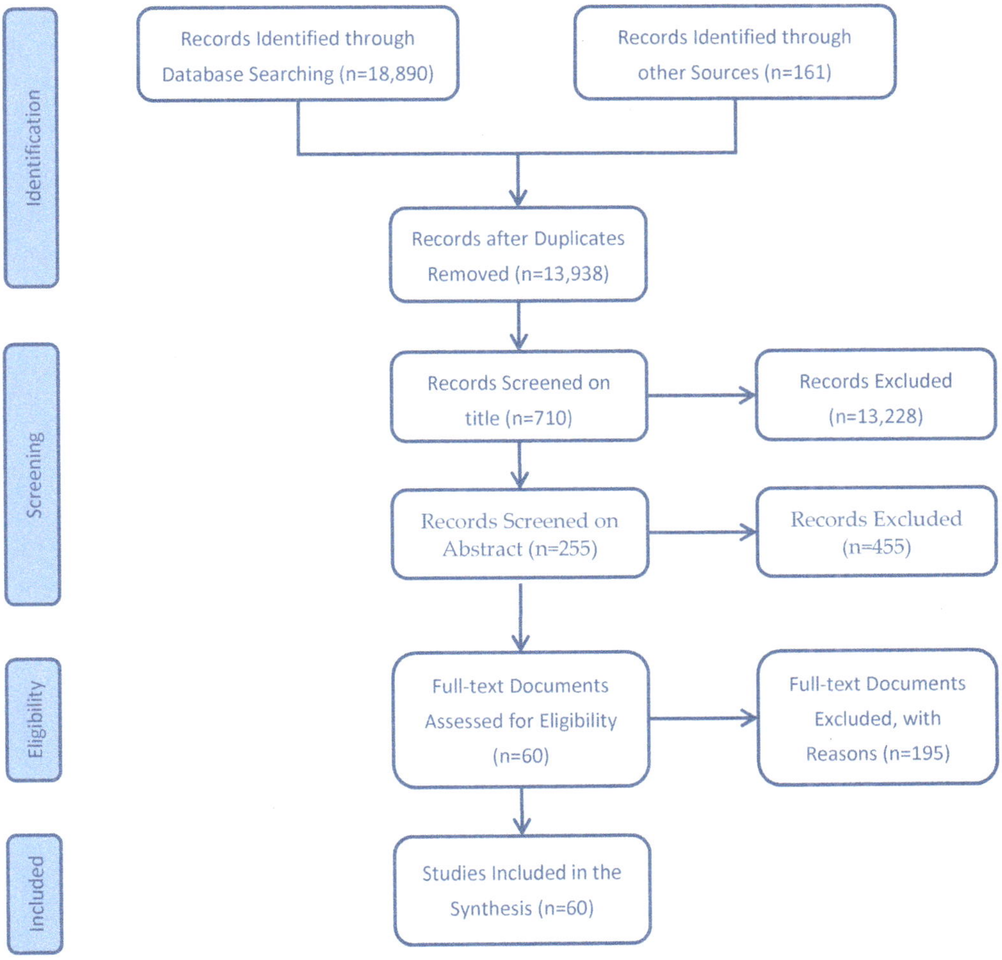

Figure 2. Flow diagram of the systematic review process (after [45]).

3. Results

The results are divided into three different sections. They will first present a general statistical analysis of the screened database, then assess the latter according to climate change processes reviewed to address the models and tools found in the literature in the final section.

3.1. Bibliometric Analysis

A total of 60 eligible studies were retrieved from the literature on modelling climate-driven aspects related to coffee production (Figure 2). The documents were comprehensively searched and categorised according to different categories: region, year of publication, model type, data used to validate the models, coffee species, climate scenarios, climate impact, and climate adaptation (Supplementary Materials). All documents showing simulation models and producing predictions for specific periods for data synthesis were considered. The number of documents were identified and classified according to publication years. Indeed, even though not constantly linear, the trend over time in publication numbers shows an increase over the last decade (Figure 3), with peaks in the number of publications in 2017 and 2022 ($n = 8$, and $n = 9$), while also 2011, 2015, 2018, and 2020 show a consistent number of publications (i.e., 5 and 6).

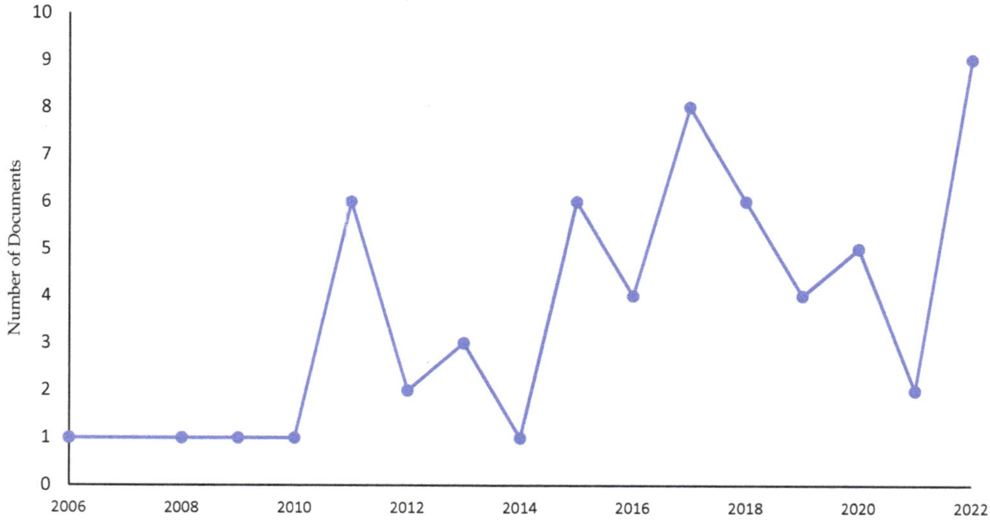

Figure 3. Documents published each year depicting modelling tools to simulate climate change impacts and adaptation on coffee production until 2022.

From another side, Figure 4 depicts the literature published on coffee production in different countries, most of which relate to Brazil ($n = 14$). The research on the coffee crop in Ethiopia is reported in seven documents, whereas we obtained a similar number of documents ($n = 2$) for Indonesia, Tanzania, Mexico, Colombia, Zimbabwe, and Costa Rica. Some articles combined studies including many countries ($n = 14$), and a few ($n = 4$) analysed climate-related aspects of coffee crops worldwide.

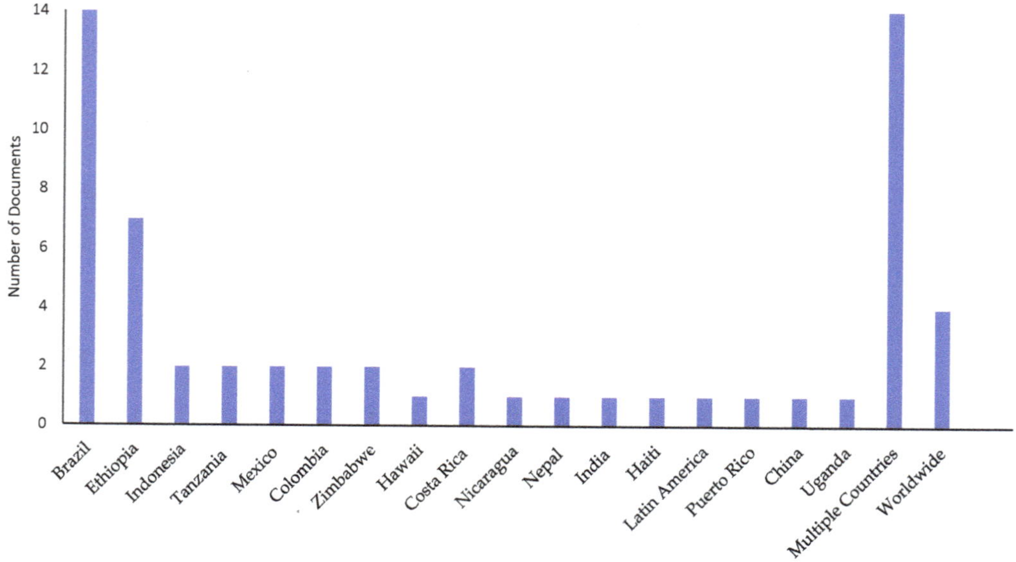

Figure 4. The number of documents published in different countries modelling impacts and adaptation to climate change in coffee production till 2022.

The available literature mainly focused on *Coffea arabica* L., as reported in Table 4. Most documents included results for the American (South America = 18, North America = 15) and the African (n = 20) continents and four papers carried out global research, focusing on *Coffea arabica* L. species. A few documents did not define the assessed species (n = 6). Among document types, the available literature is covered mainly by research articles (n = 55).

Table 4. The number of documents published in different continents, document types, and coffee species.

Continents	Document Type			Species				Total
	Research	Chapter	Report	*Coffee arabica* L.	*Coffea robusta* L.	Both Species	Not Mentioned	
North America	12	1	2	12		1	2	**15**
South America	18			15	1	1	1	**18**
Africa	18	1	1	15	2	1	2	**20**
Asia	3			1		1	1	**3**
Worldwide	4			3		1		**4**
Total	**55**	**2**	**3**	**46**	**3**	**5**	**6**	**60**

3.2. Processes Reviewed

The results identified two climate change processes using modelling tools: impacts and adaptation. The values in Figure 5 refer to the percentage of documents related to climate variability's effect on coffee production.

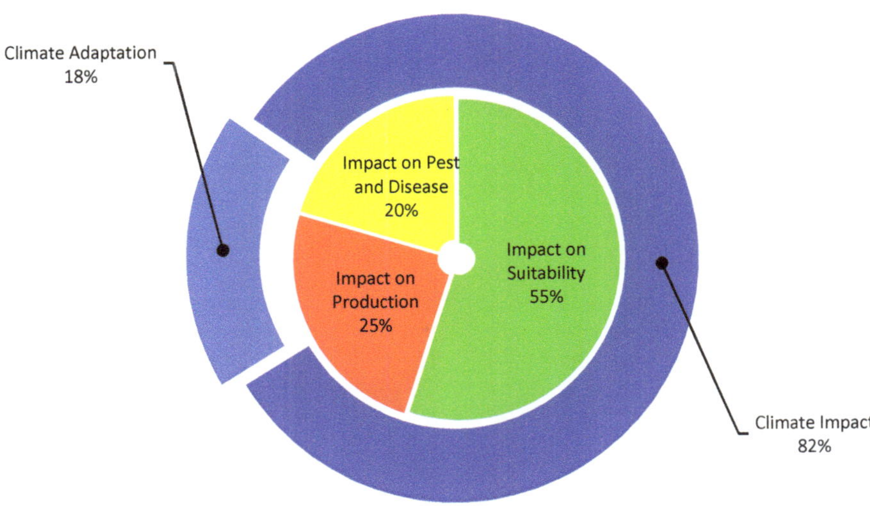

Figure 5. Percentage of documents reviewed according to climate process (climate impact, adaptation). The climate impact is further divided into three categories: impact on suitability, impact on production, and impact on pests and diseases.

The impacts of climate change on coffee production are thoroughly assessed in the literature (82%, in particular, the climate suitability (55%), increased incidence of pests and disease (20%), and decline in production (25%). A consistent number of reviewed studies proposed different adaptation strategies (18%).

3.3. Analysis of Models and Tools

A model is a simplified representation of reality though a functional scheme that allows one to investigate the properties of a system and, in some cases, predict its future outcome. Different models were developed for coffee crops to estimate current and future production and distribution, considering climate variability as a driving factor. Based on the review, the models were classified into deterministic, stochastic, and mixed stochastic/deterministic (Table 5). Deterministic models do not account for randomness in data, nor have a probability function, so a set of inputs and established relationships determines the output.

Table 5. Classification of the reviewed models.

		Model Types			**Total**
		Deterministic Models	**Stochastic Models**	**Deterministic and Stochastic**	
Model Category	Regression models	12	3		**15**
	Crop models	2	8		**10**
	Species distribution models	3	30	1	**34**
Total Models		**17**	**41**	**1**	**59**

On the contrary, a stochastic model includes a random component that uses a distribution as one of the inputs and results in a distribution as output. It presents data and predicts outcomes that account for certain levels of unpredictability or randomness. In addition, some models are stochastic and become deterministic after training. The training installs rules into a network that prescribes its behaviours, so an untrained model shows inconsistent behaviours. These models were included in the mixed model type.

The Eta model was excluded from this classification; it was used once in the literature to assess future suitability ranges, expressed in percentage, of coffee in Southeast Brazil based on annual mean water and temperature restrictions of the Arabica coffee [46].

Among the other 59 models identified in the literature, as reported in Table 5, 41 are stochastic models. Species distribution models are the most common ($n = 34$), followed by crop models ($n = 10$). Additionally, species distribution models include deterministic and stochastic models. Hence, the total number of deterministic models is 17, most of them being regression models ($n = 12$).

3.3.1. Models' Categories

In addition to the first model's classification, another conceptual modelling classification was applied to the modelling tools based on the type of mathematical function/process used to estimate climate change impacts and adaptation. This classification identified regression, crop, and species distribution models as the three main categories (Table 5). A description of each model category is reported in the following sections.

Regression Models

Regression models are simple models used to establish relationships between climate (and other environmental) variables and crop outcomes by fitting regression equations. These models are effective in detecting more general trends and projecting future scenarios. The literature comprises various regression models, including non-linear regression models, multivariate analysis, AutoRegressive Integrated Moving Average (ARIMA) models, climate-based statistical models, econometric models, Generalised Linear Model (GLM), and the Generalised Additive Model (GAM).

Regression models are extensively used to study the impact of climate change on coffee crop yields. Climate-based statistical models, such as those used by [22], predict yields of *Coffea arabica* L. and *Coffea robusta* L. in India for specific years (2010–2012) based on temperature, rainfall, and humidity variables. ARIMA models are employed by [47,48] to assess the influence of climate change on coffee yield in Tanzania and Brazil. The linear regression models predict coffee yield in Ethiopia until 2060 and investigate climate suitability for Arabica coffee until 2080 [49,50].

Non-linear regression equations are applied to study the effects of factors, such as temperature, leaf wetness, and distribution in the sun and shade-grown systems, on coffee rust and coffee crop vulnerability in Brazil and provide projections under climate change until 2080 [51–54]. Bio-economic models are used to predict the impact of the Coffee Berry Borer (CBB) on coffee crops in East Africa and examine the influence of climate on CBB in both full-sun and shade-grown systems [55]. Various studies used econometric models to investigate the correlation between coffee yield and climate variables and identify climate-vulnerable areas in different countries, such as Mexico, Brazil, and Colombia. The principal component analysis is also used to predict the vulnerability in Brazil's coffee region to climate change until 2080 [20,28,56].

Crop Process-Based Models

Crop models are process-based models that simulate the growth and development of crops in specific environmental conditions, and simulate biogeochemical processes to predict crop growth and yields and optimise crop management strategies under present and projected climatic conditions. However, these models require extensive effort in equations and parameter calibration. In the existing literature, several crop models have been identified, including mechanistic models, the yield-safe model, the DynaCof model, dynamic models, Irrigation Management System (IManSys) model, and agrometeorological models.

Several studies applying crop models (e.g., DynaCof) focus explicitly on agroforestry systems and compare them with open sun-grown systems under changing climate variables in Costa Rica, Guatemala, Nicaragua, Colombia, and Brazil [24,57–59]. The yield-safe model determines coffee yield under changing climate scenarios in Ethiopia [21]. Agrometeorological models incorporate irrigation methods to counter the effects of high temperatures and frost from 2040 to 2070 in Brazil. Other studies focus on shade levels to mitigate drought intensity in East Africa, and the IManSys model is used to calculate irrigation requirements for coffee crops under IPCC scenarios in Hawaii [60–64].

Species Distribution Models

Species distribution models identify the distribution among environmental and spatial gradients of a particular species and confirm the suitability of its niche, considering climate impact and other environmental variables. Some reviewed models driven by machine-learning algorithms [39] can investigate climate suitability and include Maximum Entropy (MaxEnt), Random Forests (RF), Boosted Regression Trees (BRT), Generalised Boosted regression Model (GBM), Support Vector Machine (SVM), and Multivariate Adaptive Regression Spline (MARS) models. Moreover, other suitability models are (i) the agro-ecological land elevation model for *Coffea arabica* L. (ALIECA), (ii) the EcoCrop Model, and (iii) the crop niche selection for tropical agriculture (CaNaStA).

Furthermore, the ensemble modelling approach, which combines different models to perform specific scientific activities, has become more common lately to ensure the projections' reliability and reduce modelling uncertainty. Other various modelling tools exist for pest species distribution and disease occurrence. They are (i) the bio-economic models, (ii) the empirical disease models, (iii) the Dinamica EGO model, (iv) the Climex model, (v) the thermal constant model, and (vi) the Insect Life Cycle Modelling Software (ILCYM). The species distribution models are more common in the literature.

The MaxEnt model is widely used to assess the climate suitability of Arabica coffee and predict its future implications worldwide. The model uses various environmental factors as explanatory variables, including temperature, precipitation, aridity, evapotranspiration, soil slope, and land cover [65], to simulate (i) climate suitability in Nepal, Indonesia and Haiti, (ii) the impact on indigenous Arabica coffee in Sudan and Ethiopia, (iii) the adaptation strategy for coffee communities in Mexico, and (iv) to assess climate vulnerability in Puerto Rico by 2099 and Mesoamerica by 2050 [66–70].

Furthermore, the MaxEnt model applied in Indonesia and Zimbabwe produced climate suitability until 2050 and extended projections for China (2060) and Ethiopia

(2070) [71–74]. It also assesses the coffee-pollinating species occurrence in Latin America in response to climate variability (temperature, precipitation, and dry season) and sets the suitable ecological zones in Costa Rica against temperature, elevation, and diurnal range [75,76]. Finally, the MaxEnt model assesses agroforestry systems for adaptation in Brazil and Mesoamerica by 2050, respectively, in response to temperature, precipitation, and bioclimatic variables [30,77].

The influence of climate variability (temperature, rainfall, and evapotranspiration) on Arabica agroclimatic zoning and coffee production was also investigated in Brazil using Eta, a regional climate modelling tool [46].

Random Forest (RF) models are run worldwide to classify the agro-ecological zones for Arabica coffee based on climate variables (temperature, precipitation, and dry months) [78]. An empirical disease model determined the incubation period of coffee rust (*Hemileia vastatrix*) in response to maximum and minimum temperatures and interpolated them to make predictions in Brazil [79]. Ecological modelling tools are also used to evaluate Brazil's phoma leaf spot distribution related to temperature and relative humidity [80]. The Dinamica EGO model produces the distribution of understorey coffee occurrences in Ethiopia [81]. Generalised Regression Models (GRM) are applied globally to assess the impact of Vapour Pressure Development (VPD) on Arabica coffee yield. In Ethiopia, GRM evaluated the influences of extreme agroclimatic indicators on Arabica coffee's overground biomass (AGB) until 2060 [13,82].

In Zimbabwe, the Coffee White Borer (CWB) occurrence probability until 2050 is assessed against temperature and precipitation factors by an ensemble of modelling approaches (BRT and GLM models) [34]. Another ensemble approach uses several machine-learning algorithms (SVM, MaxEnt, and RF) to investigate the worldwide distribution of coffee crops (Arabica and Robusta coffee) [27]. Another ensemble of modelling techniques (GLM, MaxEnt, RF, MARS, GAM, and GBM) examines the resilience potential for Arabica coffee in Ethiopia and the risk extinction of wild Arabica species in Ethiopian and Sudan while taking into account several climate variables [83,84].

The MaxEnt and CaNaStA models also use climate variables, such as temperature and precipitation, as an ensemble of models to generate climate suitability and the quality of Arabica coffee in Nicaragua and evaluate adaptation and mitigation options in Central America [19,85]. Moreover, an integrated approach of machine-learning algorithms (BFT, RF, and SVM) investigate the influence of climate variability (temperature and precipitation) and topological (elevation, soil slope angle) and soil characteristics (pH, soil Cation Exchange Capacity (CEC), apparent Bulk Density (BD), Soil Organic Carbon (SOC)) on the speciality of the coffee sector in Ethiopia under current and future scenarios [39]. Species distribution models (GAM, MaxEnt, and BRT) also predict Robusta's ecological and genomic vulnerability in its native region by 2050 [86].

The ALIECA model predicts the land suitability of Arabica coffee production using agro-ecological variables in Central America. An EcoCrop model assesses the climate suitability of a coffee-based cropping system in Uganda for the long term (2038) [87,88]. The Climex model accounts for the spatial distribution of CBB, considering the effect of environmental variables (temperature, moisture parameters, and other environmental constraints) [89]. Based on Brazil's air and soil temperature, the thermal constant model simulates the geographic distribution of coffee nematodes and leaf miners until 2080 [90]. Finally, the Insect Life Cycle Modelling software (ILCYM) predicts the coffee stink bug (*Antestiopsis thunbergii*) in Tanzania based on the Establishment of Risk Index (ERI), the Generation Index (GI), and the Activity Risk (AI) that corresponds to changes in critical factors/thresholds linked to coffee stink bug distribution based on air temperature [91].

3.3.2. Climate Change Scenarios

A total of 48 papers analyse the impact of future climate change conditions. The analyses follow the Intergovernmental Panel on Climate Change (IPCC) climate scenarios based on the emission and concentration of greenhouse gases in the atmosphere. Over the years, the IPCC has developed several scenarios, including the Special Report on Emission Scenarios (SRES) for the third and fourth Assessment Reports (AR3 in 2001 and AR4 in 2007) [92], characterised by four qualitative storyline scenarios (A1, A2, B1, and B2) representing different demographic, social, and technological advancements. Further, the IPCC has recently developed advanced sets of scenarios: the Representative Concentration Pathway (RCP) and the Shared Socioeconomic Pathway (SSP), presented in AR4 [93] and AR6 [94], respectively. Both have a valid framework for projections until the end of the current century (2100). However, the RCP provides the concentration of greenhouse gases and radiative forcing levels associated with different emission pathways. In contrast, the SSP setup has a different approach considering greenhouse gas emissions, including population growth, economic development, energy use, land use, and other factors. The RCP is further subdivided into RCP 2.6 (optimistic), RCP 4.5 and 6.0 (intermediate), and RCP 8.5 (business as usual), and the set of the SSP scenarios are SSP126 (sustainable development), SSP245 (middle of the road development), SSP370 (regional rivalry), SSP460 (inequity), and SSP585 (full fossil-fuelled development) pathways.

The available literature has been more consolidated over the last decade. Therefore, the RCP are more prevalent than the SRES scenarios, commonly used by the literature in the earlier years (2008–2019), and were gradually, but not entirely, replaced by the RCP scenarios between 2015 and 2022. In the review results, the RCP scenarios prevail (23 papers), followed by the SRES (21 manuscripts) and the SSP (4 studies) (Figure 6A). The use and frequency of different scenarios across the results indicate that the highest number of papers ($n = 7$) adopted RCP 4.5 and RCP 8.5, followed by RCP 4.5, SRES A2A, SRES A2, and SRES A2 ($n = 5$ for each scenario). The other scenarios found in the literature (RCP (2.6, 4.5, 6.0 and 8.5), SRES (A2 and B2), and SSP (126, 245, 370 and 585)) are adopted in two research papers each (Figure 6B). The remaining scenarios were used only once in the literature, including RCP (2.6 and 8.5), (2.6 and 6.0), (2.6, 4.5 and 6.0), (2.6, 6.0 and 8.5), SRES (B1 and A2), (B1, and A1F1), (A2A, and A1F1), (A1B, A2A and B2A), (A2, B2, A2A and B2A), (A1, A1B and B2), and SSP (126, 245 and 585) (Figure 6C,D).

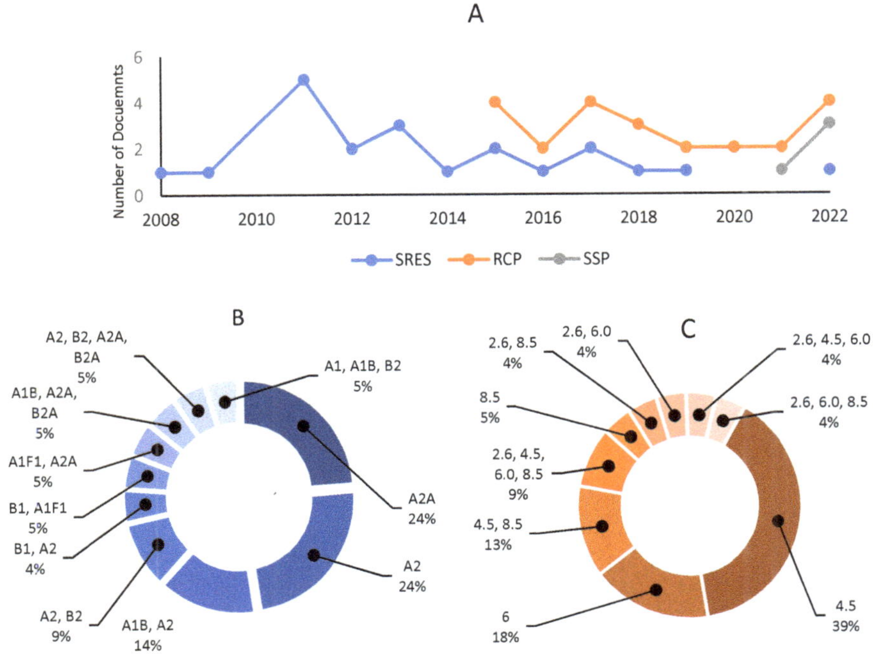

Figure 6. *Cont.*

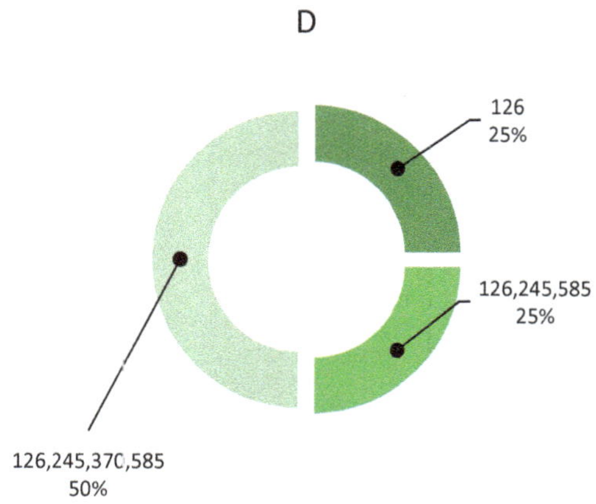

Figure 6. The coffee agrosystems are analysed using IPCC scenarios illustrating the trend of various scenarios (**A**), along with the frequency of sub-scenarios found in the literature, which are presented as SRES (**B**), RCPs (**C**), and SSPs (**D**), respectively.

4. Discussions

The literature was extensively searched to put together information on published modelling tools used to predict the impacts and adaptation of coffee agrosystems to climate change. Arabica is at a greater risk between the two major coffee species than Robusta; thus, research focused primarily on this species. Most models in the screened literature have incorporated the IPCC scenarios and evaluated climate change's impact on coffee agrosystems. Less attention is given to the adoption of adaptation practices. Most models have incorporated the IPCC scenarios, using projections until 2050, particularly the medium-emission sub-scenario RCP 4.5, either solely or combined with other sub-scenarios. The SERS' high-emission scenarios (A2) were also commonly used in the reviewed literature. The Supplementary Materials contain detailed information about each document screened for review.

Among the various regression models, econometric models based on multiple regression equations, which integrate the economic and climate variables, are commonly used to capture the effect of extreme events on coffee yield. The quadratic functional farm generates multi-collinearity, which does not affect the model's prediction but makes the estimator less accurate [28]. The AutoRegressive Integrated Moving Average (ARIMA) is particularly efficient in forecasting time series analysis, but its application with non-linear regression can compromise the model's accuracy [88]. The climate-based regression models monitor the coffee crop under critical stages, considering the effect of climate variables to analyse the coffee growth and developments at different growth cycles [22].

The MaxENT model is widely applied to determine the climate suitability of coffee crops. The model output is 1, considered the maximum probability, and 0, where species have a less suitable climate. The model calculates these values by dividing each weighted variable's sum by a scaling constant. This model is robust because it incorporates statistical and machine-learning techniques. However, the parameter selection is crucial, otherwise, the results may be biased. There is an inbuilt option to check the quality of the model using the Area Under Curve (AUC) index, as it provides a single overall measure of model accuracy [19,30].

A Random Forest (RF) model is also a popular machine learning classifier with high efficiency over large datasets without overfitting [22]. The Crop Niche Selection in Tropical Agriculture (CaNaSTA) is built on Bayesian statistics, aiming to determine a species' presence or absence and appraise the crop's performance. However, this model only works with its specific dataset format, and expertise in Bayesian statistics is also required, making it complicated and time-consuming.

Another model, EcoCrop (EC), determines the crop niche using environmental ranges, expressed in percentage, producing overall crop suitability [61]. The model gives individual suitability values for temperature and precipitation. Expert knowledge is essential for the model's accuracy in setting the crop parameters. The EcoCrop (EC) model can assess climate suitability even with limited ecological and environmental information [80].

The ensemble modelling approach is more efficient in suitability-related tasks because different machine learning and regression models perform together, highlighting modelling uncertainty and conservative choices for specific tasks. For example, integrating RF, CaNaStA, and MaxEnt has more precise results than the output produced by individual models [29]. The Dinamica's EGO model applies Weight of Evidence (WoE) to find the coffee occurrence understorey [74]. The agro-ecological land elevation model for *Coffea arabica* L. (ALIECA) is based on a Bayesian algorithm and provides information about land suitability in percentages, but does not provide data about the presence or absence of coffee crops. This model can also provide accurate results when the data are missing or uncertain [81].

The literature has applied and described several models for Identifying pests in coffee agroecosystems under climate change scenarios, including the ILCYM, climax, and thermal constant models. All these models have their strengths, but the ILCYM model stands out due to its ability to provide detailed information at a very high geographical scale, resulting in more precise results than other pest distribution models. However, the thermal constant model and ILCYM are based on temperature variables. They lack flexibility in accepting other climate variables important for pest–crop interactions, such as rainfall and relative humidity. Furthermore, neither model offers any crop or pest adaptation options. In addition, these models require daily or hourly data on a short time scale because pest–crop interactions may occur within 24 h [83,84]. The empirical disease and non-linear regression models use air temperature to determine the occurrence and intensity of diseases [72,89]. On the other hand, bioeconomic models are more flexible in considering shading, coffee berry borer (CBB) infestation, and temperature to generate information along with economic variables to estimate the shading value according to the disease infestation intensity [48].

Crop models are commonly used to assess climate impacts against adaptation strategies while modifying crop management systems. The available crop models recommend the shading level for optimal coffee yield in various regions [19,51,52]. However, few studies explored other adaptation options. As agrometeorological models, the Irrigation Management System (IManSys) model and the Yield SAFE model have been developed to enhance irrigation techniques and the efficiency of CO_2 fertilisation in a coffee production system [53,55,57]. All models obtained in each article are available in a supplementary file submitted with this manuscript.

5. Conclusions

Various modelling approaches have been applied to determine the climate change impact and adaptation of coffee agrosystems. The research adopted the systematic review approach to assess and classify the available literature according to (i) categories (regression models, crop models, species distribution models), (ii) types (deterministic, stochastic, mixed), and (iii) processes (climate impacts, climate adaptation). The results also included an assessment of the scenarios used to run different modelling tools.

In conclusion, machine learning models have complex algorithms and are stochastic, which produce predictions in situations where data include uncertainty or randomness, thereby generating more accurate results. The ILCYM model is particularly efficient in pest distribution due to its flexibility in accepting multiple variables, thus providing reliable data. The application of crop models was limited to a few studies on crop agrosystems.

Therefore, based on our results, we recommend intensifying adaptation research to explore the best options for different case studies. We advocate applying crop models to fill the gap for coffee phenology and propose adaptation strategies, for example, introducing new varieties, water conservation methods, shading management at various altitudes, and soil organic matter management. The Robusta coffee species also needs to be further investigated in the literature because it is underestimated, despite having a higher adaptation potential.

We finally address these results to decision-makers to support scientific and applied policy design and implementation in climate change resilience and adaptation.

Supplementary Materials: The following supporting information can be downloaded at: https://www.mdpi.com/article/10.3390/su151914582/s1.

Author Contributions: Data curation, M.F. and D.E.C.; Formal analysis, M.F. and D.E.C.; Investigation, M.F. and D.E.C.; Methodology, M.F. and D.E.C.; Writing—original draft, M.F. and D.E.C.; Writing—review & editing, M.F. and D.E.C., V.M., A.T., S.M. and D.S.; Supervision, V.M., A.T., S.M. and D.S.; Funding acquisition, D.S. All authors have read and agreed to the published version of the manuscript.

Funding: This work was developed within the framework of the strategic project "CAT4—Agriculture Management Analysis for Adaptation" promoted by the Euro-Mediterranean Centre on Climate Change (CMCC) Foundation.

Institutional Review Board Statement: Not applicable.

Informed Consent Statement: Not applicable.

Data Availability Statement: Not applicable.

Conflicts of Interest: The authors declare no conflict of interest

References

1. IPCC. Summary for Policymakers. In *Climate Change: Impacts, Adaptation, and Vulnerability*; Pörtner, H.-O., Roberts, D.C., Poloczanska, E.S., Mintenbeck, K., Tignor, M., Alegría, A., Craig, M., Langsdorf, S., Löschke, S., Möller, V., et al., Eds.; Contribution of Working Group II to the Sixth Assessment Report of the Intergovernmental Panel on Climate Change; Cambridge University Press: Cambridge, UK; New York, NY, USA, 2022; pp. 3–33.
2. IPCC. Climate Change: Impacts, Adaptation and Vulnerability. In *Contribution of Working Group II to the Sixth Assessment Report of the Intergovernmental Panel on Climate Change*; Pörtner, H.-O., Roberts, D.C., Poloczanska, E.S., Mintenbeck, K., Tignor, M., Alegría, A., Craig, M., Langsdorf, S., Löschke, S., Möller, V., et al., Eds.; Cambridge University Press: Cambridge, UK; New York, NY, USA, 2022; p. 3056.
3. Anjum, S.A.; Xie, X.; Wang, L.; Saleem, M.F.; Man, C.; Lei, W. Morphological, Physiological and Biochemical Responses of Plants to Drought Stress. *Afr. J. Agric. Res.* **2011**, *6*, 2026–2032. [CrossRef]
4. Shao, G. Understanding the Appeal of User-Generated Media: A Uses and Gratification Perspective. *Internet Res.* **2009**, *19*, 7–25. [CrossRef]
5. Canal Daza, D.S.; Andrade, C.H.J. Mitigation-adaptation synergies of climate change of coffee (*Coffea arabica*) production sys-tems in Tolima, Colombia [Sinergias mitigación-adaptación al cambio climático en sistemas de producción de café (*Coffea arabica*), de Tolima, Colombia]. *Rev. Biol. Trop.* **2019**, *67*, 36–46. [CrossRef]
6. Angima, S.D.; Stott, D.E.; O'Neill, M.K.; Ong, C.K.; Weesies, G.A. Soil erosion prediction using RUSLE for central Kenyan highland conditions. *Agric. Ecosyst. Environ.* **2003**, *97*, 295–308. [CrossRef]
7. Vieira, M.; Mahdi, S.; Casas-Gallego, M.; Fenton, J. Three New Paleocene Dinoflagellate Cysts from the North Sea and the Norwegian Sea. *Rev. Palaeobot. Palynol.* **2018**, *258*, 256–264. [CrossRef]
8. Fahad, S.; Bajwa, A.A.; Nazir, U.; Anjum, S.A.; Farooq, A.; Zohaib, A.; Sadia, S.; Nasim, W.; Adkins, S.; Saud, S.; et al. Crop Production under Drought and Heat Stress: Plant Responses and Management Options. *Front. Plant Sci.* **2017**, *8*, 1147. [CrossRef] [PubMed]
9. Chen, S.; Gong, B. Response and Adaptation of Agriculture to Climate Change: Evidence from China. *J. Dev. Econ.* **2021**, *148*, 102557. [CrossRef]
10. Gammans, M.; Mérel, P.; Ortiz-Bobea, A. Negative Impacts of Climate Change on Cereal Yields: Statistical Evidence from France. *Environ. Res. Lett.* **2017**, *12*, 054007. [CrossRef]
11. Jagermeyr, J.; Mueller, C.; Ruane, A.C.; Elliott, J.; Balkovic, J.; Castillo, O.; Faye, B.; Foster, I.; Folberth, C.; Franke, J.A.; et al. Climate Impacts on Global Agriculture Emerge Earlier in New Generation of Climate and Crop Models. *Nat. Food* **2021**, *2*, 873–875. [CrossRef]
12. Zhang, X.; Cai, X. Climate Change Impacts on Global Agricultural Land Availability. *Environ. Res. Lett.* **2011**, *6*, 014014. [CrossRef]

13. Kath, J.; Craparo, A.; Fong, Y.; Byrareddy, V.; Davis, A.P.; King, R.; Nguyen-Huy, T.; van Asten, P.J.A.; Marcussen, T.; Mushtaq, S.; et al. Vapour Pressure Deficit Determines Critical Thresholds for Global Coffee Production under Climate Change. *Nat. Food* **2022**, *3*, 871–880. [CrossRef]
14. Chain-Guadarrama, A.; Martínez-Salinas, A.; Aristizábal, N.; Ricketts, T.H. Ecosystem Services by Birds and Bees to Coffee in a Changing Climate: A Review of Coffee Berry Borer Control and Pollination. *Agric. Ecosyst. Environ.* **2019**, *280*, 53–67. [CrossRef]
15. Krishnan, S.; Matsumoto, T.; Nagai, C.; Falconer, J.; Shriner, S.; Long, J.; Medrano, J.F.; Vega, F.E. Vulnerability of Coffee (*Coffea* spp.) Genetic Resources in the United States. *Genet. Resour. Crop Evol.* **2021**, *68*, 2691–2710. [CrossRef]
16. Mishra, M.; Slater, A. Recent Advances in the Genetic Transformation of Coffee. *Biotechnol. Res. Int.* **2012**, *2012*, 17. [CrossRef]
17. Kouadio, L.; Byrareddy, V.M.; Sawadogo, A.; Newlands, N.K. Probabilistic Yield Forecasting of Robusta Coffee at the Farm Scale Using Agroclimatic and Remote Sensing Derived Indices. *Agric. For. Meteorol.* **2021**, *306*, 108449. [CrossRef]
18. USDA. *Coffee: World Markets and Trade*; Foreign Agricultural Service; United States Department of Agriculture (USDA): Washington, DC, USA, 2022; pp. 1–9.
19. Laderach, P.; Ramirez-Villegas, J.; Navarro-Racines, C.; Zelaya, C.; Martinez-Valle, A.; Jarvis, A. Climate Change Adaptation of Coffee Production in Space and Time. *Clim. Change* **2017**, *141*, 47–62. [CrossRef]
20. Koh, I.; Garrett, R.; Janetos, A.; Mueller, N.D. Climate Risks to Brazilian Coffee Production. *Environ. Res. Lett.* **2020**, *15*, 104015. [CrossRef]
21. Gidey, T.; Oliveira, T.S.; Crous-Duran, J.; Palma, J.H.N. Using the Yield-SAFE Model to Assess the Impacts of Climate Change on Yield of Coffee (*Coffea arabica* L.) under Agroforestry and Monoculture Systems. *Agrofor. Syst.* **2020**, *94*, 57–70. [CrossRef]
22. Jayakumar, M.; Rajavel, M. Coffee Yield Forecasting Using Climate Indices Based Agrometeorological Model in Kerala. *Mausam* **2017**, *68*, 309–316. [CrossRef]
23. Villers, L.; Arizpe, N.; Orellana, R.; Conde, C.; Hernandez, J. Impacts of Climatic Change on Coffee Flowering and Fruit Development in Veracruz, México [Impactos Del Cambio Climático En La Floración y Desarrollo Del Fruto Del Café En Veracruz, México]. *Interciencia* **2009**, *34*, 322–329.
24. Rodríguez, D.; Cure, J.R.; Cotes, J.M.; Gutierrez, A.P.; Cantor, F. A Coffee Agroecosystem Model: I. Growth and Development of the Coffee Plant. *Ecol. Modell.* **2011**, *222*, 3626–3639. [CrossRef]
25. Kath, J.; Mittahalli Byrareddy, V.; Mushtaq, S.; Craparo, A.; Porcel, M. Temperature and Rainfall Impacts on Robusta Coffee Bean Characteristics. *Clim. Risk Manag.* **2021**, *32*, 100281. [CrossRef]
26. Magrach, A.; Ghazoul, J. Climate and Pest-Driven Geographic Shifts in Global Coffee Production: Implications for Forest Cover, Biodiversity and Carbon Storage. *PLoS ONE* **2015**, *10*, e0133071. [CrossRef]
27. Bunn, C.; Läderach, P.; Ovalle Rivera, O.; Kirschke, D. A Bitter Cup: Climate Change Profile of Global Production of Arabica and Robusta Coffee. *Clim. Change* **2015**, *129*, 89–101. [CrossRef]
28. Gay, C.; Estrada, F.; Conde, C.; Eakin, H.; Villers, L. Potential Impacts of Climate Change on Agriculture: A Case of Study of Coffee Production in Veracruz, Mexico. *Clim. Change* **2006**, *79*, 259–288. [CrossRef]
29. Nab, C.; Environment, M.M.-G.G. Life cycle assessment synthesis of the carbon footprint of Arabica coffee: Case study of Brazil and Vietnam conventional and sustainable coffee production and export to the United Kingdom. *Wiley Online Libr.* **2020**, *7*, e00096. [CrossRef]
30. Gomes, L.C.; Bianchi, F.J.J.A.; Cardoso, I.M.; Fernandes, R.B.A.; Filho, E.I.F.; Schulte, R.P.O. Agroforestry Systems Can Mitigate the Impacts of Climate Change on Coffee Production: A Spatially Explicit Assessment in Brazil. *Agric. Ecosyst. Environ.* **2020**, *294*, 106858. [CrossRef]
31. Giovannucci, D.; von Hagen, O.; Wozniak, J. *Voluntary Standard Systems*; Springer: Berlin/Heidelberg, Germany, 2014; Volume 1, ISBN 978-3-642-35715-2.
32. Kath, J.; Byrareddy, V.M.; Craparo, A.; Nguyen-Huy, T.; Mushtaq, S.; Cao, L.; Bossolasco, L. Not so Robust: Robusta Coffee Production Is Highly Sensitive to Temperature. *Glob. Change Biol.* **2020**, *26*, 3677–3688. [CrossRef]
33. van der Vossen, H.; Bertrand, B.; Charrier, A. Next Generation Variety Development for Sustainable Production of Arabica Coffee (*Coffea arabica* L.): A Review. *Euphytica* **2015**, *204*, 243–256. [CrossRef]
34. Kutywayo, D.; Chemura, A.; Kusena, W.; Chidoko, P.; Mahoya, C. The Impact of Climate Change on the Potential Distribution of Agricultural Pests: The Case of the Coffee White Stem Borer (*Monochamus leuconotus* P.) in Zimbabwe. *PLoS ONE* **2013**, *8*, e73432. [CrossRef]
35. Bunn, C.; Castro, F.; Lundy, M. *The Impact of Climate Change on Coffee Production in Central America*; CCAFS: Wageningen, The Netherlands, 2017.
36. Camargo, M.B.P.D. The impact of climatic variability and climate change on Arabic coffee crop in Brazil [Impacto da variabilidade e da mudança climática na produção de café Arábica no Brasil]. *Bragantia* **2010**, *69*, 239–247. [CrossRef]
37. Andrade, H.J.; Segura, M.A.; Feria, M. Allometric Models for Estimating Belowground Biomass of Individual Coffee Bushes Growing in Monoculture and Agroforestry Systems. *Agrofor. Syst.* **2021**, *95*, 215–226. [CrossRef]
38. Sarmiento-Soler, A.; Vaast, P.; Hoffmann, M.P.; Jassogne, L.; van Asten, P.; Graefe, S.; Rötter, R.P. Effect of Cropping System, Shade Cover and Altitudinal Gradient on Coffee Yield Components at Mt. Elgon, Uganda. *Agric. Ecosyst. Environ.* **2020**, *295*, 106887. [CrossRef]
39. Chemura, A.; Mudereri, B.T.; Yalew, A.W.; Gornott, C. Climate Change and Specialty Coffee Potential in Ethiopia. *Sci. Rep.* **2021**, *11*, 8097. [CrossRef] [PubMed]

40. Bilen, C.; El Chami, D.; Mereu, V.; Trabucco, A.; Marras, S.; Spano, D. A Systematic Review on the Impacts of Climate Change on Coffee Agrosystems. *Plants* **2023**, *12*, 102. [CrossRef] [PubMed]
41. El Chami, D.; Trabucco, A.; Wong, T.; Monem, M.A.; Mereu, V. Costs and Effectiveness of Climate Change Adaptation in Agriculture: A Systematic Review from the NENA Region. *Clim. Policy* **2022**, *22*, 445–463. [CrossRef]
42. El Chami, D.; Daccache, A.; El Moujabber, M. What Are the Impacts of Sugarcane Production on Ecosystem Services and Human Well-Being? A Review. *Ann. Agric. Sci.* **2020**, *65*, 188–199. [CrossRef]
43. El Chami, D.; Daccache, A.; El Moujabber, M. How Can Sustainable Agriculture Increase Climate Resilience? A Systematic Review. *Sustainability* **2020**, *12*, 3119. [CrossRef]
44. CEE. *Guidelines for Systematic Review and Evidence Synthesis in Environmental Management*; Version 5.0. 483; Pullin, A.S., Ed.; Collaboration for Environmental Evidence (CEE): London, UK, 2018.
45. Moher, D.; Liberati, A.; Tetzlaff, J.; Altman, D.G. Preferred reporting items for systematic reviews and meta-analyses: The PRISMA statement. *Int. J. Surg.* **2010**, *8*, 336–341. [CrossRef]
46. Tavares, P.S.; Giarolla, A.; Chou, S.C.; Silva, A.J.P.; Lyra, A.A. Climate Change Impact on the Potential Yield of Arabica Coffee in Southeast Brazil. *Reg. Environ. Change* **2018**, *18*, 873–883. [CrossRef]
47. Ferreira, W.P.M.; Ribeiro Júnior, J.I.; de Fátima Souza, C. Climate Change Does Not Impact on *Coffea arabica* Yield in Brazil. *J. Sci. Food Agric.* **2019**, *99*, 5270–5282. [CrossRef]
48. Craparo, A.C.W.; Van Asten, P.J.A.; Läderach, P.; Jassogne, L.T.P.; Grab, S.W. *Coffea arabica* Yields Decline in Tanzania Due to Climate Change: Global Implications. *Agric. For. Meteorol.* **2015**, *207*, 1–10. [CrossRef]
49. Ginbo, T. Heterogeneous Impacts of Climate Change on Crop Yields across Altitudes in Ethiopia. *Clim. Change* **2022**, *170*, 12. [CrossRef]
50. Ridley, F.V. The Past and Future Climatic Suitability of Arabica Coffee (*Coffea arabica* L.) in East Africa. Ph.D. Thesis, Durham University, Durham, UK, 2011; p. 128.
51. De Carvalho Alves, M.; Sanches, L. Potential Effects of Spatio-Temporal Temperature Variation for Monitoring Coffee Leaf Rust Progress Under CMIP6 Climate Change Scenarios. *Earth Syst. Environ.* **2022**, *6*, 421–436. [CrossRef]
52. De Carvalho Alves, M.; Da Silva, F.M.; Sanches, L.; De Carvalho, L.G.; E Silva Ferraz, G.A. Geospatial Analysis of Ecological Vulnerability of Coffee Agroecosystems in Brazil. *Appl. Geomat.* **2013**, *5*, 87–97. [CrossRef]
53. Alfonsi, W.M.V.; Coltri, P.P.; Júnior, J.Z.; Patrício, F.R.A.; do Valle Gonçalves, R.R.; Shinji, K.; Alfonsi, E.L.; Koga-Vicente, A. Geographical Distribution of the Incubation Period of Coffee Leaf Rust in Climate Change Scenarios. *Pesqui. Agropecu. Bras.* **2019**, *54*. [CrossRef]
54. De Alves, M.C.; De Carvalho, L.G.; Pozza, E.A.; Sanches, L.; De Maia, J.C.S. Ecological Zoning of Soybean Rust, Coffee Rust and Banana Black Sigatoka Based on Brazilian Climate Changes. *Procedia Environ. Sci.* **2011**, *6*, 35–49. [CrossRef]
55. Atallah, S.S.; Gómez, M.I.; Jaramillo, J. A Bioeconomic Model of Ecosystem Services Provision: Coffee Berry Borer and Shade-Grown Coffee in Colombia. *Ecol. Econ.* **2018**, *144*, 129–138. [CrossRef]
56. Ceballos-Sierra, F.; Dall'Erba, S. The Effect of Climate Variability on Colombian Coffee Productivity: A Dynamic Panel Model Approach. *Agric. Syst.* **2021**, *190*, 103126. [CrossRef]
57. Ovalle-Rivera, O.; Van Oijen, M.; Laderach, P.; Roupsard, O.; de Virginio Filho, E.M.; Barrios, M.; Rapidel, B. Assessing the Accuracy and Robustness of a Process-Based Model for Coffee Agroforestry Systems in Central America. *Agrofor. Syst.* **2020**, *94*, 2033–2051. [CrossRef]
58. Vezy, R.; le Maire, G.; Christina, M.; Georgiou, S.; Imbach, P.; Hidalgo, H.G.; Alfaro, E.J.; Blitz-Frayret, C.; Charbonnier, F.; Lehner, P.; et al. DynACof: A Process-Based Model to Study Growth, Yield and Ecosystem Services of Coffee Agroforestry Systems. *Environ. Model. Softw.* **2020**, *124*, 104609. [CrossRef]
59. Van Oijen, M.; Dauzat, J.; Harmand, J.-M.; Lawson, G.; Vaast, P. Coffee Agroforestry Systems in Central America: II. Development of a Simple Process-Based Model and Preliminary Results. *Agrofor. Syst.* **2010**, *80*, 361–378. [CrossRef]
60. De Oliveira Aparecido, L.E.; Rolim, G. de S. Forecasting of the Annual Yield of Arabic Coffee Using Water Deficiency. *Pesqui. Agropecu. Bras.* **2018**, *53*, 1299–1310. [CrossRef]
61. Rahn, E.; Vaast, P.; Läderach, P.; van Asten, P.; Jassogne, L.; Ghazoul, J. Exploring Adaptation Strategies of Coffee Production to Climate Change Using a Process-Based Model. *Ecol. Modell.* **2018**, *371*, 76–89. [CrossRef]
62. De Oliveira Aparecido, L.E.; de Souza Rolim, G.; Camargo Lamparelli, R.A.; de Souza, P.S.; dos Santos, E.R. Agrometeorological Models for Forecasting Coffee Yield. *Agron. J.* **2017**, *109*, 249–258. [CrossRef]
63. Verhage, F.Y.F.; Anten, N.P.R.; Sentelhas, P.C. Carbon Dioxide Fertilization Offsets Negative Impacts of Climate Change on Arabica Coffee Yield in Brazil. *Clim. Change* **2017**, *144*, 671–685. [CrossRef]
64. Fares, A.; Awal, R.; Fares, S.; Johnson, A.B.; Valenzuela, H. Irrigation Water Requirements for Seed Corn and Coffee under Potential Climate Change Scenarios. *J. Water Clim. Change* **2016**, *7*, 39–51. [CrossRef]
65. Ovalle-Rivera, O.; Läderach, P.; Bunn, C.; Obersteiner, M.; Schroth, G. Projected Shifts in *Coffea arabica* Suitability among Major Global Producing Regions Due to Climate Change. *PLoS ONE* **2015**, *10*, e0124155. [CrossRef]
66. Purba, P.; Sukartiko, A.C.; Ainuri, M. Modeling the Plantation Area of Geographical Indication Product under Climate Change: Gayo Arabica Coffee (Coffea arabica). In *Proceedings of the IOP Conference Series: Earth and Environmental Science*; IOP Publishing: Bristol, UK, 2019; Volume 365.

67. Ranjitkar, S.; Sujakhu, N.M.; Merz, J.; Kindt, R.; Xu, J.C.; Matin, M.A.; Ali, M.; Zomer, R.J. Suitability Analysis and Projected Climate Change Impact on Banana and Coffee Production Zones in Nepal. *PLoS ONE* **2016**, *11*, e0163916. [CrossRef]
68. Eitzinger, A.; Läderach, P.; Carmona, S.; Navarro, C.; Collet, L. *Prediction of the Impact of Climate Change on Coffee and Mango Growing Areas in Haiti*; Centro Internacional de Agricultura Tropical (CIAT): Cali, Colombia, 2013.
69. Davis, A.P.; Gole, T.W.; Baena, S.; Moat, J. The Impact of Climate Change on Indigenous Arabica Coffee (*Coffea arabica*): Predicting Future Trends and Identifying Priorities. *PLoS ONE* **2012**, *7*, e47981. [CrossRef]
70. Schroth, G.; Laderach, P.; Dempewolf, J.; Philpott, S.; Haggar, J.; Eakin, H.; Castillejos, T.; Moreno, J.G.; Pinto, L.S.; Hernandez, R.; et al. Towards a Climate Change Adaptation Strategy for Coffee Communities and Ecosystems in the Sierra Madre de Chiapas, Mexico. *Mitig. Adapt. Strateg. Glob. Change* **2009**, *14*, 605–625. [CrossRef]
71. Chemura, A.; Kutywayo, D.; Chidoko, P.; Mahoya, C. Bioclimatic Modelling of Current and Projected Climatic Suitability of Coffee (*Coffea arabica*) Production in Zimbabwe. *Reg. Environ. Change* **2016**, *16*, 473–485. [CrossRef]
72. Schroth, G.; Laderach, P.; Cuero, D.S.B.; Neilson, J.; Bunn, C. Winner or Loser of Climate Change? A Modeling Study of Current and Future Climatic Suitability of Arabica Coffee in Indonesia. *Reg. Environ. Change* **2015**, *15*, 1473–1482. [CrossRef]
73. Benti, F.; Diga, G.M.; Feyisa, G.L.; Tolesa, A.R. Modeling Coffee (*Coffea arabica* L.) Climate Suitability under Current and Future Scenario in Jimma Zone, Ethiopia. *Environ. Monit. Assess.* **2022**, *194*, 271. [CrossRef]
74. Zhang, S.; Liu, B.; Liu, X.; Yuan, Q.; Xiao, X.; Zhou, T. Maximum Entropy Modeling for the Prediction of Potential Plantation Distribution of Arabica Coffee under the CMIP6 Mode in Yunnan, Southwest China. *Atmosphere* **2022**, *13*, 1773. [CrossRef]
75. Coto-Fonseca, A.; Rojas, C.; Molina-Murillo, S. Climate Change-Based Modeling of Potential Land Use Arrangements for Coffee (*Coffea arabica*) and Forest in Costa Rica. *Agric. Eng. Int. CIGR J.* **2017**, *19*, 224–229.
76. Imbach, P.; Fung, E.; Hannah, L.; Navarro-Racines, C.E.; Roubik, D.W.; Ricketts, T.H.; Harvey, C.A.; Donatti, C.I.; Läderach, P.; Locatelli, B.; et al. Coupling of Pollination Services and Coffee Suitability under Climate Change. *Proc. Natl. Acad. Sci. USA* **2017**, *114*, 10438–10442. [CrossRef]
77. Quiroz-Guerrero, I.; Pérez-Vázquez, A.; Landeros-Sánchez, C.; Gallardo-López, F.; Velasco-Velasco, J.; Benítez-Badillo, G. Resilience of coffee agroecosystems in light of climate chang [resiliencia del agroecosistema café ante el cambio climático]. *Trop. Subtrop. Agroecosyst.* **2022**, *25*, 3. [CrossRef]
78. Bunn, C.; Läderach, P.; Jimenez, J.G.P.; Montagnon, C.; Schilling, T. Multiclass Classification of Agro-Ecological Zones for Arabica Coffee: An Improved Understanding of the Impacts of Climate Change. *PLoS ONE* **2015**, *10*, e0140490. [CrossRef]
79. Ghini, R.; Hamada, E.; Pedro, M.J., Jr.; Gonçalves, R.R. V Incubation Period of Hemileia Vastatrix in Coffee Plants in Brazil Simulated under Climate Change [Simulação Dos Efeitos Das Mudanças Climáticas Sobre o Período de Incubação de Hemileia Vastatrix Em Cafeeiro No Brasil]. *Summa Phytopathol.* **2011**, *37*, 85–93. [CrossRef]
80. Moraes, W.B.; De Jesus Junior, W.C.; De Azevedo Peixoto, L.; Moraes, W.B.; Coser, S.M.; Cecílio, R.A. Impact of Climate Change on the Phoma Leaf Spot of Coffee in Brazil. *Interciencia* **2012**, *37*, 272–278.
81. Hailu, B.T.; Maeda, E.E.; Pellikka, P.; Pfeifer, M. Identifying Potential Areas of Understorey Coffee in Ethiopia's Highlands Using Predictive Modelling. *Int. J. Remote Sens.* **2015**, *36*, 2898–2919. [CrossRef]
82. Chalchissa, F.B.; Diga, G.M.; Feyisa, G.L.; Tolossa, A.R. Impacts of Extreme Agroclimatic Indicators on the Performance of Coffee (*Coffea arabica* L.) Aboveground Biomass in Jimma Zone, Ethiopia. *Heliyon* **2022**, *8*, e10136. [CrossRef] [PubMed]
83. Moat, J.; Williams, J.; Baena, S.; Wilkinson, T.; Gole, T.W.; Challa, Z.K.; Demissew, S.; Davis, A.P. Resilience Potential of the Ethiopian Coffee Sector under Climate Change. *Nat. Plants* **2017**, *3*, 17081. [CrossRef]
84. Moat, J.; Gole, T.W.; Davis, A.P. Least Concern to Endangered: Applying Climate Change Projections Profoundly Influences the Extinction Risk Assessment for Wild Arabica Coffee. *Glob. Change Biol.* **2019**, *25*, 390–403. [CrossRef]
85. Laderach, P.; Lundy, M.; Jarvis, A.; Ramirez, J.; Portilla, E.P.; Schepp, K.; Eitzinger, A. Predicted Impact of Climate Change on Coffee Supply Chains. In *Climate Change Management*; Springer: Berlin/Heidelberg, Germany, 2011; pp. 703–723. [CrossRef]
86. Tournebize, R.; Borner, L.; Manel, S.; Meynard, C.N.; Vigouroux, Y.; Crouzillat, D.; Fournier, C.; Kassam, M.; Descombes, P.; Tranchant-Dubreuil, C.; et al. Ecological and Genomic Vulnerability to Climate Change across Native Populations of Robusta Coffee (*Coffea canephora*). *Glob. Change Biol.* **2022**, *28*, 4124–4142. [CrossRef]
87. Mulinde, C.; Majaliwa, J.G.M.; Twinomuhangi, R.; Mfitumukiza, D.; Waiswa, D.; Tumwine, F.; Kato, E.; Asiimwe, J.; Nakyagaba, W.N.; Mukasa, D. Projected Climate in Coffee-Based Farming Systems: Implications for Crop Suitability in Uganda. *Reg. Environ. Change* **2022**, *22*, 83. [CrossRef]
88. Lara Estrada, L.; Rasche, L.; Schneider, U.A. Modeling Land Suitability for *Coffea arabica* L. in Central America. *Environ. Model. Softw.* **2017**, *95*, 196–209. [CrossRef]
89. Jaramillo, J.; Muchugu, E.; Vega, F.E.; Davis, A.; Borgemeister, C.; Chabi-Olaye, A. Some like It Hot: The Influence and Implications of Climate Change on Coffee Berry Borer (*Hypothenemus hampei*) and Coffee Production in East Africa. *PLoS ONE* **2011**, *6*, e24528. [CrossRef] [PubMed]
90. Ghini, R.; Hamada, E.; Pedro, M.J., Jr.; Marengo, J.A.; Gonçalves, R.R.D.V. Risk Analysis of Climate Change on Coffee Nematodes and Leaf Miner in Brazil. *Pesqui. Agropecu. Bras.* **2008**, *43*, 187–194. [CrossRef]
91. Azrag, A.G.A.; Pirk, C.W.W.; Yusuf, A.A.; Pinard, F.; Niassy, S.; Mosomtai, G.; Babin, R. Prediction of Insect Pest Distribution as Influenced by Elevation: Combining Field Observations and Temperature-Dependent Development Models for the Coffee Stink Bug, Antestiopsis Thunbergii (Gmelin). *PLoS ONE* **2018**, *13*, e0199569. [CrossRef]

92. IPCC. Glossary of terms. In *Managing the Risks of Extreme Events and Disasters to Advance Climate Change Adaptation*; Field, C.B., Barros, V., Stocker, T.F., Qin, D., Dokken, D.J., Ebi, K.L., Mastrandrea, M.D., Mach, K.J., Plattner, G.-K., Allen, S.K., et al., Eds.; A Special Report of Working Groups I and II of the Intergovernmental Panel on Climate Change; Cambridge University Press: Cambridge, UK; New York, NY, USA, 2012; pp. 555–564.
93. Pachauri, R.K. *Climate Change 2014 Synthesis Report*; IPCC: Geneva, Szwitzerland, 2014; ISBN 9789291691432.
94. Lee, J.-Y.J.; Marotzke, J.; Bala, G.; Cao, L.; Corti, S.; Dunne, J.P.; Engelbrecht, F.; Fischer, E.; Fyfe, J.C.; Jones, C.; et al. *Future Global Climate: Scenario-Based Projections and Near-Term Information*; Cambridge University Press: Cambridge, UK, 2021; ISBN 9781009157896.

sustainability

MDPI

Article

Bibliometric Network Analysis of "Water Systems' Adaptation to Climate Change Uncertainties": Concepts, Approaches, Gaps, and Opportunities

Aymen Sawassi [1,2] and Roula Khadra [1,*]

1 International Center for Advanced Mediterranean Agronomic Studies, Mediterranean Agronomic Institute of Bari, Via Ceglie 9, 70010 Valenzano, Italy; aymen.sawassi@uniparthenope.it
2 Department of Science and Technology, Parthenope University of Naples, 80133 Naples, Italy
* Correspondence: Khadra@iamb.it

Abstract: In response to the impact of climate change and to the uncertainties associated with the various dimensions of hydrologic variability, water systems' adaptation has risen to the top of global agendas. In accordance, identifying the additional science needed to improve our understanding of climate change and its impacts, including the scientific advances needed to improve the effectiveness of actions taken to adapt water systems, is of the utmost importance. To this aim, this research draws on a systematic bibliometric study of data, generated from the Web of Science research engine between 1990 and 2019, combined with a statistical analysis, to explore academic publication trends, and identify the strategic gaps and opportunities in global scientific research. The analysis shows the consistent level of national and international collaboration among authors, institutions, and countries, and highlights the substantial contribution of the USA and the UK to this research field. The statistical examination shows that the adaptation-informed literature on water systems remains fragmented, and predominantly centred on the framing of water resource planning and management, in addition to water engineering and infrastructure. The analysis also revealed a relatively skewed understanding of various important dimensions, such as governance, integrated water resources management, and stakeholder engagement, which are crucial for planning and implementing an efficient adaptation process. Observations reflect on the need to build water-related adaptive approaches based on a thorough understanding of potential climate uncertainties, rather than to generically address all the uncertainties in one scenario analysis. These approaches are required to combine short and longterm actions rather than considering only current and short-term measures, and to similarly associate policy and engineering, and equally consider the robustness, flexibility, reliability, and vulnerability during the planning phase.

Keywords: climate change; adaptation; water; flexible design; uncertainty; bibliometrics

Citation: Sawassi, A.; Khadra, R. Bibliometric Network Analysis of "Water Systems' Adaptation to Climate Change Uncertainties": Concepts, Approaches, Gaps, and Opportunities. *Sustainability* **2021**, *13*, 6738. https://doi.org/10.3390/su13126738

Academic Editors: Daniel El Chami and Maroun El Moujabber

Received: 19 May 2021
Accepted: 10 June 2021
Published: 14 June 2021

1. Introduction

Climate change is likely to aggravate threats on social, environmental, economic, and political dimensions [1,2]. These threats are illustrated either directly through the impact of thermal stress (drought, heatwaves), floods, and storms, or indirectly through changes in the ranges of disease vectors (e.g., mosquitoes), water quantity and quality, air quality, and food availability and quality, as well as the migration of populations [3,4].

The expected changes in climatic variables such as the amount, intensity, frequency, type of precipitation (e.g., snow vs. rain), temperature fluctuations, and other related hydro-climatic shocks such as a decrease in dry season flows (drought) or anticipated high-intensity precipitation (floods) are most likely to cause extreme events at the scale relevant to impact studies [5,6]. An estimated 821 million people, approximately one out of every nine people in the world, are classified as undernourished partly due to drought [7]. More than 1600 deaths may be caused annually by drought and heatwaves [7]. More than

35 million people are affected by floods [8], and more than 2 million people are displaced from their homes partly due to climate-related disasters [9]. Besides, some models predict that an increase of 40% in the number of people living in conditions of absolute water scarcity, with less than 500 cubic meters per capita per year, will be associated with only 2 °C warmings above current temperature levels [10].

Currently, around one-third of the world's population lives under physical water scarcity for at least two months of the year [7,11,12]. However, if a population faces a water scarcity induced by policies, socioeconomic relations, and/or political power, this becomes a socio-water scarcity, which is the second order of scarcity that threatens another third of the world population [11,13,14]. Consequently, societal well-being development and economic prosperity are always dependent on the constraint of water availability [10,11]. Studies have proven that agricultural production and food security would be most sensitive to water availability constraints, especially in many developing countries where the climate is already changing and economies are heavily dependent on agriculture [12–14]. Notably, arid, semi-arid, and sub-humid areas present observational evidence for changes, where all the above-mentioned threats can dramatically endanger agriculture, i.e., crop yields and livestock [15].

Since the 1970s, human actions have focused on the adaptability to these expected threats and limitations as pre-conditions for social justice, human well-being, economic development, and socio-environmental sustainability [16]. Hence, the concept of system resilience (flexibility) was introduced as the capacity to sustain human well-being and progress throughout disturbance events, uncertainties, societal changes [17,18], and to adapt to external perturbations [19–21]. New approaches were developed, advocating a more flexible, reliable, and robust water infrastructure design and planning to be able to deal with those expected risks associated with climate change [22].

As a matter of fact, flexible water infrastructure planning has the potential to manage uncertainty at a reduced cost by building less infrastructure upfront, yet enabling expansion in the future if needed [23]. However, enabling flexibility often requires substantial proactive quantification research, based on several studies and sources, to weigh the risks of rigid design, and to understand the benefits of flexible infrastructures in responding to climate change uncertainties [24].

Despite the general consensus about climate change's impact on water requirements, agricultural production, and food security, there is still a lack of knowledge about the magnitude of that impact on the users' communities. The current systems' flexibility is still unknown, the capacity of these systems to adapt to stresses induced by climate change and to absorb drastic shocks of some sudden events, i.e., floods and droughts is doubted, and the amount of research done on this issue is still unquantified.

So far, the bibliometric analysis showed to be an appropriate tool that addresses the global scientific production of academic research. The approach has been widely used in several specific fields of science: Yi-Ming Guo et al. (2019) studied the research productivity of smart cities [25]; Pauna et al. (2019) focused on microplastic in the marine ecosystem [26]; Zeshui Xu (2019) carried out a bibliometric analysis to understand the development status and trends in the field of big data [27]; Wei Li et al. (2015) assessed the global environmental assessment in a 20-year period in terms of trends of growth, international collaboration, geographic distribution of publications, and scientific research issues [28]; and Cuiqian Huai et al. (2016) used the bibliometric analysis as a tool for assessing the performance and pattern dynamics of water security research [29]. However, no attempt has been made, to the best of the authors' knowledge, to consider water systems' adaptation to climate change uncertainties.

This research draws on a systematic bibliometric study and presents a quali-quantitative analysis of data generated from the Web of Science research engine. It explores academic publication trends with the aim of identifying the strategic gaps and opportunities in the global scientific research concerning the adaptation of water systems to climate change by looking for answers to the following questions: To what extent is research addressing water systems' adaptation to climate change? What are the dimensions and concepts that have been framed? How have water management adaptive approaches been outlined?

2. Data Source and Methodology

2.1. Data Acquisition

The search covers the English language academic literature retrieved from the Web of Science search engine, Web of Science Core Collection Database (WSCCD), within a time frame set from 1990 to 2019. Specifically, the authors searched for the terms "adaptive planning", "uncertainty", "resilient design", and "climate change" using the "Related Record" option in the Web of Science database. The assumption behind Related Records searching is that articles that cite the same works have a subject relationship, regardless of whether their titles, abstracts, or keywords contain the same terms. The greater the number of cited references shared by two articles, the closer the subject relationship [30].

The resulting bibliographic database files consist of books, ordinary articles, and scientific reports. The search produced 8338 documents, a large portion of which were not relevant to our investigation due to the rise of interdisciplinary work that implies a multidimensional application of the concept of adaptation to climate change, such as Environmental Science, Water Resources Science, Meteorology, Atmospheric Sciences, and so forth.

The subject of adaptive planning under climate change uncertainties attracted several researchers working in the field of environmental science, where the investigation focused on impact assessments and strategic environmental assessments. The water engineering domain tackled adaptive water policies, adaptive planning, and the design of hydraulic infrastructures, in addition to applications in real case studies. The Meteorology Atmospheric Science focused on climate projections based on a variety of scenarios to adapt to climate change. Models and simulations were performed to assess future climate change and adaptation patterns.

In order to narrow the study to articles that explicitly connect adaptation to some aspects of water systems, the number of documents was refined to 3645 (44% of total records) by filtering the search to those within the "water resources category" of the Web of S cience database.

Subsequently, the bibliographic database files were exported as Tab-delimited files-(Mac-OS or Windows) that contain "Full Record" and "Full Record and Cited References" files. The derived information from these full record files included the following: authors, titles, year of publication, keywords, subject categories, journals, organizations, and the number of citations for each article.

The analysis was performed using VOS viewer software (version 1.6.13) widely used in the literature and particularly in bibliometrics (Section 2.3).

2.2. The Bibliometric Analysis

The Bibliometric measurement methods are the proper tools to interpret and represent temporal trends and growth regarding subject categories and journals, national and international collaborations between authors, the geographic distribution of publications in terms of authors' affiliations and their citations, and some other related scientific research issues. The scientific network is based on three key analyses: the co-authorship analysis, the citation analysis, and the co-occurrence analysis (Table 1) [31].

The co-authorship analysis addresses all the links between the authors, countries, and scientific institutions associated with water-related adaptation to climate change. This analysis aims to show the level of collaboration between items and assesses the national and international cooperation between authors and countries regarding the adaptation of water systems to climate change uncertainties.

Table 1. Key bibliometric analyses applied in this study.

Types of Analyses	Description
(Authors and countries) Co-authorship	In co-authorship networks, researchers or countries are linked to each other based on the number of publications they have authored jointly.
Journals Citation	In citation networks, two items are linked if at least one cites the other.
Author's Keywords Co-occurrence	The number of co-occurrences of two keywords is the number of publications in which both keywords occur together in the title, abstract, or keywords list.

The citation analysis explores the journals that are cited for scientific papers on water-related climate change adaptation. This analysis aims to inquire to what extent the topic was discussed in various journals from diverse disciplines.

The analysis of the keywords (co-occurrence analysis) is intended to ascertain the tendency of the water systems' adaptation to climate change uncertainties research, as well as the boundaries that limit the scope of the investigated subject and involve other disciplines. In this analysis, the focus is on the keywords provided by the authors in their scientific papers, to simplify the field, the subfield, the topic, and the research issue.

2.3. The Software

VOSviewer uses bibliometric database files to construct a scientific network of items (e.g., publications, scientific journals, researchers, countries, and research organizations). Items are graphically mapped through the application of mathematical modeling. The resulting maps display the possible links or connections among the different objects of interest and classify them into clusters or groups of similar concerns. The software is compatible with dataset files exported from Web of Science, Scopus, PubMed, RIS, or Crossref JSON [31].

Basically, a link is a connection or relation between two items. Examples of links are bibliographic coupling links between publications, co-authorship links between researchers, and co-occurrence links between terms (keywords). Each link has a strength, represented by a positive numerical value. The higher this value, the stronger the link. The strength of a link may indicate the number of cited references that two publications share, the number of publications that two researchers co-authored, and the number of publications in which two terms occurred together [31].

For a given item, "the links" and "the total link strength" attributes indicate, respectively, the number of links of an item with other items and the total strength of the links of an item with other items.

An attribute value of strength, expressed with a positive numerical value, is assigned to each item and link. Items with high weights are regarded as more important than items with lower weights, while a high weight linkage reflects a strong relationship between the two concerned items [31]. Table 2 lists the terms used in this publication and reports the related definitions.

The software uses the fractional counting method in calculating the attribute value of strength. The strength of a co-authorship link is determined not only by the number of documents co-authored, but also by the total number of authors of each of the co-authored documents, and the number of citations of the document. For example, when an author has co-authored a document with n authors, each of the n co-authorship links has a strength of 1, resulting in a total strength of the n co-authorship links [31].

Table 2. The terminology involved in this study.

Term	Description
Item	The object of interest (e.g., publication, researcher, keyword, author).
Link	Connection or relation between two items (e.g., co-occurrence of keywords).
Link strength	The attribute of each link, expressed by a positive numerical value. In the case of the co-authorship link, the higher the value, the higher the number of publications the two researchers have co-authored.
Network	Set of items connected by their links.
Cluster	Sets of items included in a map. One item can belong only to one cluster.
Number of links	The number of links between an item with other items.
Total link strength	The cumulative strength of the links of an item with other items.

3. Results and Discussion

3.1. Trend Analysis

Overall, the results show a steep rise, in the past 25 years, in the number of papers dealing with water adaption to climate change, be it water systems and/or water management, in a variety of aspects. The records started to considerably grow between 2009 and 2011, with a significant jump from 77 to 145 publications. The number of documents per year, exponentially displayed in Figure 1, exhibits a sustained growth, throughout the period between 1996 and 2019, that reaches 580 documents only in the year 2019. This research shows that scientific concern is expressed as significant interest in the subject of water systems' adaptation to climate change uncertainties.

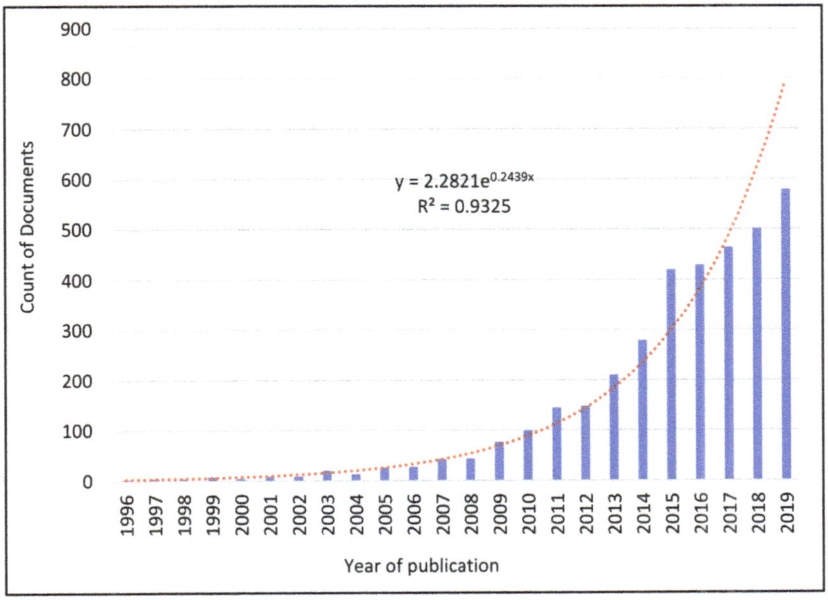

Figure 1. The annual number of publications during the period 1990–2019: the red curve represents the trendline of the growth pattern (the minimum record count appears in 1996 with 3 publications).

3.2. Citation Analysis of Journals

This analysis revealed 172 journals, of which only 68 journals eventuated in fundamental research advances with important contributions to water systems' adaptation to climate change uncertainties (Figure 2). The analysis clearly shows that the topic has been undertaken by journals of different disciplines.

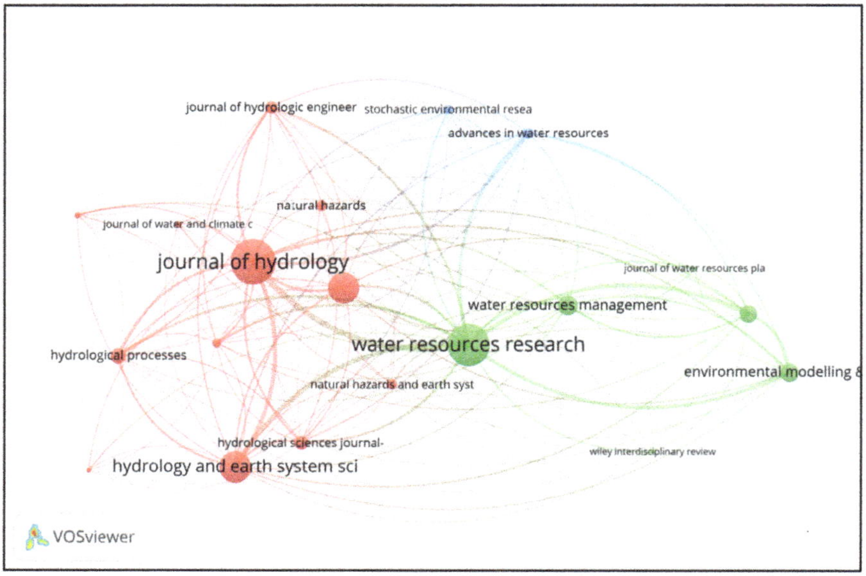

Figure 2. Network map of top 20 cited journals based on their total link strength.

The Water Resources Research Journal lead the search ranking with 388 documents and 20,187 citations, followed by the Journal of Hydrology (with 421 documents and 18,218 citations) and the Hydrology and Earth System Science Journal (with 289 documents, and 10,343 citations), and so forth. Furthermore, the analysis shows that despite the highest number of documents published in the Journal of Hydrology (421 documents), the journal is ranked only second according to its total link strength. This position might be related to the good quality of publications of the Water Resources Research Journal that are cited more.

3.3. *Co-Authorship Analysis*

The co-authorship analysis resulted in a network of 27,524 authors. This number was reduced to 788 by excluding documents with a large number of authors (by default, the maximum allowed number of authors per document is 25) and authors with low relevance scores. Only authors having a minimum of five publications on the topic of climate change adaptation in water systems were included. At first sight, the large number of authors who perform research related to water systems' adaptation to climate change announces a high interest in this field. Figure 3 illustrates the network map of co-authorship, which reflects the consistency level of collaboration between authors within their cluster and among other clusters. The topic has been widely tackled by means of interdisciplinary authors such as hydrologists, water engineers, geoscientists, and environmental scientists (Table 3).

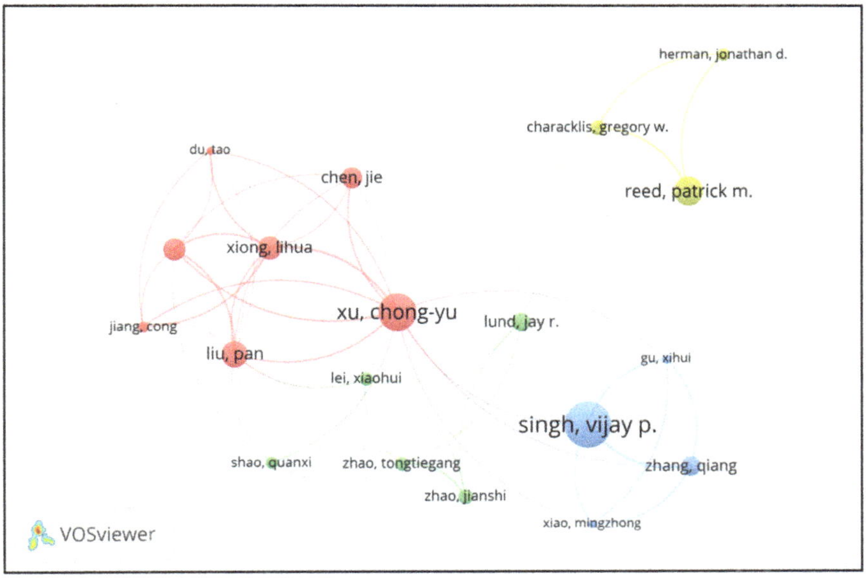

Figure 3. Network map of top 20 co-authorships based on the total link strength.

Table 3. Top 5 authors, affiliations, and research fields.

Authors	Documents	University	Department
Singh, Vijay p	36	Texas A&M University, USA	Water Engineering department
Xu, Chong-yu	29	University of Oslo, Japan	Dept of GeosciencesHydrology
Liu, Pan	20	Southeast University, China	Dept of Transportation Engineering
Xiong, Lihua	18	University of Ireland, Ireland	Dept of Hydrology and Water Resources Engineering
Guo, Shengalian	17	Wuhan University, China	Laboratory of Water Resources and Hydropower Engineering Science

3.4. Co-Occurrence Keywords Analysis

The analysis yielded 15,385 keywords. After excluding the general keywords with a low relevance score (e.g., 'New Approach', 'Results', etc.), and those with low occurrence (by default, a minimum of five occurrences of a keyword is selected, to strengthen the co-occurence results), 961 items were finally identified. Based on the total link strength, each resulting keyword is sketched in a node, creating a network map of all keywords. Figure 4 shows the network map of the top 20 authors' keyword co-occurrence. The size of the node reflects the keyword's degree of importance. However, its proximity to other nodes indicates some related subjects that are frequently occurring, along with the investigated ones.

Figure 4 reports the top 20 keywords most likely representing the research hot spots over the study period (1990–2019) where: the keyword "climate change" was ranked first, followed by "uncertainty", indicating that the climate issue is reflected in deep uncertainty about future changes [32,33]. Along with temporal evolution, the third most frequent keyword was "adaptation", showing that adaptability to the changing future is progressively viewed as a valuable design approach concerning strategic planning in water systems [24,34].

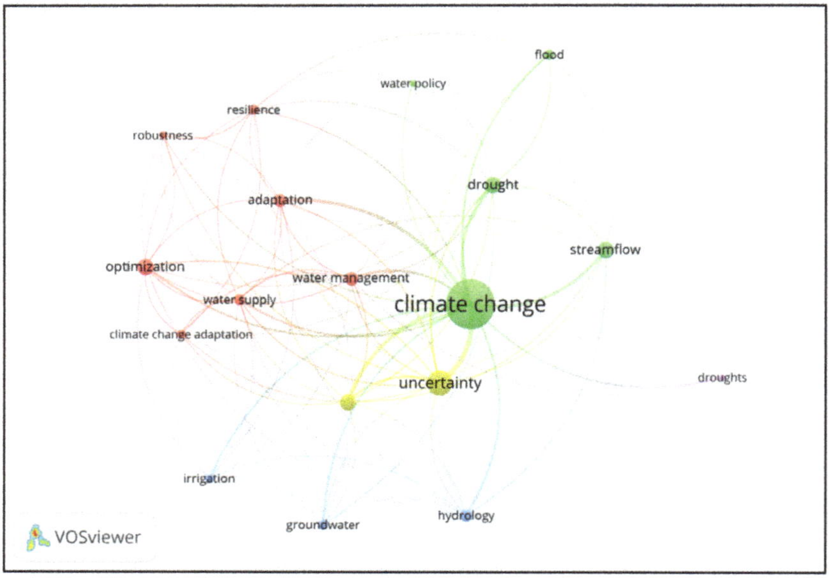

Figure 4. Top 20 frequently used author keywords during the period 1990–2019.

The increasing occurrence of keywords such as "mitigation" and "climate change adaptation" over the last two decades demonstrates the continued emphasis on these approaches. On the contrary, approaches such as "resilience" and "flexibility" are given less attention (52 and 31 documents respectively). From this perspective, there is a risk that the current adaptive water management strategies can effectively stifle potentially transformative shifts towards more infrastructure-centered thinking about water adaptation to climate change, and, at last, more resilient and flexible forms of water management.

3.4.1. The Water Dimensions, Concepts, and Scales

The quantitative analysis shows that the most prominent water dimensions integrated with the context of adaptation to climate change are "water resources management" with 1544 documents representing 43% of the total count, followed by "water planning" (537 documents), "groundwater management" (361 documents), "water policy" (335 documents), and other water-related dimensions, such as "drainage", "water security" and "wastewater management" (with 152, 110, and 40 documents, respectively).

Note that important key emerging water concepts related to the adaptation to climate change were poorly addressed (Table 4) despite being critical dimensions of an efficient adaptation planning process, such as water governance (31 documents), stakeholders' participation (10 documents), stakeholders' engagement (08 documents), and integrated water management (06 documents).

Table 4. Poorly articulated concepts in the literature search.

Concepts	Number of Documents
Management practices	51
Water governance	31
Stakeholders' participation	10
Stakeholders' engagement	08
Integrated water management	06

This lack continues to cast doubt on how the water sector is to adapt to climate change, not only in terms of physical adaptation (modern technology and long-lived infrastructure), but also in terms of institutional adaptation, which includes the social dimension and the water communities [35,36]. In the water sector, site-specific solutions need to be considered within the broader context of integrated water management approaches. A lack of regard for particular contexts, alongside poor planning and overemphasis on short-term outcomes, or failure to account for possible climatic consequences and adaptation limits, can result in maladaptation or "an adaptation that does not succeed in reducing vulnerability but increases it instead" [3].

The scales of water systems at which the adaptation is applied varies from ecological scales, being the most mentioned (including rivers with 1526 documents, watersheds with 360 documents, and stream flows with 80 documents, etc.), followed by the water distribution systems (including water supply systems with 326 documents and water infrastructure systems with 292 documents), and others. However, despite the multiplicity of scales that are relevant to the adaptive pathways to climate change, the irrigation systems are receiving less attention (only 72 documents), notwithstanding the large share of water consumed by the agricultural sector. This might be related to the fact that adaptation is a newly highlighted challenge in irrigation systems as compared to other scales. Yet, this does not exclude that the current formal processes and institutional organizations are ill-equipped to deal with building adaptive irrigation systems. Nevertheless, this fact highlights the compelling need to address these critical research questions, and once again underlines the importance of considering the institutional and organizational arrangements in the adaptation processes.

3.4.2. Water-Related Adaptive Approaches

The search revealed two pathways of water-related adaptative approaches: (1) static adaptiveness and (2) dynamic adaptiveness. Static adaptiveness aims to concentrate all efforts on protecting the basic existing plan from failure via proactive interventions and actions. It has been used in several design approaches, such as (i) Assumption Based Planning (ABP), (ii) Adaptive Policy Making (APM), and (iii) Robust Decision Making (RDM). With ABP, the challenges are to identify important assumptions and to select the vulnerable ones within the planning time horizon to define signposts that indicate the drawbacks of an assumption, and to enable hedging actions that preserve important options in light of the plausibility of that assumption's failing at some point [37]. The APM suggests the formulation and implementation of new policies that combine time-urgent actions with other actions that maintain necessary flexibility in the future. With an explicit provision for learning, this adaptive approach implies fundamental changes in policy-making by creating policies that respond to changes over time [38]. RDM is a quantitative decision–analytic approach based on the predict-then-act framework. The approach rests on a full understanding of the ranges of deep future uncertainty, suggests a set of robust strategies, and incorporates a vulnerability-and-response-option framework as part of a management strategy [39].

On the other hand, dynamic adaptiveness aims to alternatively develop parallel actions to be addressed when required. This approach includes the Multi-Objective Robust Decision Making (MORDM) [40], the Engineering Options Analysis (EAO) [41], and the Dynamic Adaptive Policy Pathways (DAPP) [32].

The current study highlights the main limitations of the above literature on water-related adaptive approaches. First, the RDM, being very generic, addresses all the future uncertainties together, including climate change, using the same scenario analysis rather than classifying the uncertainties and tackling each of them with the appropriate analysis tool. The RDM assumes that the current changing conditions are to be extended over the entire planning horizon, rather than characterizing and understanding the future climate and hydrological uncertainties. Moreover, static adaptiveness, being so limited, considers only short-term options instead of combining short-term options and long-term actions, and considering the upcoming changing circumstances. Besides, the MORDM considers robustness as a key criterion for evaluating alternative decisions under conditions of deep uncertainty, without considering the flexibility, reliability, and vulnerability options. Finally, the DAPP approach relies only on policy rather than combining it with the engineering aspect.

Although there remains a persistent lack of specificity around the boundaries and the limitations of these approaches, a limited number of authors have tackled these gaps. Haasnoot (2013) suggested that a new paradigm for planning under uncertainty should be developed by combining short and long-term options, following a dynamic adaptation over time to meet changing circumstances [32]. Maier (2016) proposed a multidisciplinary approach to dealing with the uncertain future that includes: (i) simulating uncertainties according to different sets of hypotheses by representing future coherent pathways; (ii) understanding the levels of insensitivity to changes in future conditions in terms of water system performance; (iii) developing adaptive strategies alongside with the commonly used static counterparts [42]. Haasnoot (2018) developed a design framework using a monitoring plan as part of the Dynamic Adaptive Policy Pathways to support water-related infrastructure investments under uncertain climate change. The approach suggests criteria to evaluate the monitoring signposts as measurability, timeliness, reliability, and institutional connectivity [43].

3.5. Co-authorship Country Analysis

The co-authorship country analysis revealed 118 countries, but only 74 met the minimum requirement of five publications related to the topic of adaptation in water systems to climate change. This analysis (Figure 5) shows that most documents originate from the USA (37% contribution), followed by the People's Republic of China (15% contribution), England (9%), and Australia (9%). On the other hand, Germany and the Netherlands, with a relatively low number of publications, yielded a high number of citations, which hints at the excellent quality of the publications (Table 5).

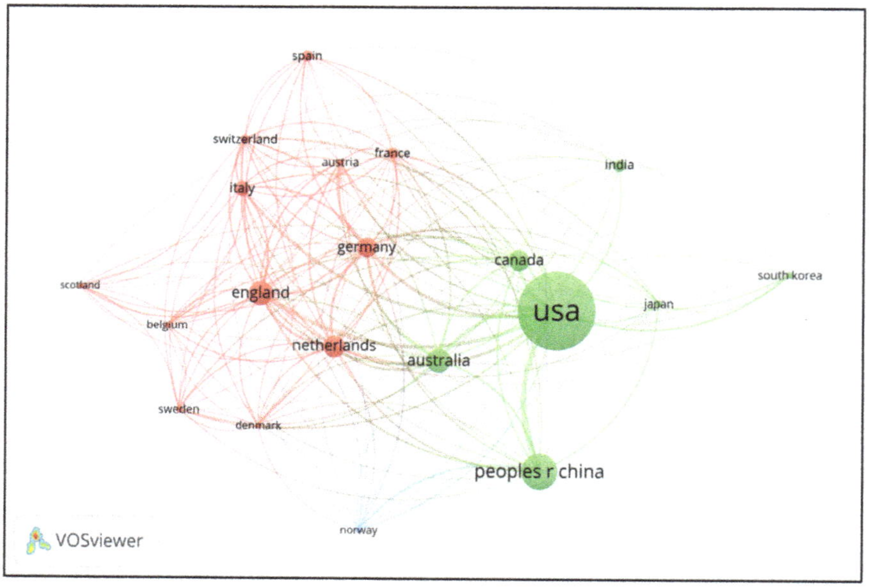

Figure 5. Network map of top 20 countries by co-authorship based on the total link strength.

Table 5. Top 5 countries ranked according to their total link strength.

Country	Documents	Citations	Total Link Strength
USA	1334	42,237	983
England	341	14,900	550
Germany	268	12,316	502
Netherlands	307	12,643	494
China	541	10,351	465

The top 20 ranked countries include the traditionally defined developed countries in Europe (Switzerland, Italy, France, and Spain) and North America (Canada). This might be mainly related to the fact that these countries house a large number of highly-rated research centers and universities.

Although there were still no developing countries ranked in the top 20 countries, the topic of water systems adaptability to climatic change is receiving more attention in Africa (Tunisia and South Africa), South America (Columbia, Peru, and Chile), South Asia (Nepal), and Western Asia (Turkey, Jordan, and Lebanon). Al-Omari et al. (2014) proposed an adaptation approach to climate change in Jordan: the Case of the Red Dead Canal (RDC) project [44]. Almazroui et al. (2018) developed a new approach to the design of water infrastructure based on a hydro-meteorological model [45]. Amamou et al. (2018) studied adaptation strategies as perceived by Tunisian farmers to cope with attributable climate change variability [46]. However, these countries still lack the experience that the developed countries have gained over their long history with climate change issues: a traditionally earned experience illustrated through their management strategies, new policies, and adaptive technologies.

4. Conclusions and Recommendations

The analysis of water systems' adaptation to climate change uncertainties would ideally require a deep evidence-based assessment across all the dimensions (i.e., water, climate, governance, policy, and decision making), as well as agreements on how significantly these dimensions are tackled, related, and defined.

Given the multiple dimensions of this work, it is important to note that the data used in this framework were obtained from a broad range of studies, developed using different methods and tackling the topic from different perspectives. This broad analysis may increase the uncertainty level of the results. However, at some point, it is difficult to establish commonality across all the studies covering a particular topic.

Data collection was limited to the core collection of WoS and refinements that were applied, such as "document categories", "documents types", and "languages", while other international databases (e.g., PubMed or Scopus) may have been combined. However, it was decided to use the WoS database since it is perceived to be more accurate and reproducible, as well as being the world's most trusted citation index for research and being used as the standard by official organizations.

The study considered "climate change" as the sole main driver of uncertainty that defines the adaptation strategies and measures that are largely contextual. This may lead to underestimating the other drivers, such as population growth, increased demand, and migration, etc., and might have missed some papers that do not specifically use the search terms that were used, while still referring to similar framings and subjects, e.g., flexibility, adjustment, or conversion of water systems to climate change. Nonetheless, this research investigated the bulk of the global scientific production of academic literature from 1990 to 2019. The combined use of statistical methods and bibliometrics resulted in quantitative statements that captured the interdisciplinarity of research topics, crossing the boundary of specific disciplines, and has provided a valuable and seminal reference for researchers and practitioners in the adaptation of water to climate change research.

The trend analysis showed exponential growth in the research interest concerning the subject of adaptation in water systems to climate change uncertainties during the study period. The research showed that different disciplines were concerned with the subject, while the majority of research was mainly derived from the Environmental and Water Management fields. This search shows, through different analyses (co-authorship, journals citation, and country analyses), that the scientific literature is significantly interested in the approach to adaptation in water systems to climate change uncertainties, where there are countries, such as the USA and the UK, that have contributed to the largest number of national and international collaborative documents.

The analysis shows that a large proportion of the literature concerning the adaptation of water systems to climate change tends to focus primarily on water resources planning and management (implied by investors, water managers, and the government), and also on building adaptive water supply infrastructure (implied by designers, and engineers). This infrastructure-centered management can be highly problematic, as there remains a relatively fractured understanding of the governance principles and the engagement of stakeholder's options in implementing an efficient adaptation planning process.

This search also addresses the main limitations of adaptive water approaches confronted in the literature, and suggests the following opportunities. Adaptive approaches should be built on a good understanding of the future climate and hydrological uncertainties. To this aim, uncertainties should be classified, mapped, and modelled with the appropriate analysis tool. An effective adaptive approach should be built on a set of combinations involving short-term options and long-term actions, policies, and engineering processes, and should equally consider robustness, flexibility, reliability, and vulnerability options within the planning process.

It is strongly recommended to address the short-term concepts, such as "resilience", which includes reconsidering the communities' resilience to drought, developing climate-resilient infrastructures, and also evolving some new strategies that guarantee an efficient and equitable service to the water users under limitations.

Author Contributions: Conceptualization, R.K.; Methodology, A.S.; Software, A.S. Validation, R.K.; Formal Analysis, A.S.; Investigation, R.K.; Resources, R.K.; Data Curation, A.S.; Writing—Original Draft Preparation, A.S.; Writing—Review & Editing, R.K.; Visualization, A.S.; Supervision, R.K. All authors have read and agreed to the published version of the manuscript.

Funding: This research received no external funding.

Informed Consent Statement: Not applicable.

Data Availability Statement: The data presented in this study are openly available in [Mendeley data] at [doi: 10.17632/h3rswz6hz5.1].

Acknowledgments: The authors greatly appreciate the English Language editing and review services supplied by Lea Arida, language consultant (linkedin.com/in/leaarida). They wish to thank the three anonymous reviewers for their constructive comments and suggestions.

Conflicts of Interest: The authors declare no conflict of interest.

References

1. Gosling, S.N.; Arnell, N.W. A global assessment of the impact of climate change on water scarcity. *Clim. Change* **2016**. [CrossRef]
2. Osman Elasha, B. Mapping of Climate Change Threats and Human Development Impacts in the Arab Region. Arab Human Development Report. *UNDP Arab Dev. Rep. Res. Pap.* **2010**.
3. Albritton, D.L.; Dokken, D.J. World Bank Climate Change 2001: Synthesis report. *Choice Rev. Online* **2001**, 409. [CrossRef]
4. Tol, R.S.J. The economic impacts of climate change. *Rev. Environ. Econ. Policy* **2018**. [CrossRef]
5. Trenberth, K.E. The Impact of Climate Change and Variability on Heavy Precipitation, Floods, and Droughts. *Encycl. Hydrol. Sci.* **2008**. [CrossRef]
6. Pradhan, P.; Parajuli, U.N.; Khanal, R.C. Framework for effectiveness and resilience of small-and medium-scale irrigation systems in Nepal. *Gov. Nepal* **2017**.
7. FAO; IFAD; UNICEF; WFP; WHO. *Food Security and Nutrition in the World the State of Building Climate Resilience for Food Security and Nutrition*; FAO: Rome, Italy, 2018; ISBN 9789251305713.

8. Below, R.; Wallemacq., P. *Annual Disaster Statistical Review 2018*; Centre for Research on the Epidemiology of Disasters: Brussels, Belgium, 2018.
9. Migration, I.O. World Migration Report 2020 (Full Report). International Organization for Migration: Geneva, Switzerland, 2019; ISBN 1561-5502.
10. Schewe, J.; Heinke, J.; Gerten, D.; Haddeland, I.; Arnell, N.W.; Clark, D.B.; Dankers, R.; Eisner, S.; Fekete, B.M.; Colón-González, F.J.; et al. Multimodel assessment of water scarcity under climate change. *Proc. Natl. Acad. Sci. USA* **2014**. [CrossRef]
11. Oki, T.; Agata, Y.; Kanae, S.; Saruhashi, T.; Musiake, K. Global water resources assessment under climatic change in 2050 using TRIP. *IAHS Publ.* **2003**, *280*, 10.
12. Mendelsohn, R. The impact of climate change on agriculture in developing countries. *J. Nat. Resour. Policy Res.* **2008**. [CrossRef]
13. Dinar, A.; Hassan, R.; Mendelsohn, R.; Benhin, J. *Climate Change and Agriculture in Africa: Impact Assessment and Adaptation Strategies*; Earthscan: London, UK, 2012; ISBN 9781849770767.
14. Khadra, R.; Sagardoy, J.A. *Irrigation Governance Challenges in the Mediterranean Region: Learning from Experiences and Promoting Sustainable Performance*; Springer International Publishing: New York City, NY, USA, 2019.
15. FAO. *World Agriculture: Towards 2015/2030*; Earthscan Publications Ltd: London, UK, 2002; ISBN 9251047618.
16. Folke, C.; Biggs, R.; Norström, A.V.; Reyers, B.; Rockström, J. Social-ecological resilience and biosphere-based sustainability science. *Ecol. Soc.* **2016**. [CrossRef]
17. Walker, B.; Holling, C.S.; Carpenter, S.R.; Kinzig, A. Resilience, adaptability and transformability in social-ecological systems. *Ecol. Soc.* **2004**. [CrossRef]
18. Boltz, F.; LeRoy Poff, N.; Folke, C.; Kete, N.; Brown, C.M.; St. George Freeman, S.; Matthews, J.H.; Martinez, A.; Rockström, J. Water is a master variable: Solving for resilience in the modern era. *Water Secur.* **2019**, *8*, 100048. [CrossRef]
19. Rockström, J.; Steffen, W.; Noone, K.; Persson, Å.; Chapin, F.S.; Lambin, E.F.; Lenton, T.M.; Scheffer, M.; Folke, C.; Schellnhuber, H.J.; et al. A safe operating space for humanity. *Nature* **2009**, *461*, 472–475. [CrossRef]
20. Folke, C. Resilience: The emergence of a perspective for social-ecological systems analyses. *Glob. Environ. Chang.* **2006**. [CrossRef]
21. Folke, C.; Carpenter, S.R.; Walker, B.; Scheffer, M.; Chapin, T.; Rockström, J. Resilience thinking: Integrating resilience, adaptability and transformability. *Ecol. Soc.* **2010**. [CrossRef]
22. Stakhiv, E.Z. Pragmatic approaches for water management under climate change uncertainty. *J. Am. Water Resour. Assoc.* **2011**. [CrossRef]
23. Sawassi, A.; Khadra, R.; Lamaddalena, N. Assessing the correlation between service flexibility and the cost of modernized large-scale pressurized irrigation systems: A perspective of resilience. *Irrig. Sci.* **2021**. [CrossRef]
24. Fletcher, S.; Lickley, M.; Strzepek, K. Learning about climate change uncertainty enables flexible water infrastructure planning. *Nat. Commun.* **2019**. [CrossRef]
25. Guo, Y.M.; Huang, Z.L.; Guo, J.; Li, H.; Guo, X.R.; Nkeli, M.J. Bibliometric analysis on smart cities research. *Sustainability* **2019**, *11*, 3606. [CrossRef]
26. Pauna, V.H.; Buonocore, E.; Renzi, M.; Russo, G.F.; Franzese, P.P. The issue of microplastics in marine ecosystems: A bibliometric network analysis. *Mar. Pollut. Bull.* **2019**, *149*, 110612. [CrossRef]
27. Xu, Z.; Yu, D. A Bibliometrics analysis on big data research (2009–2018). *J. Data Inf. Manag.* **2019**. [CrossRef]
28. Li, W.; Zhao, Y. Bibliometric analysis of global environmental assessment research in a 20-year period. *Environ. Impact Assess. Rev.* **2015**. [CrossRef]
29. Huai, C.; Chai, L. A bibliometric analysis on the performance and underlying dynamic patterns of water security research. *Scientometrics* **2016**. [CrossRef]
30. Clarivate Analytics. Available online: https://images.webofknowledge.com//WOKRS535R111/help/WOS/hp_related_records.html (accessed on 12 January 2021).
31. Van Eck, N.J.; Waltman, L. VOSviwer Manual version 1.6.10. CWTS Meaningful Metrics: Leiden, The Netherlands, 2019.
32. Haasnoot, M.; Kwakkel, J.H.; Walker, W.E.; ter Maat, J. Dynamic adaptive policy pathways: A method for crafting robust decisions for a deeply uncertain world. *Glob. Environ. Chang.* **2013**. [CrossRef]
33. Erfani, T.; Pachos, K.; Harou, J.J. Real-Options Water Supply Planning: Multistage Scenario Trees for Adaptive and Flexible Capacity Expansion Under Probabilistic Climate Change Uncertainty. *Water Resour. Res.* **2018**. [CrossRef]
34. Wall, T.A.; Walker, W.E.; Marchau, V.A.W.J.; Bertolini, L. Dynamic adaptive approach to transportation-infrastructure planning for climate change: San-Francisco-Bay-Area case study. *J. Infrastruct. Syst.* **2015**. [CrossRef]
35. Christiansen, L.; Olhoff, A.; Trærup, S. *Technologies for Adaptation: Perspectives and Practical Experiences*; UNEP Risoe Centre: Roskilde, Denmark, 2011; Volume 53, ISBN 9788578110796.
36. Technology Excecutive Committee (TEC). In *Technologies for Adaptation in the Water Sector*; UN: San Francisco, CA, USA, 2014; p. 12.
37. Dewar, J.A.; Builder, C.H.; Hix, W.M.; Levin, M.H. Assumption-Based Planning; A Planning Tool for Very Uncertain Times. RAND CORP SANTA MONICA: Santa Monica, CA, USA, 1993.
38. Walker, W.E.; Rahman, S.A.; Cave, J. Adaptive policies, policy analysis, and policy-making. *Eur. J. Oper. Res.* **2001**. [CrossRef]
39. Lempert, R.J.; Bankes, S.C.; Popper, S.W. *Shaping the Next One Hundred Years*; Rand Publishing: Santa Monica, CA, USA, 2003; ISBN 0833032755.

40. Kasprzyk, J.R.; Nataraj, S.; Reed, P.M.; Lempert, R.J. Many objective robust decision making for complex environmental systems undergoing change. *Environ. Model. Softw.* **2013**. [CrossRef]
41. Smet, K.S.M. *Engineering Options: A Proactive Planning Approach for Aging Water Resource Infrastructure under Uncertainty*; Harvard University Library, Office for Scholarly Communication: Cambridge, MA, USA, 2017.
42. Maier, H.R.; Guillaume, J.H.A.; van Delden, H.; Riddell, G.A.; Haasnoot, M.; Kwakkel, J.H. An uncertain future, deep uncertainty, scenarios, robustness and adaptation: How do they fit together? *Environ. Model. Softw.* **2016**. [CrossRef]
43. Haasnoot, M.; van 't Klooster, S.; van Alphen, J. Designing a monitoring system to detect signals to adapt to uncertain climate change. *Glob. Environ. Chang.* **2018**. [CrossRef]
44. Al-Omari, A.; Salman, A.; Karablieh, E. The Red Dead Canal project: An adaptation option to climate change in Jordan. *Desalin. Water Treat.* **2014**, *52*, 2833–2840. [CrossRef]
45. Almazroui, M.; Şen, Z.; Mohorji, A.M.; Islam, M.N. Impacts of Climate Change on Water Engineering Structures in Arid Regions: Case Studies in Turkey and Saudi Arabia. *Earth Syst. Environ.* **2019**, *3*, 43–57. [CrossRef]
46. Amamou, H.; Sassi, M.B.; Aouadi, H.; Khemiri, H.; Mahouachi, M.; Beckers, Y.; Hammami, H. Climate change-related risks and adaptation strategies as perceived in dairy cattle farming systems in Tunisia. *Clim. Risk Manag.* **2018**, *20*, 38–49. [CrossRef]

sustainability

MDPI

Article

Proximities and Logics of Sustainable Development of the Territorial Resource: The Case of the Localised Agro-Food System of Kalâat M'gouna in Morocco

Mohamed Zahidi [1], Jamila Ayegou [1] and Mohamed Ait Hou [2,*]

[1] Management, Information and Governance Research Laboratory (LARMIG), Hassan II University of Casablanca, Casablanca 20000, Morocco
[2] Polydisciplinary Faculty of Errachidia, Moulay Ismail University, Meknes 50050, Morocco
* Correspondence: m.aithou@umi.ac.ma; Tel.: +212-670-898-735

Abstract: Sustainable agriculture is fundamental to strengthening the resilience of rural communities to climate risks. This article aims to integrate traditional and scientific knowledge by identifying the types of proximity and the modes of valorisation of the territorial resource "rose" activated by the territorial actors of a localised agri-food system (LAFS), specialised in the production and processing of fresh rose in Kalâat M'gouna (K-M), Province of Tinghir in Morocco. The objective is to show to what extent the logics of proximity activated allow for the inclusion/understanding of the mechanisms of resource valorisation implemented by the actors. In other words, we seek to see whether the activation of several types of interaction between actors can promote the emergence of new modes of valuing the rose resource. This will be done by analysing the level of valorisation of the "rose" resource by the actors of this system while proposing tracks of improvement to take part in the sustainable development of this traditional system through its orientation towards the cognitive and innovative aspects. Our methodology is based mainly on a case study of a localised agri-food system (LAFS) using a set of semi-directive interviews conducted with 47 different actors of the system. Thus, we relied on SPSS software to study the possible correlations between the types of proximities existing and the modes of resource valuation activated. In terms of results, we were able to verify the domination of traditional proximities (geographical, organisational, relational, etc.) at the level of interactions between actors. On the other hand, there is a kind of insufficiency of the relations which are based on the knowledge and the (cognitive) knowledge.

Keywords: proximity; sustainable valorisation; territorial resource; actors; localised agri-food system; resilience of ecosystems

check for updates

Citation: Zahidi, M.; Ayegou, J.; Ait Hou, M. Proximities and Logics of Sustainable Development of the Territorial Resource: The Case of the Localised Agro-Food System of Kalâat M'gouna in Morocco. *Sustainability* **2022**, *14*, 15842. https://doi.org/10.3390/su142315842

Academic Editors: Daniel El Chami and Maroun El Moujabber

Received: 21 October 2022
Accepted: 21 November 2022
Published: 28 November 2022

Publisher's Note: MDPI stays neutral with regard to jurisdictional claims in published maps and institutional affiliations.

1. Introduction

In a given territory, there is a set of socio-economic actors that are geographically close to each other, but other types of ties or relationships that bring these actors together may exist, such as family, membership in an organisation, obedience to the same norms and values, etc. These elements constitute traditional knowledge that is essential to the construction of sustainable and climate-change-resilient agrosystems if integrated with scientific knowledge [1,2]. In this context, we use the notion of proximity, whose basic definition, according to the National Center for Textual and Lexical Resources (NCTLR), is: the situation of something or someone being within a short distance of someone or something. The notion of proximity is the basis of the proximity economy, a model for organising the economy around a set of relationships between actors. This notion aims to improve well-being through the valorisation of the territory and its resources by local actors for a sustainable territorial development for resilient rural communities.

Indeed, proximity has a plural meaning (meaning that it takes different forms) and has several facets. Proximity refers first of all to the geographical or spatial aspect, but it is not only on the basis of a spatial connection that actors or individuals interact. There are other supports that favour the interaction of actors in a territory, such as belonging to the same organisation, belonging to the same value system, etc. Despite their differences, there are many classifications and approaches in the literature; they share many common points. Indeed, where we find a classification of proximity in two types, "geographical" and "organised" [3–5], there are two other classifications based on five types: the first distinguishes "geographical" and "socio-economic" proximities insofar as the latter is divided into two segments; proximity of resources (cognitive and material), and proximity of coordination (relational and mediation resources) [4,6], and the second classification distinguishes between proximities; cognitive, organisational, institutional, geographical and social [4,5,7].

The development of means of communication (car, plane, train ...), as well as of Information and Communication Technologies, would sign the end of geography: space will no longer be of interest, as soon as it is possible to abolish it either by travelling or, more and more, via technologies that allow us to discuss but also to interact in a more and more detailed way at a distance (social networks, internet, etc.). Internet-based diagnostics, interventions and consultations are one example. However, opponents of these approaches argue that the local or the territories have never mattered as much as today. Local or original products are in the spotlight, with terroir values, and traceability of goods is sought after while local systems and clusters are the alpha and omega of innovation policies [8].

The purpose of this research is twofold. First, it seeks to identify the different types of proximity existing between the actors of the localised agri-food system operating in the exploitation of the fresh rose at the level of the territory of Kalâat M'gouna (K-M rose ecosystem), as well as the different modes of valorisation of this resource adopted by the local actors (private essentially). Secondly, it aims to analyse the extent to which the logics of proximity activated make it possible to include or understand the mechanisms of valorisation of the resource implemented by the actors. In other words, this research seeks to see how the activation of several types of interaction (proximities) between actors can promote the emergence of new modes of valuing the pink resource. The aim is therefore to investigate the possible adequacy between the proximities of the actors and the modes of valorisation of the territorial resource "rose" (the engine of the Localised Agri-food System of K-M) using the correlation coefficients between the different types of proximities and valorisation. This is to detect the gaps in terms of proximity and valorisation of the resource and propose recommendations to develop the exploitation of fresh rose perfume at the level of the territory of Kalâat M'gouna.

The present study aims to fill the existing gap in this sense, because the majority of studies on the rose agri-food system of M'gouna focuses mainly on traditional aspects, such as the natural potential of the territory, the problems of exploitation of resources, marketing of production, etc. However, no study has previously dealt with the relationship between the proximity of actors and the valorisation of the rose. However, no study has previously dealt with the relationship between the proximity of actors and the development of the rose. For this, we seek to exploit the existing resource of "proximity" between actors in order to develop new ways of valuing the rose based on knowledge, research and development. From there, we present our problem as follows: to what extent can the logics of proximity between the actors of the LAFS rose in Kalâat M'gouna in Morocco explain and promote the mechanisms of valorisation of the territorial resource "rose"? In this context, the combination of the two logics "proximity of actors versus sustainable valorisation of the pink resource" is at the centre of this research. In short, we focus on the logic of proximity to explain and analyse the logic of valorisation.

2. Proximity: A Plural Meaning and Many Classifications

In terms of typology of proximity, there is a great debate between the "tri-type" approaches and the five-type approaches. However, in general, the different approaches can be considered as complementary or even substitutable from our point of view, given the similarities they present. To this end, we will try to present all the types of proximity existing in the literature in order to broaden the analysis, while distinguishing between "classical" and "neoclassical" proximities.

2.1. Tri-Type Proximity

At this level, we present three types of proximity which constitute the foundation or basis of all classifications of proximity. The first is geographical, the second is organisational and the third is institutional (see Figure 1 below). In our treatment, we will refer to the last two types of proximity (organisational and institutional) under the name of "organised" proximity.

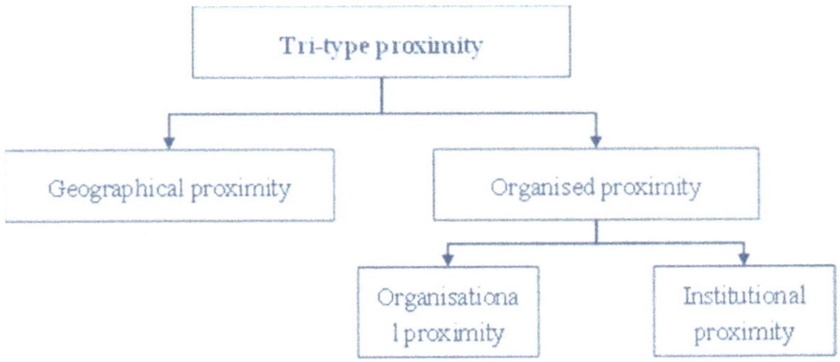

Figure 1. Tri-type proximitis. Source: our elaboration (theory information).

2.1.1. Geographical Proximity

This type does not pose any problems for the authors since there is a kind of homogeneity. Thus, as its name indicates, geographical proximity is defined as "the distance in kilometres between two entities (individuals, organisations, cities) weighted by the temporal and monetary cost of crossing it" [9]. This proximity is therefore spatial in nature and refers to two properties: firstly, it is binary (determines whether one is close or far); secondly, it is doubly relative (in relation to a set of criteria such as subjectivity and means of transport) [10,11]. Thus, it determines whether actors are geographically close or dispersed.

In other words, this type of proximity is defined as "the travel distance, functionally expressed in terms of cost and/or time, and dependent on the transport infrastructures and services" [12]. Thus, according to geographical proximity is translated by "the respective positioning of located agents" [4,5]. In this framework, two units (organisation or individual) are considered as close if they can meet or exchange in a given space at the lowest cost and/or in the shortest time and in a simple way [11]. Geographical proximity refers to face-to-face relations between actors [13] facilitating "exchanges, the circulation of new ideas and the diffusion of innovations" [14]. This means that direct and daily relations in a given territory allow for better communication and sharing of experiences, knowledge (tacit and codified), norms and values (e.g., trust), innovations, etc. between the different territorial actors, as well as the generation of financial externalities (economies of scale and minimisation of transaction costs) [14]. In this context, territorial anchorage through geographical proximity "participates largely in the deepening of cognitive processes and in the durability-stability of coordination relations between actors" [13].

However, it can be assumed that the advantages of geographical proximity in terms of innovation and spatial coordination between actors are recognised in numerous studies. In this case, this type of proximity has three limitations [15]. Firstly, geographical proximity only allows the transfer of a certain type of tacit knowledge. Secondly, the trust that could be established through co-location can be dangerous. Third, because many successful innovative networks (or territorial organisations) result from actors who are not spatially close [15].

Thus, we can announce that the analysis of territorial organisations, in terms of geographical proximity, is instrumental to apprehend the relations of inter-actor coordination (inter-individual and collective). Thus, it is essential to understand the processes of creation, transfer and exchange of crucial knowledge in the dynamics of territorial construction. However, this geographical proximity is not a sine qua non condition to coordinate and cooperate between actors. It sometimes constitutes a "constraint" according to the work of Lamara [16] on the coordination of territorial actors.

2.1.2. Organised or "Socio-Economic" Proximity

Additionally, called socio-economic proximity [6], organised proximity has a relational vocation, unlike geographical proximity. This type of proximity is of a different nature "it results from a social link" [12]. It therefore goes beyond the physical framework of the territory to focus on the social links between the actors, from which point on, organised proximity "translates the respective positioning of the agents in terms of coordination potential" [4]. Two entities (individuals or organisations) can be considered as close if they "share values, rely on identical coordination rules, share a precise knowledge base, speak the same language, regularly exchange e-mails ... " [15]. All these links can be grouped under the term "organised proximity". Thus, beyond geographical concentration or physical proximity, there are other possibilities for the actors of a network or a territorial organisation to coordinate effectively.

Additionally, organised (relational) proximity reflects "the capacity that an organisation offers to make its members interact" [10], and thus facilitates interactions between actors according to two logics: the logic of belonging to the same organisation, which necessarily produces interdependencies between its actors, that is to say, "the sharing of common objectives and rules" [13], and the logic of similarity, which increases the possibilities of interaction between members of territorial organisations is "the adhesion to representations, rules of action and models of thought" [13]. From these two logics emerge other types of proximity, we speak here of organisational proximity (belonging) and institutional proximity (similarity) which constitute the two pillars of organised proximity:

Organisational proximity (or the first pillar of organised proximity) refers to a logic of belonging of an individual or an actor to the same organisation or to the same territorial structure. It can therefore be deployed "within organisations (companies, establishments, etc.) and, if necessary, between organisations linked by a relationship of dependence or economic or financial interdependence (between companies that are members of an industrial or financial group, within a network, etc.)" [17]. It therefore seems that this proximity is not identified by direct or indirect relations, but by belonging to the same group [7]. This proximity favours collective learning processes and collective projects.

This type of proximity "refers to the complementary resources held by actors potentially capable of participating in the same finalised activity of a meso-economic type, within the same organisation (large group, etc.) or a set of organisations (network of co-operations, sector of activity, local productive system, etc.)" [18]. This dimension of proximity refers to the belonging of individuals or a group of actors to the same organisation (such as the belonging of a group of students to the same faculty). This type of proximity based on belonging is criticised by a large number of authors because belonging to the same organisation is not a condition for identifying interactions, but the majority of authors defend the idea that belonging to the same structure facilitates and favours interactions and relations between actors. In fact, these interactions are facilitated by the actions, routines and rules according to which they behave.

In this case, this type of proximity is presented as a "material proximity" [6], because they have the same foundations (interactionist logics and belonging to the same organisation or network). Bouba-Olga and Grossetti consider that individuals are similar or complementary from the point of view of the resources they possess (assets, income, diplomas, social status, etc.) [6]. Thus, actors interact because they belong to the same organisation, the same social structure or the same territory.

As for institutional proximity (institutional means organisations, firms, networks, etc. and behaviours [19]), the second pillar of organised proximity, it "reflects the fact that a group of individuals share and conform to the same set of institutions" [20]. Thus, as long as institutions exert a certain influence on individuals, it is obvious that these institutions exert the same or even greater influence on organisations and territorial configurations. In an increasingly uncertain global and local context, institutions intervene to reduce this uncertainty by offering the necessary and precise information and analyses to succeed in any collective or individual action. The institutional sphere therefore intervenes as "structures that frame behaviour, particularly collective behaviour, and are therefore the basis of social relations and therefore of a form of proximity" [21]. From there, the institutional sphere constitutes an important support and a dimension of proximity.

This type of proximity thus refers to "the adherence of the actors to common rules of action, explicit or implicit (habitus), and, in certain situations, to a common system of representations, even of values" [18]. This proximity thus refers to a logic of similarity, which means that all the actors of the territorial system or organisation adhere to or share "the same system of representations, or set of beliefs, and the same knowledge" [10]. In other words, this proximity is based on tacit or immaterial links by nature that intervene so that all the actors of the system resemble each other or become similar on the basis of an infinity of criteria (for example, language, beliefs, rules, etc.). We can say, for example, that two companies in a territory are institutionally close if they obey the same labour code. In the following figure, we summarise the classical proximities mentioned in this axis.

2.2. Five-Types Proximity

Following the logic of similarity, organisations and individuals can be similar in knowledge, and this similarity creates a kind of cognitive proximity. The latter "refers to the degree of similarity in the knowledge bases of organisations (. . .) it is a crucial issue for communication and knowledge transfer. Effective knowledge transfer and collaboration requires the ability to identify, interpret and exploit new knowledge" [4,7]. This means that when individuals, organisations or actors in a given territory share the same knowledge or expertise, they are cognitively close and can therefore interact and establish cooperative and coordinative relationships regardless of the physical distance between them.

In this sense, Pecqueur emphasises that "the common reference allowing coordination cannot be reduced to an accumulation of common knowledge" [22], but must integrate "common social representations" [13], i.e., opinions, beliefs, etc. Because, according to this perspective, the only accumulation of common knowledge by the actors of a given territory does not allow the construction of interrelations and the emergence of a cognitive proximity, the activation of this type of proximity must be supported by a social or identity proximity, that is to say, by a network of social and cultural relations.

Additionally, consequently, cognitive proximity refers to the idea of "a similarity or complementarity of values, goings-on, projects, routines, conventions, referents, etc. (all of which can be grouped under the term "cognitive resources" [6]. In this logic, Dupuy [20] defines cognitive proximity as "the capacity of actors to learn from others. Actors are cognitively close when they share the same knowledge base and expertise". Then, according to this logic, most cognitive resources are shareable, such as languages, norms, values, representations, cultures, etc. [6]. Consequently, they facilitate the coordination of actors. This makes possible the development of collective learning processes between actors, the construction of new resources, the valorisation of the territory's resources and assets, and the discovery of new opportunities for the development of the territory.

The second type of this category has a social vocation—social proximity—because it challenges the micro-economic level in terms of interactions and coordination (agents and individuals), contrary to the previous proximities which challenge the meso- and macro-economic levels (organisations and institutions). Indeed, social proximity highlights "the role of social relations, based on trust, friendship and family relations between individuals" [20]. Thus, we can say that this proximity is based, in terms of the creation of interdependencies, on links of a personal or family nature, developed over time or in the environment of the family and acquaintances.

Thus, social proximity can be defined as "socially anchored relationships between agents at the micro level" [7]. Thus, these relationships that have emerged a social proximity between individuals in a territory, are anchored and rooted in the social structure of the territory, that is, in the history and identity of the territory. Thus, between organisations, social proximity is defined as "the degree of overlap between the personal networks of the individuals who make up these two organisations" [7]. Indeed, the social structure of organisations is composed of a group of people, so it is normal that there are no virtual boundaries between the social networks of individuals and the relationships with the organisation.

Thirdly, we find two other types of proximity, a relational type which refers to the "relational proximity", and another type of mediation, which refers to the "mediation resource" proximity. This distinction comes down to the fact that the links, coordinations and interactions between actors can take place in two ways: the first is direct or without mediation resources and the second is indirect or with mediation. Thus, at the level of relational proximity, the interactions take place in a network between the actors and in a direct way, in fact, this proximity is defined by "the position of the different actors in the networks" [6]. Concerning the proximity of mediation resources, it is based on "devices that allow exchange without mobilising relational chains" [6]; therefore, at the level of this type, exchanges or interrelations cannot exist without mediation or without means that facilitate the relationship between actors, knowing that mediation resources can be material or immaterial.

Finally, there is another proximity that is widely used and studied in the technological or innovation field, namely the "electronic" proximity, which makes it possible to overcome the geographical or physical constraint [15]. This proximity is based almost on the same logic of similarity already mentioned, so that organisations and companies that possess the same technologies or a similar level of innovation are likely to coordinate and collaborate through collective projects. This is the activation of electronic or technological proximity. In the following Figure 2, we summarize the "five-types" proximities mentioned in this axis.

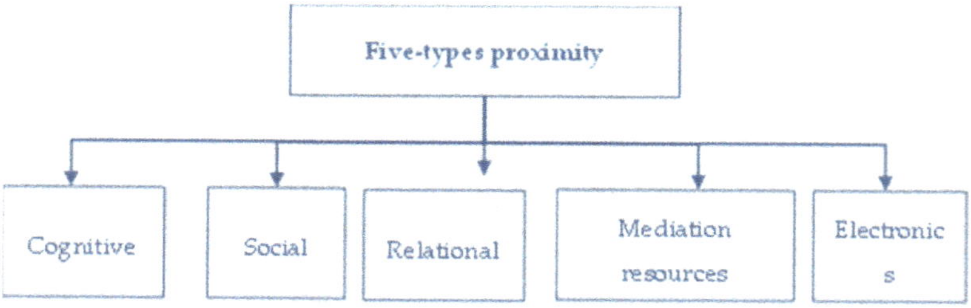

Figure 2. The five-types proximities. Source: Our elaboration from the definitions of [5–7].

3. Geographical, Socio-Economic and Climatic Conditions of the Study Area

The cultivation of the perfume rose (Rosa damascena) is the main agricultural activity in the Dades and M'gouna valleys in the province of Tinghir in Morocco. This activity currently occupies 4200 linear km in the form of hedges or fences around agricultural plots, i.e., about 1000 ha (10% of the cultivated agricultural area). Rose production is often compromised by frosts and cold weather which affect the flowering period and the industrial quality of the rose. It is very variable from one year to the next with an average of 4000 t of fresh roses per year.

The name "Valley of Roses" comes from one of the agricultural specificities of these two parts of the valley: the cultivation of the Damascus rose (Rosa damascena), which, even if it remains discreet, is a real marker of the landscape. The "rose of M'Goun" is dried and distilled on site. This local industry is shared between three large factories, several cooperatives (37), and a few private distillers. Distillation produces mainly rose water, which is exported to Marrakech and the major cities, and is not very open to the market. Its function in the agrarian system was limited to uses that did not make it a cash crop, before the colonial era. This rose water and the products derived from it (soaps, cosmetics, dried roses), which are marketed nationally and worldwide, but mainly locally in the many shops located along the main road, constitute a marker of the economy of this region. The rose has moreover been labelled through a protected designation of origin (PDO) "Rose de Kelâat M'Gouna Dadès" in 2011 [22]. The following Figure 3 shows the geographical location of the LAFS of M'gouna rose.

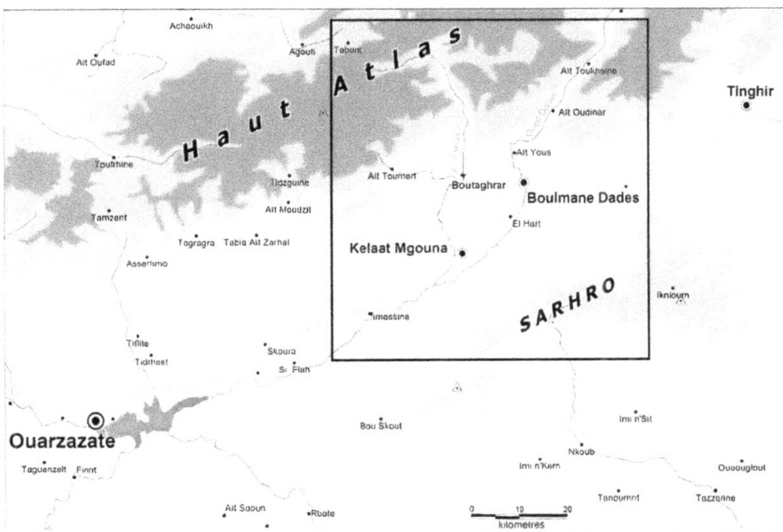

Figure 3. Map of the study area (the M'Gouna and Dadés valleys). Source: [23].

The rose is generally harvested in the first half of May, in the morning, at sunrise. It is used to make several products such as dried flower buds, rose water, rose essence and rose concrete and rose absolute, it is also used in its dry state as a cosmetic product. It takes 4 to 5 kg of fresh roses to make 1 kg of dry rose and 5000 kg of fresh roses to make 1 kg of rose oil. Pale rose petals contain 0.03–0.04% of an essential oil composed of geraniol, nerol, citronellol and phenylethylaldehyde (soluble in water, hence the fragrance given to rose water).

Bioclimatically, the perfume rose production area is arid to Saharan (the average temperature over the year is 18.0 °C) with very irregular rainfall from one year to the next (the average rainfall is 171 mm per year) and with a marked continental aspect. Rainfall is often intense and concentrated in time in the form of storms, causing violent floods. In fact, the Dades belongs to the cool pre-Saharan bioclimatic stage. The cultivated soils consist of deep silts, differentiated by their colloidal element content. Their slightly alkaline pH generally varies between 7.4 and 7.8. The organic matter content is low. The total nitrogen content is medium to low.

This rose-growing region, through its oasis vocation, constitutes an ecological bulwark against the advance of desertification and a source of income for the populations established in these areas. They contain a variety of plant and animal resources in an environment marked by low annual rainfall. This fragile ecosystem is increasingly degraded, due to a combination of factors, including climatic disturbances, which are already having an impact on farmers.

The various aromatic rose extracts are produced by two companies: Bioland and Arômes du Maroc, which have modern processing units in Kelaa M'gouna. Both companies also have processing units in the town of Khémisset (Bioland) and Tiddas (Arômes du Maroc). In addition to these two major players in the "rose" sector, the study area has 27 processing units with a capacity of 920 T/year, and the organisation of the profession includes more than 35 agricultural cooperatives operating in the field of rose production and an economic interest grouping (EIG). These different actors have helped to improve the productivity and cultivated area of rose in the study area, as shown in Figure 4 below.

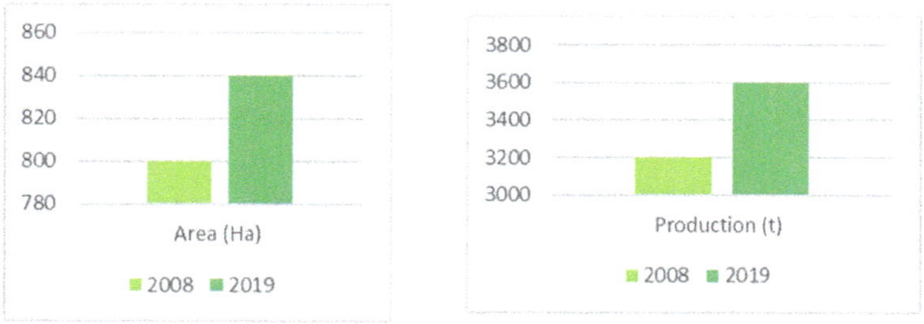

Figure 4. Evolution of cultivated area and rose production in the Dades and M'gouna area between 2008 and 2019. Source: [24].

4. Methodology

In order to analyse and explain the relationship between the level of proximity of the territorial actors of the rose LAFS of Kalâat M'gouna (K-M) and the power of the mechanisms of valorisation of the "rose" resource by these actors on the said territory, and thus to verify the existence or not of a coherence between these two variables, we divide the article into two parts. The first is theoretical, in which we present the meaning and classifications of the notion of proximity in the literature. The second is empirical, in which we propose our methodological framework, the results of the field survey, the analysis and discussion of these results, and we conclude by proposing avenues for improving the exploitation of the fresh rose in the province of Tinghir. In terms of the empirical part, we adopt the K-M rose LAFS case study approach in three steps.

The first step consists of contacting the actors of the rose ecosystem of the KM. To do this, we administer a questionnaire to the various local actors in the private and public system, with the aim of gathering as much information as possible about the actors, the types of proximity that link them, and the practices of valuing the "rose" resource adopted by the actors.

The second step consists of classifying and processing the data collected from the respondents (actors of the LAFS rose), in order to analyse, on the one hand, the dominant inter-actor proximities in the study area, and to identify, on the other hand, the modes of promotion and development of the "rose" resource adopted by the actors of the LAFS of M'gouna. This with the aim of correlating the types of proximity and the existing modes of valorisation, with the help of the SPSS software (v23), in order to lead, in fine, to a conclusion on the nature of the relations linking the locally activated proximities between the actors and the logics of valorisation of the territorial resource "rose".

Thirdly, we will detect the anomalies that prevent a better valorisation of the "rose" resource by the actors of the ecosystem, while proposing ways to improve the conditions and processes of valorisation of the resource by the actors.

In fact, we relied on a representative sample of 47 actors, the majority of whom are private actors (farmers, cooperatives and agricultural enterprises in the rose sector), to which we added a few public actors responsible for the governance and organisation of the sector at the local level. The survey was conducted over a temporary interval of three months (from February 2020 to April 2020).

Our sample is composed of different profiles: presidents of cooperatives, administrative executives (executives of public institutions such as the chamber of commerce, the chamber of agriculture, the agricultural development agency, the regional office of agricultural development of Ouarzazate—ORMVAO, etc.), members of cooperatives or cooperants, treasurers, etc. Our questionnaire also aims to identify the proximities activated between the actors of the ecosystem and the modes of valorisation of the rose resource adopted by the actors at the local level, in order to correlate the results, with a view to analysing the links between proximities and modes of valorisation and to identify the lack of proximity or valorisation at the level of the M'gouna rose EPA. To do this, we have constructed the questionnaire taking into account the following themes:

The proximities activated by the actor with other actors, or how he communicates and reacts with other actors in the ecosystem (face-to-face, in an organisational framework—such as an association, a cooperative, unions, etc.—via digital tools, via mediation resources, etc.), as well as the level of use of each means.

The modes of valorisation of the rose resource adopted by the actor: how the actor participates in the valorisation of the M'goun rose ... during the phases of cultivation, production, processing, packaging, transformation, marketing and communication, etc.

The insufficiencies or points of weakness perceived by the actor in terms of the proximity process or the valorisation of the rose resource at the LAFS rose de M'gouna and suggestions for improvement.

To calculate our sample size, we used the G*power 3.1 software [25,26], with the following parameters: F2 = 0.02 (small), $\alpha = 5\%$, the number of predictors = 2 (two variables (proximity and valuation)), and the power was set to 80% [27]. The required (significant) sample size set by the software, to test our model is 43. The following table shows some characteristics of our study sample (status of the respondents, gender, number ...). More than 70% of the respondents belong to the co-operative world (presidents, directors, cooperators) and almost 60% of the respondents are women. The characteristics of our study sample are explained in Table 1 below.

Table 1. Status and Gender of Respondents to the Questionnaire.

Status	Number of observations	Frequency (%)	Type		Frequency (%)	
			Male	Woman	Male	Woman
President of cooperatives	17	46.17	19	28	40.42	59.58
Plant managers/distillation unit	3	6.38				
Treasurer	2	4.26				
Members/Cooperators	12	25.53				
Administrative framework (public institutions)	09	19.15				
Association/federation	4	8.51				
Total	**47**	**100.0**	**47**		**100.0**	

Source: Own elaboration.

In order to process and analyse the results obtained in the field, with a view to understanding the mechanisms of valorisation of the territorial resource "rose" through the proximities activated by the territorial actors of the K-M rose ecosystem, we follow two logics: the first is analytical, and aims to identify the types of proximity as well as the modes of valorisation of the resource experienced or implemented by the actors. The second logic is synthetic, and aims to correlate the forms of proximity identified among the actors with the modes or processes of valorisation of the rose resource implemented by the latter, in order to explain whether there is an adequacy or a direct relation between proximity in the plural sense and the processes of valorisation of the "rose" territorial resource of the LAFS of K-M.

In order to analyse the correlation or the level of interdependence between the two variables (proximity and modes of valorisation of the territorial resource), we opted for Pearson's correlation (and relying on the results collected from our sample of actors of the ecosystem (47) via the administered questionnaire). The use of this technique tests our main hypothesis of this research, which stipulates that the proximity (all forms) between actors has a direct impact on the processes of valorisation of the territorial resources of this system, and conditions the local innovation processes.

Thus, from the answers of the respondents to the questionnaire, concerning the activated proximities between the actors and the operational processes of valorisation of the rose (each respondent selects the proximities and the processes of valorisation that he thinks are active within the LAFS of the K-M rose), we were able to set up the dynamic crossing of these two variables.

5. Results

In this section, we will present the results of our field survey. First of all, we will start by presenting the territorial dynamics of the M'gouna rose ecosystem, then its actors, its structure and its history. Then, we will present the results obtained in relation to the proximities and modes of valorisation of the "rose" resource identified among the actors of this productive territorial configuration.

5.1. The Rose LAFS of Kalâat M'gouna: A Territorial Dynamic for the Development of the Rose

For a long time, the Dades Valley has been known for the cultivation of the perfume rose with the aim of separating the agricultural plots and not in a purely agricultural and economic logic. With time, this culture has developed more and more to cover today an area of more than 3250 km, or 800 ha, in parallel with the development of the entrepreneurial and cooperative spirit that relies on economic, social and cultural logic aimed at the development of the production and development of the perfume rose. This culture is mainly concentrated in the communes of Kelâat Mgouna, Aït Sedrate, Aït Ouassif, Souk Lekhmis and Ighil N'oumgoun. In the following Table 2, we present the main actors of this activity, who have participated throughout the time in the construction of the rose LAFS of M'gouna.

The cooperation of these actors has enabled the creation of a large structure in Dadès (30 km from Kalâat M'gouna) with the aim of organising the sector and creating a space for the exchange of expertise between the actors of this sector, to supervise the producers in the operations of production and valorisation and to establish the bases of a partnership and to reinforce the capacities of negotiation in the field of the marketing of the products at the local, national and international level As well as the creation of a rose and its derivatives exchange by the determination of a reference price and the coordination of the operations of the designations of origin of the rose and its derivatives, without forgetting the insurance of the respect of the quality standards in accordance with the specifications of the designation of origin. This structure is named Maison de la Rose Parfumée, and it includes a museum of the perfumed rose, a laboratory, an exhibition hall, administrative buildings, a conference room and another for meetings.

The interaction of this amalgam of local private, public and community actors, and the implementation of action plans and common strategies for the development of this sector in the Dades valley, have enabled the emergence of a Localised Agri-food System (SAL) on the scale of the M'gouna and Dades territory. This SALA, marked by the diversity and territoriality of the actors, performed unsatisfactorily before 2011, when the average level of production of fresh roses was 2500 t per year, due to a number of reasons such as drought, the poor structuring of the sector, poor communication and coordination between the actors in the sector, etc. However, since 2011, with the implementation of Pillar I of the Green Morocco Plan in 2008 [28].)—the programme that aimed to develop the rose sector in the Tinghir area—and the signing of the programme-contract between the government and the professionals of the sector in 2012, the average production level of this system has increased to 3350 t annually, i.e., an increase in production of 30%. This result is due to efforts to structure and energise the sector, as well as to the collaboration of the main actors in this LAFS. Additionally, despite the fact that exports from this ecosystem remain mediocre (63 t/year) [24], but we are witnessing a great movement of industrialisation and transformation of fresh roses into innovative derivative products. In the following Figure 5 we show the evolution of the sector's performance over the last ten years.

Table 2. Main actors of the localised agri-food system of Kalâat M'gouna rose.

	Actors	Legal Form
Actors Main	Ministry of Agriculture, Maritime Fishing, Rural Development and Water and Forests	Public
	Agricultural Development Agency (ADA)	Public
	Regional Office of Agricultural Development of Ouarzazate (ORMVAO)	Public
	National Agency for the Development of Oasis and Argan Zones (ANDZOA)	Public
	National Initiative for Human Development (INDH)	Public
	Group of municipalities AL WARDA	Association
	The Moroccan Interprofessional Federation of the Perfume Rose (FIMAROSE)	Association
	Rose production and processing entities (80 local entities including 2 companies)	Private (cooperatives and companies)
	Farmers or peasants	Private
	The National Office of Sanitary and Food Safety (ONSSA)	Public
	National laboratories: OARC—EAFLabo—Charles Nicoles	Private
	International laboratories: Pyrenessences, Intertek, Sofia	Private
Supporting structures	Equipment suppliers: Ura Industrie, etc.	Private
	Cosmetic laboratories: Cerra SAS, Azbane	Private
	The cosmetic experts: A. Y. Cosmetic Conseil	Private
	The German Agency for International Development Cooperation—GIZ	Non-governmental organisation (NGO)
	The Belgian Cooperation and the APEFE program (Association for the Promotion of Education and Training Abroad)	NGO
	The UNDP	NGO

Source: Our elaboration.

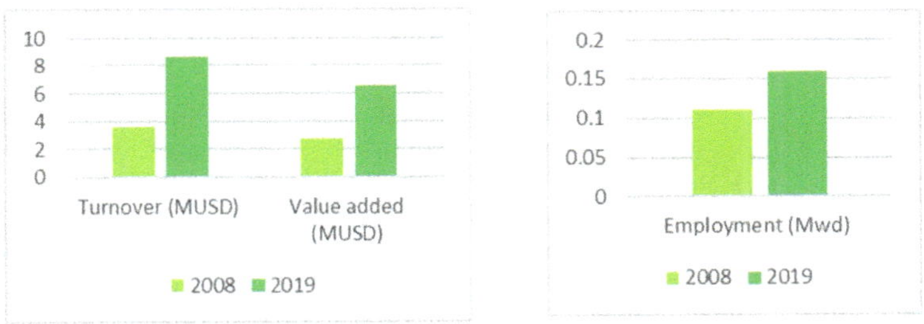

Figure 5. Impact indicators of the rose sector in the study area. Source: [24].

5.2. The K-M Rose LAFS: From Proximities to the Mechanisms of Valorisation of the "Rose" Territorial Resource

According to the results obtained from the actors of the Kalaât M'gouna rose ecosystem, we identified the existence of many links between the different actors of the system, however, these links are not of the same level and vary according to a set of variables such as the objectives of the organisation, the size of the organisation, the entrepreneurial culture of the entrepreneurs, etc. These different links allowed us to raise the existence of eight types of proximity between these actors, as shown in the figure below. These different links have allowed us to raise the existence of eight types of proximity between these actors, as shown in Figure 6 below.

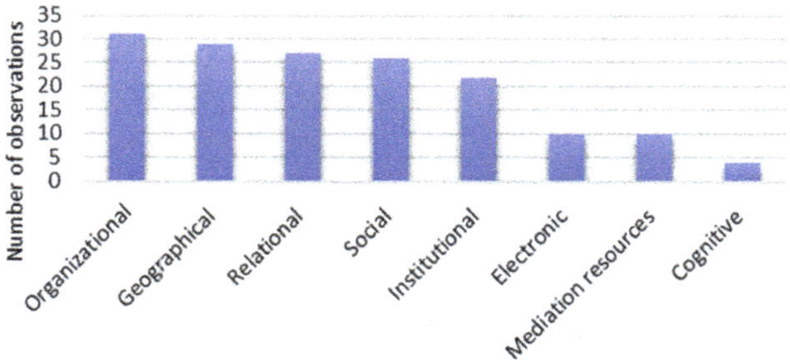

Figure 6. Types and frequencies of proximity identified among the actors of K-M LAFS rose. Source: Our elaboration.

From this Figure 6, it emerges that the links that dominate the interdependencies between the actors of the M'gouna rose sector are essentially geographical (88.6%), organisational (82.9%), social (74.3%), relational (77.1%) and institutional (62.9%). It follows that the territorial actors of the KM rose ecosystem weave their relations on the basis of geographical proximity mainly, as the majority of actors are anchored in the province of Tinghir. Thus, their relations are of an organisational nature as the majority of actors are members of the Moroccan Federation of Rose Professionals (FIMAROSE). The last two types of relations are, on the one hand, social and/or relational, that is to say the family, school and friendship ties that exist between the actors. On the other hand, they are institutional in nature, as the actors obey the same system of norms and values and are governed by the same public institutions.

On the other hand, other types of proximity (cognitive, electronic and mediation resources) do not represent a significant share and remain marginalised. This shows the weakness or absence of cognitive-based relationships and links (joint research projects, training initiatives, etc.) between the actors of the system. It also reflects the lack of use of digital and electronic solutions by the actors to foster their integration into the rose ecosystem.

Additionally, the field survey enabled us to identify the different modes of valorisation of the "rose" resource by the actors of the K-M rose ecosystem. In Figure 7 below, we have summarized the different modes of valorization identified in the field. In fact, we were able to identify five dominant modes of valorisation of the resource; recovery during the cultivation phase (85.7%), recovery during the harvesting phase (91.4%), productive recovery or transformation of the rose into by-products (77.1%), recovery through packaging (54.3%), and commercial recovery or sale (51.4%). Whereas the other modes of valorisation of the resource such as valorisation by innovation, commercialisation, historical and cultural valorisation are very weak or not implemented by the actors of the K-M rose ecosystem.

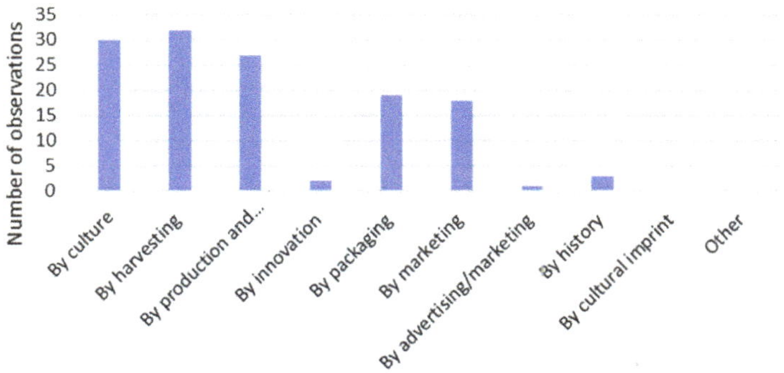

Figure 7. Typology and level of rose resource development processes in the K-M rose LAFS. Source: Our elaboration.

From this figure, it can be seen that the lowest levels in terms of valorisation of the "rose" resource, are those that focus on innovation, advertising or marketing valorisation, and the presence of the historical and cultural touch in the products of the K-M rose ecosystem. This is due to the insufficiency or even the non-existence of local structures responsible for research and development and the training of the actors of the localised agro-food system in the areas of creativity and productive and valorising innovation, new techniques and production processes, as well as new channels and alternatives for the commercialisation of production.

Thus, the presence of these local training structures would have allowed the development of local knowledge of operators, farmers and traders of the perfume rose on the whole territory of the K-M rose ecosystem. Their presence would have made it possible to improve the strategic competences of the cooperatives and companies of the system concerning essentially the mastery of the value chain of rose cultivation to the mastery of the processes of valorisation and marketing of the production, while passing by the good mastery of the processes of production and innovative transformation which goes beyond the simple transformation of the rose into traditional products. However, the absence of training institutions and actors at the local level in the field of the exploitation of the perfume rose, shows the absence of cognitive links between the actors of the rose ecosystem of K-M.

The inexistence of these links between actors (cognitive, knowledge, training and technology) therefore refers to the absence of proximities that relate to these links; in this case, we speak of proximities of cognitive, electronic and mediation resources. Additionally, as a logical consequence of the absence of proximities, we find a great insufficiency of the modes of valorisation of resources based on cognitive, innovative and electronic skills.

After analysing the links of interdependence and the types of proximity that link the actors of the K-M rose ecosystem, and after analysing the different modes of valorisation of this resource among the actors, it is necessary to analyse the correlation or the possible interdependence between these two variables (proximities and modes of valorisation of the resource). In the following Table 3, we show the results of the correlation between the types of proximities and the ways of valuing the pink resource in the LAFS-KM.

From the table that relates the forms of proximity and the logics of valuing the rose resource in the LAFS rose of M'gouna, we see that the significant values are expressed at the level of the following proximities: geographical, organisational, institutional, relational and social, while the significant values that concern the rose valorisation processes are expressed at the level of the following processes: cultivation/production, harvesting, transformation, packaging, marketing and sale. The cognitive, electronic and social proximities, as well as the valorisation processes (innovation, marketing and valorisation through the historical and cultural imprint of the M'gouna territory) express non-significant values.

Table 3. Results of the correlation between types of inter-actor proximities and rose valuation patterns in LAFS-KM.

			The Rose Recovery Process								
			Production	**Harvest**	**Transformation**	**Innovation**	**Packaging**	**Sale**	**Marketing**	**Historical Footprint**	**Cultural Footprint**
Types of proximity	**Geo**	C.P	**0.540** ***	0.114	−0.037	−0.1	0.038	−0.199	−0.011	0.052	−0.175
		Sig.	**0**	0.445	0.805	0.503	0.801	0.18	0.94	0.73	0.238
	Orga	C.P	0.204	**0.763** ***	**0.714** ***	0.09	0.079	0.189	0.042	0.09	0.141
		Sig.	0.169	**0**	**0**	0.547	0.595	0.202	0.78	0.547	0.343
	Relat	C.P	0.183	0.032	0.088	−0.003	−0.184	−0.07	0.237	−0.244	−0.019
		Sig.	0.217	0.829	0.556	0.986	0.215	0.642	0.108	0.098	0.901
	Soc	C.P	0.258	0.194	0.134	−0.256	−0.07	**0.631** ***	−0.154	−0.256	**−0.329** *
		Sig.	0.08	0.192	0.37	0.083	0.642	**0**	0.303	0.083	**0.024**
	Institu	C.P	−0.083	−0.129	−0.129	0.163	**0.326** *	0.123	0.029	0.042	0.253
		Sig.	0.585	0.394	0.394	0.278	**0.025**	0.414	0.849	0.781	0.09
	Elec	C.P	−0.18	−0.151	−0.254	0.129	−0.076	0.171	0.18	−0.03	0.067
		Sig.	0.227	0.312	0.085	0.389	0.61	0.25	0.227	0.839	0.653
	Cog	C.P	0.011	−0.135	0.044	0.004	−0.077	0.074	0.253	−0.144	0.183
		Sig.	0.943	0.365	0.77	0.979	0.607	0.622	0.09	0.333	0.219
	RDM	C.P	**−0.292** *	−0.022	−0.105	−0.245	0.168	−0.259	−0.185	−0.245	−0.285
		Sig.	**0.047**	0.883	0.483	0.097	0.259	0.079	0.213	0.097	0.052

Types of proximity: Cog = Cognitive, Geo = Geographic, Instit = Institutional, Orga = Organisational, Relat = Relational, RDM = Mediation Resources, Soc = Social, Elec = Electronic. C.P = Corrélation of Pearson. Sig = Signification. Source: Our elaboration. *** significant at the 1% level, * significant at the 5% level.

On the other hand, when we asked the ecosystem actors what is missing at the local level or what needs to be improved to better valorize the rose resource at the local, national and international levels, the answers were very different. However, after processing and analyzing the responses, we found that most respondents emphasized the following components: training of actors, technical support, know-how, and cooperation and coordination of actors, as shown below in Table 4. This explains well the important lack of links and proximities between the K-M rose LAFS actors in the areas of knowledge, know-how, new technologies, technical support and cooperation in local collective projects. These ad-vances therefore logically support our analysis.

Table 4. Gaps expressed by the actors of the K-M rose LAFS and their level.

Gaps	Number	Frequency
Coordination of actors	6	9%
Training of actors	9	13%
Governance of the system	3	4%
Technical assistance	10	14%
Financial support	7	10%
The spirit of creativity	5	7%
The customers—the demand	0	0%
Simplification of the export scheme	1	1%
Trust	0	0%
Technology equipment—computer	4	6%
Cooperation of actors	8	12%
Know-how	15	22%
Other	1	1%
Total	**69**	**100%**

Source: our elaboration.

6. Discussion

Indeed, according to the results of our empirical study (notably the correlation table), we note the existence of a sort of homogeneity between the valorisation processes in place or operational and the nature of the relations linking the actors of the local system. We can see that the modes of valorisation that dominate in the M'gouna rose LAFS are consistent with the existing proximities. According to the data in the table, it can be seen that the processes of valorisation of the rose resource effectively identified by the actors in the system (cultivation/production, harvesting, processing, packaging and marketing) can be explained essentially by the existence of five forms of proximity (geographical, organisational, relational, social and institutional).

These proximities are the effective concretisation of the relations that link the actors to each other in this productive territorial configuration. However, we note the insufficiency or even the fragility of the valorisation processes which are based on knowledge, know-how and new technologies, which allows us to understand the insufficiency of the cognitive and electronic links and mediation resources between the actors of the system.

In fact, the processes of valorisation of the rose resource revealed by the actors of the KM rose ecosystem are those attached to the classical operations of the rose value chain in this territory, i.e., production (harvesting), harvesting, processing, packaging, and marketing operations.

They are based on traditional interdependencies governed by physical and geographical links justified by the geographical proximity between the actors in the system, due to their location in a small geographical area (Dades and Kalaât M'gouna). Additionally, the actors have family and social links that allow them to intervene on the quantity, the nature of the by-products, the packaging and the points of sale of the production, etc. These personal and social links have been developed over several decades, since the emergence of this activity in the territory. They are now well maintained thanks to the trust established between the actors.

Consequently, we can say that the logic of proximity expressed in the territory of Kalâat M'gouna allows us to understand the mechanisms of valorisation of the rose resource in which the private actors of the rose ecosystem of K-M participate in this territory. Thus, in the case study, we only find the modes of valorisation that already have an adequate relational base among the territorial actors.

As a result, the modes of valorisation of the "rose" resource that are revealed in M'gouna are based on direct contact between actors, informal cooperation and competition, and membership of common territorial structures that reinforce the social, relational and institutional interdependencies of the actors (in this sense, we are talking about the agricultural cooperatives of the sector operating in the territory). On the other hand, the modes of valorisation based on cognitive, creative and innovative interdependencies are very little present in this ecosystem. These types of proximity are not revealed or established by the actors until now because of the lack of coordination of the actors in terms of designing common research projects, collective training actions, knowledge and technique sharing processes or financing of innovative projects, etc.

Consequently, we can affirm that each mode or process of resource valorisation is in adequacy with a type of proximity implemented by the actors of this local agri-food system. Geographical proximity is correlated with the harvesting process. This is mainly due to the existence of direct links and face-to-face relations between the actors of the system on a limited territory, these direct links favour inter-actor cooperation relations (mainly producers) in terms of harvesting and collection operations, which allows the development of techniques attached to these processes. Consequently, it can be said that the geographical proximity of the actors improves the process of harvesting and collecting the system's production each year.

Additionally, we notice that the organisational proximity is linked to the production process or rose cultivation. This situation shows that the membership of LAFS actors to the agricultural cooperatives and the interprofessional federation of rose producers in the study area makes it possible to improve more and more the techniques and tools for rose cultivation in Kalâat M'gouna. The improvement of cultivation techniques undoubtedly favours the increase in production from one agricultural company to another.

We also note that relational proximity is correlated with the transformation process. This can be explained by the existence of direct relations between LAFS actors, supported by membership of the same territory and common professional organisations, which reinforces the pooling of knowledge, techniques and equipment. As a result, there has been an improvement in the processes of transformation of the rose (LAFS products) into other derived products, such as perfumed water, soaps, creams, etc.

In terms of institutional proximity, it is noted that it is linked to the rose conditioning process (processing, packaging and conditioning). This observation can be explained by the activation of institutional relations between the actors of the system and the public actors in charge of the governance and promotion of the rose sector at local level. The local public actors (the Regional Office for Agricultural Development of Ouarzazate, the Agricultural Development Agency, the Chamber of Agriculture, etc.) are therefore involved, through technical and financial support programmes (training, technical packaging equipment, value-adding units, etc.), in the improvement of operations and packaging techniques for the products of the KM rose LAFS.

As much, we notice that social proximity is attached to the marketing process of the system's products. As a result, it can be said that the marketing operations of the KM rose LAFS products are largely based on the personal and family ties activated between the actors of the system. Actors use their family or acquaintance networks to sell their products mainly locally, but also in other major cities in Morocco (Agadir, Casablanca, Rabat, etc.).

For the other three types of proximity, which have insignificant levels (cognitive, electronic, and mediation resources), we notice that they are linked, although at a mediocre level, with the processes of innovation, marketing and valorisation through the historical or cultural imprint of the territory. Therefore, the weakness of the innovation and marketing processes can be explained by the insufficiency or inactivation of the said proximities (or cognitive). In fact, the innovation or invention of new rose-based products by the actors of the LAFS, which reflects the historical and cultural specificities of the M'gouna territory, as well as the marketing and commercialisation of these wide-ranging products, requires innovation networks, research and development structures and links based on knowledge on the one hand, and on new technologies on the other hand, which is not the case within the framework of the rose LAFS of KM. Consequently, the overall level of resource valorisation by the actors of the ecosystem is linked to the actual level of proximity implemented by the said actors.

Indeed, as the valorisation of territorial resources depends on local actors, their collective actions and their coordination on a given territory, the valorisation of the territorial resource also depends on the proximities set up by the actors and their degree of implementation on the ground. Thus, to improve the level of valorisation of the rose resource by the actors of the K-M rose LAFS, it is judicious to develop and promote the presence of links of research, training, collective learning, collective action, etc., between the actors of this productive territorial organisation. This requires the establishment of other actors at the local level responsible for the training of professionals in the sector, the setting up of joint research projects between actors, the financing of these projects and the coordination of the actors' actions.

Indeed, apart from the efforts invested by GIZ in this field in the province of Tinghir, we do not find any other actor who deals with the connection between rose professionals in M'gouna and their training in topics related to the control of the production chain, stock and quality, etc. Thus, the great lack at the level of this ecosystem is concretised by the absence of links and actions that promote cognitive and electronic proximity between actors. Thus, the great lack at the level of this ecosystem is concretised by the absence of links and actions that favour cognitive and electronic proximity between the actors. This is due to a set of reasons such as:

- The lack of establishments and actors (private or public) in charge of training in the field of the development of the rose sector in the territory (from production, to marketing, through the development of a local industry for the transformation of the rose into innovative by-products: perfumes, cosmetics, food, etc.). Thus, for the other training organisations present in the territory (OFPPT, mainly), they do not offer adequate training to the actors concerning the rose value chain;

- The resistance of a group of private actors in the sector to join the cooperatives, associations and federations of LAFS, given that the objectives of these organisations are to provide professionals in the sector with supervision, monitoring, awareness-raising, training, support in the search for opportunities, etc. However, the low membership rate of the actors in these territorial structures shows that the professionals of the sector are reluctant to carry out collective actions and that their awareness of collective work is not yet developed;
- This can be explained by the low level of computerisation and digitalisation of the sector at the local level, as the production, transformation and marketing processes of the K-M rose LAFS products remain faithful to the traditional approach, following classic methods, which does not favour a better valorisation of this resource with high potential.

7. Conclusions

In conclusion, we can note that the rose sector in the territory of Kalâat M'gouna (the LAFS of K-M rose) is experiencing numerous difficulties on several levels, notably that of the quantitative and qualitative valorisation of this territorial resource, considered as the driving force of the economic activity in this marginalised geographical area. The production of fresh rose and the processes of transformation of the rose into derived products (oils, perfumes, cosmetics, soaps, creams, etc.) still suffer from a set of constraints such as:

- The non-modernisation of the practices of operators, farmers and even local cooperatives in the cultivation and harvesting of their produce;
- Poor control of the processing of fresh rose into by-products, which leads to many failures in the distillation process;
- The absence of structures and actors responsible for training field actors to adopt good operating practices;
- Weaknesses in the processes of valorisation of production, particularly in the cognitive, historical, cultural, commercial and technological fields;
- Weak coordination, cooperation and sharing of knowledge, techniques and tools;
- Very heterogeneous sales prices and many similarities in the packaging and labelling used;
- Non-compliance with texts and regulations on labelling and mandatory inscriptions, etc.

For that, and to restructure the activity at the level of this space and for a better valorisation of the local production in fresh rose, it proves necessary to put an end to the disorganisation of the sector which takes advantage of the normative and legal vacuum, and to set up a legal and normative framework which organises this production that it is at the quantitative or qualitative level. Thus, it is a question of reviving the governance bodies of this activity. We are talking about FIMAROSE—leader of the rose sector—which federates the vast majority of producers and operators of fresh rose in Morocco. Thus, FIMAROSE is called upon to put in place new tools for the coordination of actors, particularly in the cognitive aspect (training, collective research projects, mutualisation of knowledge and techniques, seminars related to the activity, international presence in scientific events related to the fresh rose sector, etc.).

Similarly, other public actors, at the local level, are called upon to participate in this transition towards another mode of management and governance of this sector, through the creation of a local committee encompassing all the institutions at the level of the province of Tinghir, in order to coordinate efforts and implement a program for the development of the sector, in which each organisation presents what it can offer to this activity and to the deadlines, while promoting the cognitive sides. However, the problem of enhancing production in the K-M rose ecosystem is essentially linked to knowledge, new techniques, and adaptation to environmental changes.

Lastly, it is impossible to sustainably develop the territorial resource without the human factor, the heart of any project or system. Additionally, as long as the human factor of the K-M rose ecosystem is not at the level of international good practices in terms of management, exploitation, cultivation and production, this factor is likely to remain an obstacle to the development of the resource in this sector. It is therefore time for the actors in this system, including the governance institutions, to focus on the skills and know-how of the human factor in this system, with a view to creating cognitive, technological and mediation resource proximities between the actors, and to continuously improve the strategic capacities of the actors along the value chain of the "rose" resource. It is a question of going beyond the productive issue at the level of this ecosystem and talking about an innovative local dynamic oriented towards industrialisation and based on sustainability.

Author Contributions: Conceptualization, M.Z., J.A. and M.A.H.; methodology, M.Z., J.A. and M.A.H.; validation, M.Z., J.A. and M.A.H.; formal analysis, M.Z., J.A. and M.A.H.; investigation, M.Z.; resources, M.Z.; data curation, M.Z.; writing—original draft preparation, M.Z., J.A. and M.A.H.; writing—review and editing, M.Z., J.A. and M.A.H.; visualization, M.Z., J.A. and M.A.H.; supervision, M.Z., J.A. and M.A.H.; project administration, M.Z. and M.A.H. All authors have read and agreed to the published version of the manuscript.

Funding: This research received no external funding.

Institutional Review Board Statement: Not applicable.

Informed Consent Statement: Informed consent was obtained from all subjects involved in the study.

Data Availability Statement: The data used in this study come mainly from a field survey (the rose production system in Kalâat M'gouna), through a paper questionnaire distributed to the actors, and also with the help of an interview guide. The collected data are stored in the form of an Excel file, so they are processed using the SPSS software (V 23). we can of course share with you the Excel file containing the data base For the other data (system indicators) we mentioned that they are the data of the Ministry and the Regional Agricultural Development Board. Here are the links: https://www.fellah-trade.com/fr/filiere-vegetale/chiffres-cles-rose-a-parfum?filiere=filiere_vegetale and https://docs.google.com/spreadsheets/d/1U_zC4mZrp-gTiwnZ52qlr9_CmcsBM7JjAKa1Uq_ARW4/edit?usp=sharing (all accessed on 20 October 2022).

Conflicts of Interest: The authors declare no conflict of interest.

References

1. El Chami, D. Towards Sustainable Organic Farming Systems. *Sustainability* **2020**, *12*, 9832. [CrossRef]
2. El Chami, D.; Daccache, A.; El Moujabber, E.M. How Can Sustainable Agriculture Increase Climate Resilience? A Systematic Review. *Sustainability* **2020**, *12*, 3119. [CrossRef]
3. Gilly, J.-P.; Torre, A. (Eds.) *Dynamiques de Proximité*; Harmattan: Paris, France, 2000; p. 306.
4. Pecqueur, B.; Zimmermann, J.-B. (Eds.) *Économie de Proximités*; Hermès Science Publications: Paris, France, 2004.
5. Pecqueur, B.; Zimmermann, J.-B. Les fondements d'une économie de proximités. In *Economie de Proximités*; Hermès Science Publications: Paris, France, 2004; pp. 13–41.
6. Bouba-Olga, O.; Grossetti, M. Socio-économie de proximité. *Rev. D'Écon. Régionale Urbaine* **2008**, *10*, 311–328. [CrossRef]
7. Boschma, R.; Balland, P.A.; de Vaan, M. The formation of economic networks: A proximity approach. In Regional development and proximity relations. *Dév. Régional Relat. Prox.* **2014**, *39*, 61–74.
8. Torre, A. The different approaches to proximity. *Exec. Care* **2019**, *28*, 18–20.
9. Rallet, A.; Torre, A. Proximité et localisation. *Écon. Rural.* **2004**, *280*, 25–41. [CrossRef]
10. Torre, A.; Caron, A. Réflexions sur les dimensions négatives de la proximité: Le cas des conflits d'usage et de voisinage. *Économie Inst.* **2005**, *6–7*, 183–219. [CrossRef]
11. Filippi, M.; Torre, A. *Proximités et Changements Socio-Économiques Dans les Mondes Ruraux*; INRA Editions—Collection Un point sur: Paris, France, 2005. Available online: https://hal.archives-ouvertes.fr/hal-00388487 (accessed on 20 October 2022).
12. Rallet, A. L'économie de proximités: Propos d'étape. *Études Rech. Sys. Agraires Dév.* **2002**, *33*, 11–25.
13. Colletis, G.; Gianfaldoni, P.; Richez-Battesti, N. Économie sociale et solidaire, territoires et proximité. *Rev. Int. Lécon. Soc. Recma* **2005**, *296*, 8–25. [CrossRef]
14. Courlet, C.; Pecqueur, B.; Soulage, B. Industrie et dynamiques de territoires. *Rev. D'écon. Ind.* **1993**, *64*, 7–21. [CrossRef]
15. Loilier, T. Innovation et territoire. Le rôle de la proximité géographique ne doit pas être surestimés. *Rev. Fr. Gest.* **2010**, *36*, 15–35. [CrossRef]
16. Lamara, H. Les deux piliers de la construction territoriale: Coordination des acteurs et ressources territoriales. *Dév. Durable Territ.* **2009**, *13*, 8208. [CrossRef]

17. Tremblay, D.-G.; Fontan, J.-M.; Klein, J.-L.; Rousseau, S. Proximité territoriale et innovation: Une enquête sur la région de Montréal. *Rev. D'Écon. Reg. Urbaine* **2003**, *12*, 835–852. [CrossRef]
18. Gilly, J.-P.; Lung, Y. Proximités, secteurs et territoires. In Proceedings of the 4th Proximity Congress Proximité, Réseaux et Coordination, Marseille, France, 17–18 June 2004; p. 21.
19. Kirat, T.; Lung, Y. *Innovations et proximités: Le territoire, lieu de déploiement des processus d'apprentissage. Coordination économique et apprentissage des firmes*; Economica: Paris, France, 1995; pp. 206–227.
20. Dupuy, S.R. *Le Capital Social: Un Déterminant des Coopérations Inter-Organisationnelles Territorialisées, le Thermalisme dans les Landes*; Université de Pau: Pau, France, 2017.
21. Talbot, D. Les institutions créatrices de proximités: Institutions as creators of proximities. *Rev. D'Écon. Reg. Urbaine* **2008**, *10*, 289–310. [CrossRef]
22. Pecqueur, B.; Zimmermann, J.B. Processus cognitifs et construction des territoires économiques. In *Economie de la Connaissance et Organisations*; L'Harmattan: Paris, France, 1997; 482p.
23. Michon, G.; Aderghal, M.; Berriane, M.; Landel, P.A. La rose du M'Goun. Un bon exemple pour réfléchir les relations entre ancrage, patrimonialisation et banalisation. In *L'émergence des Spécificités Locales dans les Arrières-Pays Méditerranéens*; Les Impromptus du LPED: Marseille, France, 2019; Volume 5, pp. 114–130.
24. ORMVAO. Rapport sur les Filières Agricoles. Available online: http://www.ormva-ouarzazate.ma/ (accessed on 23 November 2019).
25. Faul, F.; Erdfelder, E.; Lang, A.G.; Buchner, A. G* Power 3: A flexible statistical power analysis program for the social, behavioral, and biomedical sciences. *Behav. Res. Methods* **2007**, *39*, 175–191. [CrossRef] [PubMed]
26. Faul, F.; Erdfelder, E.; Buchner, A.; Lang, A.-G. Statistical power analyses using G*Power 3.1: Tests for correlation and regression analyses. *Behav. Res. Methods* **2009**, *41*, 1149–1160. [CrossRef] [PubMed]
27. Gefen, Rigdon, et Straub. Editor's Comments: An Update and Extension to SEM Guidelines for Administrative and Social Science Research. *MIS Q.* **2011**, *35*, iii. [CrossRef]
28. Ministry of Agriculture, Fisheries, Rural Development, Water and Forests (MAFRDWF), Key Figures for Agricultural Sectors, la Rose. Available online: https://www.agriculture.gov.ma/fr/accueil (accessed on 22 November 2019).

sustainability

Article

Food Gap Optimization for Sustainability Concerns, the Case of Egypt

Mohannad Alobid [1], Bilal Derardja [2,3,*] and István Szűcs [1]

[1] Faculty of Economics and Business, Institute of Applied Economic Sciences, University of Debrecen, H-4032 Debrecen, Hungary; mohannad.alobid@econ.unideb.hu (M.A.); szucs.istvan@econ.unideb.hu (I.S.)
[2] Department of Civil, Environmental, Land, Building Engineering and Chemistry, Polytechnic of Bari, 70126 Bari, Italy
[3] Land and Water Resources Management Department, Centre International de Hautes Etudes Agronomiques Méditerranéennes (CIHEAM)—Mediterranean Agronomic Institute of Bari, 70010 Valenzano, Italy
* Correspondence: bilal.derardja@poliba.it

Abstract: Nowadays, even with the growth and progress of the agricultural sector, the food gap (FG) is still wide, particularly for strategic crops, affecting the national economy and compromising the food security. The realization of self-sufficiency can be fulfilled only by achieving the highest production efficiency along with preserving the natural resources currently available, especially arable land and irrigation water. In this analysis, the FG in Egypt was modeled for 13 crops between the years 2000 and 2018. The linear model applied suggested a redistribution of crops in terms of production, food demand and land reallocation, in order to find the best solution to minimize the FG on the basis of crop value and under a set of constraints. It was found that the value of the modelled FG increased steadily from 2005 to 2017, then it started to decline slightly, probably due to the steady increase in the population growth rate which is a crucial factor in enlarging the FG. Furthermore, important water loss was noticed through the analysis period. In fact, there was a huge difference, reaching around 25 billion m^3 between the water consumed for the studied crops and the total amount of renewable water. The main reason for this loss can be linked to the traditional irrigation methods used, such as surface irrigation. Moreover, the calculation of food demand with the estimated production and the redistribution of crop land reallocations were performed to achieve the best model fit between the crops in terms of minimizing the FG in Egypt. So far, the current agricultural policy has reaped limited gains and a steep decline of food economic balance. Hence, significant interest on rising productivity should be given by the government to achieve the food self-sufficiency in Egypt.

Keywords: food gap; self-sufficiency; water consumption; crops land reallocation; food security

check for updates

Citation: Alobid, M.; Derardja, B.; Szűcs, I. Food Gap Optimization for Sustainability Concerns, the Case of Egypt. *Sustainability* **2021**, *13*, 2999. https://doi.org/10.3390/su13052999

Academic Editor: Daniel El Chami

Received: 23 January 2021
Accepted: 1 March 2021
Published: 9 March 2021

1. Introduction

Food is considered the primary necessitate for human needs. This need is based around the work of economic sectors such as agriculture, industry and trade, thus, food security is a significant issue that represents the usefulness and efficiency of sustainable development policies (economically, socially and environmentally), especially in light of the demand increase in food products and the continuous rise in their prices, which is accompanied by the increase in the number of people and their change in consumption patterns [1,2].

Food security is of paramount importance in times of economic, natural and human crisis due to shifts in food supply and demand, scarcity of food, lack of consumer stability and loss of income, all of which are among the factors affecting the essence and content of food security [3,4].

On the other hand, food security is a complex and interrelated issue [5]. It is a holistic issue whose dimensions overlap in most economic, social and environmental sectors, and it is defined as one of the main goals of any development process, as it is reflected in important development goals, related to living the human, health and expected age [6,7]. Consequently, the dimensions of food security are related to the dimensions of economic development (agriculture, industry, trade, prices, etc.), social (poverty and human development, especially the health dimension, etc.), the environmental dimension related to climate and desertification and its impact on agricultural production, and the food security related to population stability and commodity stability [6–8].

According to the World Food Security Summit in 2009, food security is achieved when all people, at all times, have physical, social and economic opportunities [9].

As for food insecurity, it is the situation in which people lack access to adequate quantities of safe and nutritious foods, to ensure a normal and healthy life [9].

Herein, it is necessary to distinguish between two levels of food security (absolute and relative):

- Absolute food security means the production of food within a single country equivalent to or greater than local demand, this level is synonymous with complete self-sufficiency, or complete food security. It is clear that this absolute definition is not realistic, as the state is missing out on the ability to benefit from international trade based on specialization, division of labor and the exploitation of comparative advantages [10,11].
- Concerning relative food security, this means the ability of a country, or a group of countries, to provide commodities and foodstuffs partially or totally. This is in the sense of providing the society's needs of basic food commodities, partially or completely at a minimum, as it does not mean the production of all the basic food needs; rather, it is to provide the necessary materials to secure these food needs, in cooperation with other countries [12,13].

Indeed, the components of food security are present in a country that owns favourable geographic and climatic characteristics, an abundance of water resources, human resources, agricultural lands, pastures, forests, livestock, and a modern technology [14,15].

The reality of the Arab countries requires that they provide food security to their citizens, at least. If we take Egypt, it was once considered the food basket of the Arab world and is a historical model of food self-sufficiency, due to the availability of large agricultural areas and the abundance of labor [16]. Egypt relied on beans and wheat during the historical crises, but today, it has become one of the largest importers of these two commodities [17,18].

The definition of a food gap (FG) is the disproportion between the required food quantities and the population, which leads the concerned country to import food from abroad, thus any shortage of food resources is matched by an increase in the population [19].

Many research papers have been published on the study of the FG and ways to reduce it in Egypt; most of these studies discuss the main crops such as wheat and maize, which are highly consumed in Egypt and their production does not cover the local market [20]. Other studies explain the factors that contribute to widening the FG, for example, the limited investment in agricultural and food projects [21], and the impact of climate change and water scarcity on the FG [22,23].

In the fifties and sixties of the last century, a large number of simulation and optimization models have been used for the appropriate planning and management of water use in irrigated agriculture and FG [24,25]. Soon after the simplex algorithm was found by Dantzig in 1947 [26], agricultural economists started to use linear programming for farm planning. Early publications related linear programming in agriculture as one of the best tools for the optimal allocation of land and water resources [27,28] either aimed at disseminating the mathematical knowledge by explaining the characteristics of the procedure [29,30] or at pointing out its possible applications and general potential for farm management [31,32] and applied linear programming to the hypothetical agricultural holding in order to find optimal production plans by minimizing the FG in a country [33].

In this research, we will calculate and analyze the FG during 2000–2018, as well as the water consumption for the crops studied and food demands for the country. Additionally, we will reallocate the land cultivations under a set of constraints to reduce the FG.

The problem of the study is as follows:

- Egypt as a developing country is facing a crucial challenge in providing the citizens' needs of basic food commodities to keep pace with the steady population growth. To that end, it should work to raise the productivity of the agricultural sector in order to achieve self-sufficiency and to lessen food imports. The FG that affects the staple commodities is a major problem for the state that bears the burden of providing a high budget to stipend the importations usually being paid for with foreign currency. This requires studying FG for the most important crops that represent the biggest portion in the contribution of the national food security in Egypt.

The objectives of the study are to:

1. Determine and calculate the FG during 2000–2018 for the most important food crops in Egypt;
2. Determine the water consumption and the food demand for the same study period;
3. Build a mathematical model to find the optimum land reallocation and production distribution for minimizing the FG.

2. Materials and Methods

The data used for this study describe several variables in relation to thirteen crops for the period between 2000 and 2018. The data were collected and used in all calculations (area, crop yield, crop exports, and crop imports,) from the on-line database of the Food and Agriculture Organization of the United Nations (FAOSTAT), World Bank Open Data, Central Agency for Public Mobilization and Statistics (CAPMAS) of Egypt. In addition, the average world market prices USD/ton was collected from the World Bank Open Data and "Water footprints of nations Volume 2: Appendices" [34].

Regarding the statistical analysis, it was performed by using Stata software. In this part, the linear relationship between the water consumption and the FG will be assessed and modelled for a period of nineteen years (2000–2018).

2.1. Crop Water Requirements

The crop water use for each crop included in the study were calculated with the following equation (Equation (1)):

$$\mathrm{CWU(c)} = \mathrm{CWR(c)} \times \frac{\text{production (c)}}{\text{Yeild (c)}} \quad (1)$$

where:

CWR: the amount of water requirements for each crop (c) measured in the field in m^3/hectare (ha) Table 1. It is defined as the amount of water required for evapotranspiration from the planting until the harvest for a specific crop that grows in soil containing sufficient water for it., The crop evapotranspiration (ET_c) under standard conditions was calculated following the crop evapotranspiration guidelines for computing crop water requirements [35]. No limitations are placed on crop growth or evapotranspiration from soil water and salinity stress, crop density, pests, and diseases, weed infestation or low fertility. ET_c is determined by the crop coefficient approach whereby the effect of the various weather conditions are incorporated into ET_0 and the crop characteristics into the K_c coefficient". The water requirement is assumed to be constant for each crop.

Table 1. Crop water requirements (CWR) for the studied crops.

Crop	Crop Water Requirements CWR m^3/Hectare
Wheat	4912
Maize	8312
Barley	4562
Potatoes	8487
Legumes	18,723
Vegetables	10,481
Fruits	12,446
Sugar beets	8460
Oil seeds	9770
Cotton	9667
Nuts	15,503
Aromatic plants	8412
Rice	10,346

Resource: own calculating.

Production: production of crop (c) measured in tons.

Yield: yield of crop (c) per unit area, measured in tons/ha.

The quantity of water requirements of crop (c) is calculated from the following relationship Equation (2):

$$CWR = 10 \times \sum_{d=1}^{lp} ET_c\,(c, d) \quad (2)$$

where:

The factor 10 is meant to convert mm into m^3/ha, and the summation is carried out over the period from the first to the final day of the growing period.

lp: represents the length of growth, measured in days.

Et_c: is the amount of daily evapotranspiration of the crop (c) and it is measured in mm. This evapotranspiration is obtained by the process of multiplying the reference evapotranspiration amount ET_0 by the coefficient of the crop K_c. The crop coefficient is taken from four stages of the crop growth; initial, crop development, mid-season and late season, that is the stage where the crop is ready for harvest [35,36].

$$ET_c = K_c \times ET_0 \quad (3)$$

ET_c (c) is the crop evapotranspiration [mm d^{-1}].

K_c (c) crop coefficient indicates the relationship between the evapotranspiration of the crop (ET_c) and the reference evapotranspiration (ET_0).

ET_0, the amount of reference evapotranspiration, which is the percentage of evapotranspiration from the grass in specific growth conditions, is affected by climatic conditions only [37] [mm d^{-1}].

The water consumption of crops varies according to their growth stage. In fact, the water consumption is at a low rate at the beginning of the crop growing season, when it is mostly in form of evaporation from the soil surface, then it increases with the plant growth as a result of the leaf mass surface increase and is consumed in the form of transpiration from the leaves up to the maximum growth stages. K_c differs according to the type of the crop, the growth phase, the growing season, and the associated climatic conditions. K_c expresses the effect of the properties that distinguish the field crop from the reference grass, whose appearance is stable and covers the entire ground, and therefore different crops have different K_c.

Several weather condition effects have been incorporated into the ET_0, that then represent an indicator of the atmospheric requirements needed for the process of evapotranspiration from green grass surfaces. Accordingly, the K_c varies greatly with the characteristics of the crop and to a small extent with the climate. Thing that can explain the possible transfer of the K_c values that were calculated at one of the irrigation research stations for generalization between sites and climatic regions by Hargreaves and Merkley (1998) [38] to the ET_0 reference. The ET_0 represents the evaporating potential of the atmosphere at a specific spatio-temporal point and disregards the crop features and soil factors [35].The ET_0 was calculated as an average of all the months of the year through the CROPWAT program [39,40]. As for the K_c values, they were derived from a previous study performed by the researchers Allen et al. (1998) [35] and Champagain and Hoekstra (2004) [41].

2.2. *Describing the Mathematical Model and the Resources Constraints*

Mathematical modelling is an approach allowing one to assess and to understand any system interactions, as well to solve optimization problems along with different constraints. In such problems, the decision maker aims to optimize the solution according to some criteria and limitations.

In other words, a mathematical model can be used for optimizing f(x) under the constraint g(x). Particularly, if f(x) and g(x) are linear functions, the problem will be linear. Correspondingly, the agricultural activity problems are usually evaluated through linear programming [42]. In the present study, the optimization of the system studied seeks to minimize the FG (Equations (4) and (5)) under four different constraints.

Considering that we are assessing several key crops of economic importance, we aim to reduce the FG that is expressed by the disproportion between the required food amounts of the population and the local production, which leads the concerned country to import food from abroad [19]. The FG optimization means the production optimization according to the limited resources and the population need increase. In the model created in the scope of this study, we tried to reorganize the factors that contribute to the widening of the FG, such as the crop area that can be re-allocated on the basis of different aspects in relation to the crop value and the population needs.

The objective of our model is to determine the so-called "objective function", that is a function of unknown crop area reallocation (ha) A_j, which is expressed mathematically as the following:

$$\text{Min}: \sum_{j=1}^{NC} (D_j - S_j) * Pri_j \tag{4}$$

$$\text{Min}: \sum_{j=1}^{NC} (D_j - Pro_j A_j) * Pri_j \tag{5}$$

D_j: The amount of food demand for each crop j (ton). It can be calculated following the relationship Equation (6):

$$D_j = S_j + I_j - E_j \tag{6}$$

I_j, E_j, S_j are, respectively, imports, exports and productions of each crop j (ton).

Pri_j: International crop price (USD per ton) from the World Bank Open Data and [41].

Pro_j: Productivity of each crop j (ton per hectare).

NC: Number of crops.

The minimization of FG is modelled, taking into consideration a set of constraints, that are as follows:

1. The area allocation set for each crop should be positive, this constraint is known as the non-negative variable (Equation (7)):

$$A_j \geq 0 \qquad \forall\ j \tag{7}$$

2. The total crops land allocation should not exceed the maximum exploitable land ($Land_{const}$ = 3.5 million hectares) (Equation (8)):

$$\sum_{j=1}^{NC} A_j \leq Land_{const} \tag{8}$$

3. The total crops water consumption should be less than the renewable water volume ($Water_{const}$ = 45 billion m^3) (Equation (9)):

$$\sum_{j=1}^{NC} CWR_j \times A_j \leq Water_{max} \tag{9}$$

CWR_j: Crop water requirement for each crop j (m^3 per hectare).

4. The production of the allocated area should not exceed the required amount for each crop (the demand) (Equation (10)). Actually, this constraint is necessary, especially for the strategic crops in order to determine the volume of production for each crop which have a high economic return whose production may exceed the market need in Egypt [43]:

$$Pro_j \times A_j \leq D_j \tag{10}$$

As mentioned previously, the problem consists of finding the crops' area reallocation A_j that minimizes the FG per year. We say that a constraint is saturated when the optimal solution uses the total available resources.

The first algorithm used for solving linear programs was presented by Dantzig in 1947. We used the so-called simplex algorithm even though there are many competitive alternative algorithms at the present time. By the means of the programming language named "Octave" and the package "GLPK" (GNU Linear Programming Kit), our linear programming (LP) problem was successfully solved, and the model output expresses the optimum land reallocation for minimizing the FG.

The model was applied on two types of data (average and annual) only for the production and food demand. A temporal evaluation of the model outputs was established to follow the fluctuation of the FG, the water consumption, and the crop land reallocation throughout the study period.

3. Results and Discussion

As it was disclosed previously, our model outputs express the optimal propositions in terms of crop area reallocation, in such a manner to reduce the FG as much as possible depending on several constraints. Mentioning that, the constraint for the total cultivated land was saturated in all the proposed scenarios.

3.1. Crops Land Reallocation for the Studied Crops during 2000–2018

As mentioned previously, the problem consists of setting a new crops' area al-location (Aj) that minimize the FG per year. The area for the whole crops should not exceed the total available area in the country which is equal to 3.5 million hectares. Figure 1 represents the real allocation of all crops through the studied period while Figure 2 illustrates the model output, indeed, all the crops except rice, wheat, potatoes, sugar beet, and maize, are reallocated fairly close to the real situation. Mostly, the crops that are similarly distributed converge to the maximum allowed area (for non-exceeding the crop demand). The excepted crops are critical, in which their allocation is affected by the policy followed by the government as well as other factors, some of which were mentioned heretofore.

As it is known, land allocation refers to assigning land to be used in a certain manner, however, it may not mean that the actual use of the land reflects the initial plan of its allocation [44]. That is what happened with rice, wheat, potatoes, sugar beet and maize crops in Figure 1.

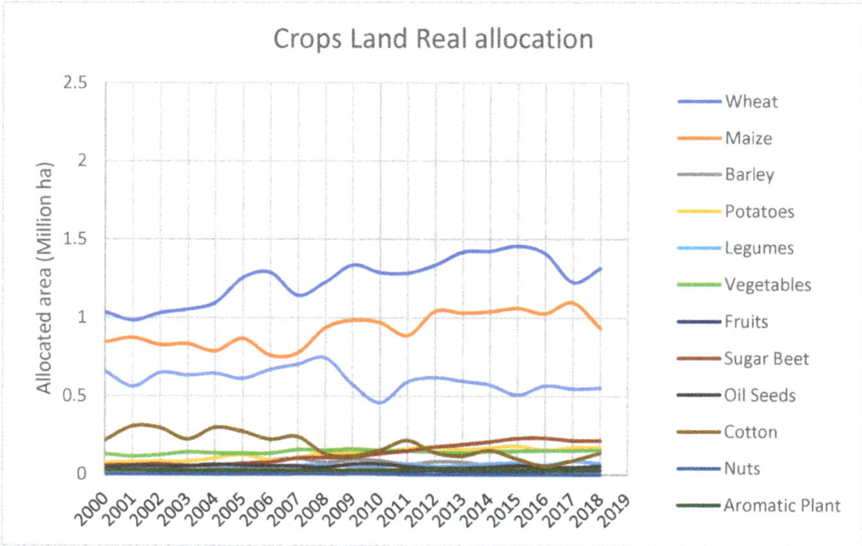

Figure 1. Crops land real allocations for the studied crops.

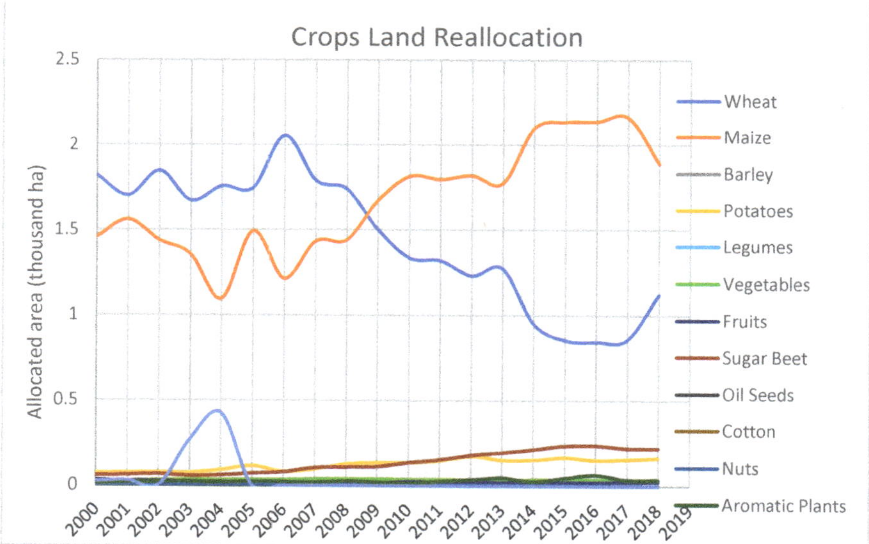

Figure 2. Crops land reallocations for the studied crops. Resource: own calculating.

3.2. Food Gap

From Figure 3, it is clear that the FG has increased with the years, starting from 2005 to 2017. This that can be explained by several reasons in relation to different aspects. In fact, the steady increase in the population growth rate is an important factor amongst others in the widening of the FG, particularly for the period from 2000–2018, during which it had reached 2.56% according to the Central Agency for Public Mobilization and Statistics in Egypt, ensuing an increase in the food demand notably for strategic crops such as wheat, maize and rice. Furthermore, the global financial crisis in 2008 engendered volatile global prices for these strategic crops. Additionally, the technical integration of the production procedures is still low and does not cover the whole cultivated area [45]. Over and above, there are two important aspects of the widening FG in Egypt for these crops:

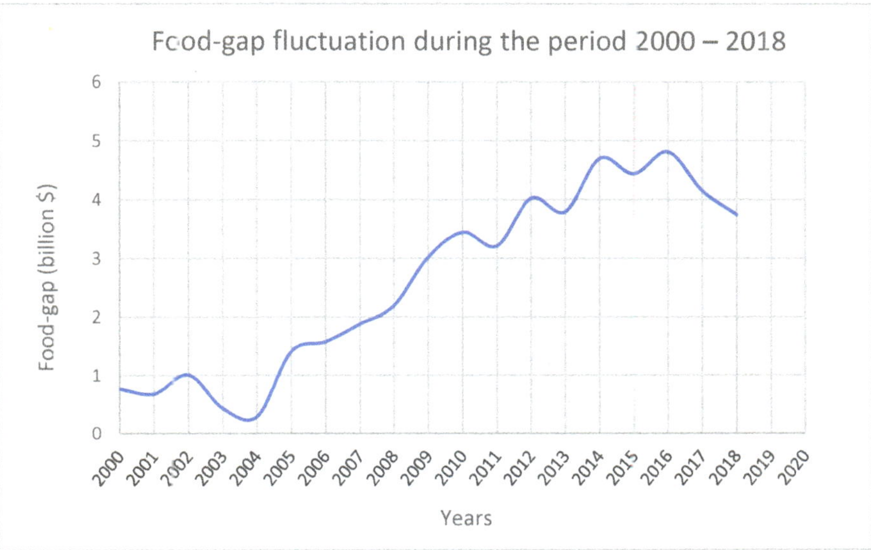

Figure 3. Food gap fluctuation of some important crops for the period between "2000–2018". Resource: own calculating.

Firstly, the supply in the food market (the adopted agricultural policies, the loss of agricultural production and its impact on public consumption, the availability of agricultural production requirements, the agricultural sector's share of investments, environmental problems and climate change, the costs and prices of crop production, and government support provided to farmers) [46,47].

Secondly, demand in the food market (population increase and growth as mentioned above, per capita national income, price policies adopted, customs, traditions and consumption patterns, migration and economic openness of the country) [48,49]. All these reasons have led to the widening of the FG in Egypt recently.

3.3. *The Water Consumption for the Studied Crops*

The mathematical model applied under different constraints gave us an estimation of the water consumption of the studied crops considering their CWR. One of the model constraints is that the water consumption of the crops should not exceed the renewable water resource for the agricultural sector in the country and that is about 45 billion m^3. Through the analysis period, a huge difference, reaching around 25 billion m^3, between the water consumed for the studied crops and the total amount of renewable water ($Water_{const}$) was observed (Figure 4). Even though this water amount difference is supposed to be exploited to irrigate other crops, it is clear that there is an important water loss. This can be explained by several reasons, including the traditional irrigation methods as well as the used technique such as surface irrigation in many nearby areas of the Nile river, where the water use efficiency is very low [50]. In addition, the lack of use of modern technology skills in irrigation and the lack of guidance for the farmers through the establishment of introductory courses that contribute to teaching them some methods about using modern irrigation that reduce the water waste, such as sprinkler and drip irrigation [51]. This is as well as the environmental impact (high temperature, low precipitation), especially for the crops which need high amounts of water.

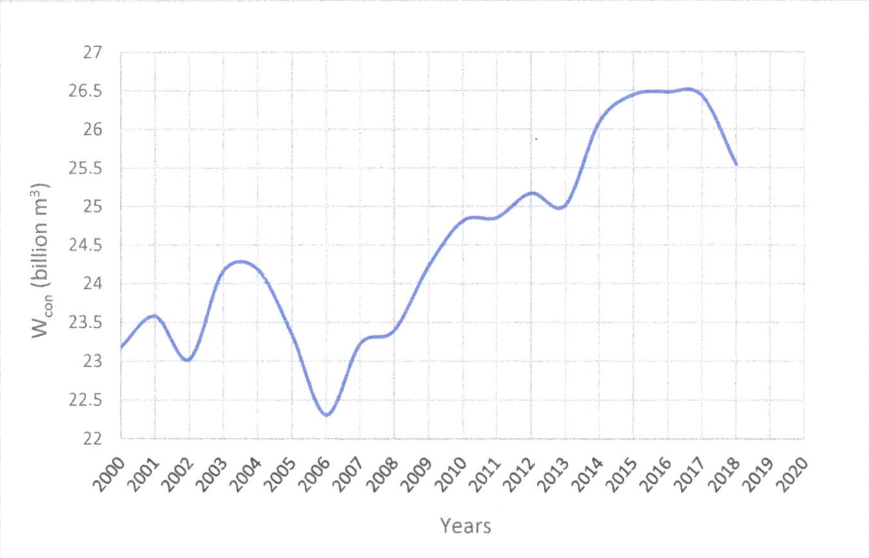

Figure 4. The water consumption for the studied crops. Resource: own calculating.

3.4. *Statistical Analysis for the Food Gap and the Water Consumption*

Figure 5 reveals in a better way the trend between the FG and the water consumptions during 19 years for the studied crops. We noticed that the variation is a direct proportion between water consumption and the FG, and this can also be proved statistically by modelling them through SPSS program using the simple linear regression (LR) model.

- The response variable (dependent variable) is the FG;
- The covariate (independent variable) is the water consumption (Wcon);
- The observations (OBS) from 2000 to 2018.

The question was to check whether the water consumption was a significant variable to explain the FG changes, or not.

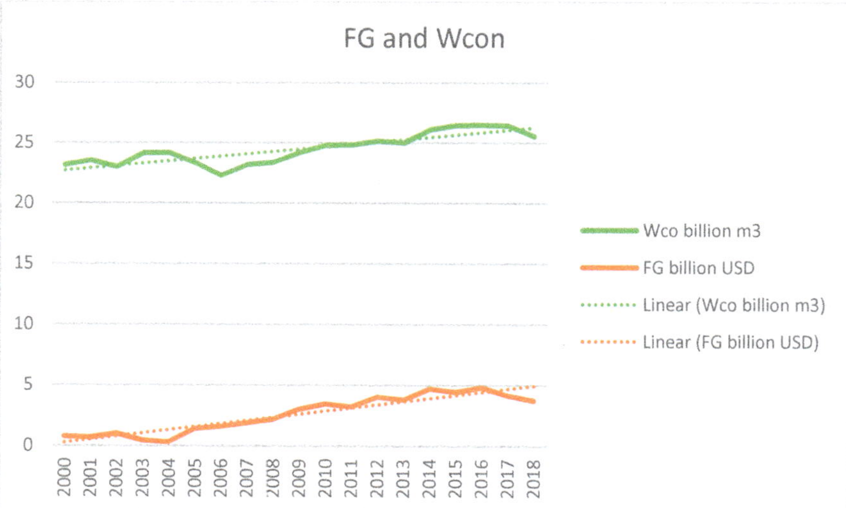

Figure 5. The FG and the water consumption trend for 19 years' period. Resource: own calculating.

Simple linear regression was used in order to determine the influence of the independent variable Wcon on a dependent variable FG. In addition, it was necessary to transform the data by Square Root Transformation (SQRT), since there is a noticeable moderate negative skewness in the studied data [52]. The results are reported in the Table 2 and they can be expressed in the following regression Equation (11):

$$FG = 8986.700 + 26{,}877.114\ Wcon + e \quad (11)$$

The intercept value shows the influence of Wcon on FG. It means that if all the independent variables are zero, FG as the dependent variable is predicted to be 8986.700. Furthermore, the value 26,877.114 represents the coefficient of Wcon, obviously meaning that if Wcon increases by one unit then FG is predicted to increase by 26,877.114.

Table 2. The LR model applied on the studied data.

Variable	B
Constant	8986.700
Wcon	26,877.114

Afterwards, the test of goodness to fit was used Table 3. Based on the analysis, the correlation coefficient (R) is equal to 0.875 indicating the strong relationship of independent variable to the dependent variable. The coefficient of determination (R^2) is equal to 0.766, which means that the independent variable affected the dependent variable with 76.6% in this model.

Table 3. Test of goodness of the LR model applied on the studied data.

R	R^2
0.875 [a]	0.766

[a] Dependent Variable: SQRT(k-FG).

A classical assumption test was needed to ensure that there was no impediment to use the regression analysis, verifying the homoscedasticity, the normality and the autocorrelation.

Figure 6 shows that the pattern of points is spread from both sides of the zero line of the ordinate, revealing that there is no heteroscedasticity in this regression.

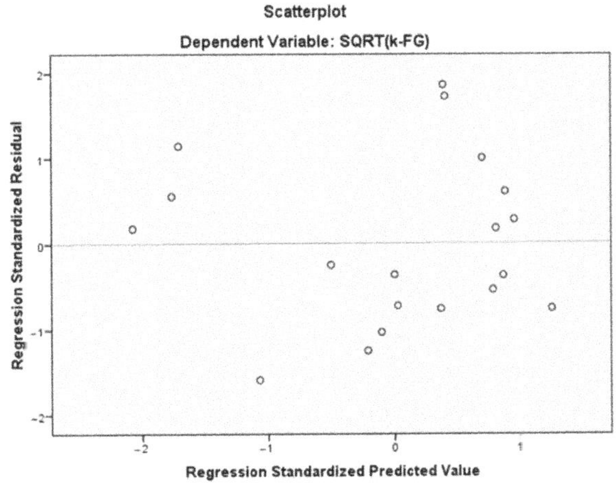

Figure 6. The heteroskedasticity of the residuals resulted from the LR model applied on the studied data.

Later, the normality test was used, and the resulted graph Probability Plot plot (Figure 7) shows that the expected values are strongly correlated to the observed values of the FG.

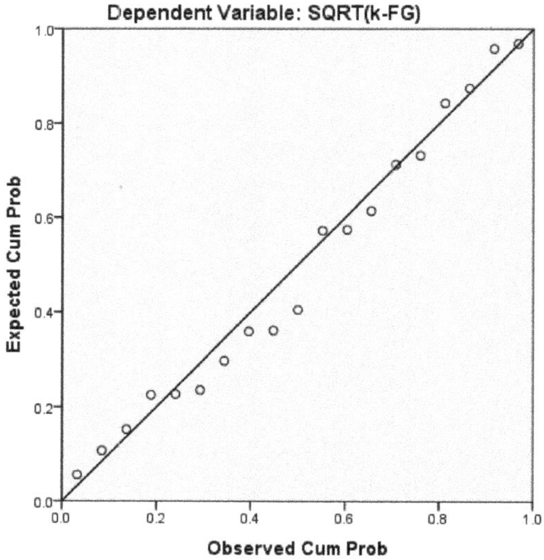

Figure 7. Normal probability plot of regression standardized residual displaying expected vs. observed values of the studied data using LR model.

Additionally, we used the Shapiro–Wilk test since the data observations are less than 50 [53]. In this test, we set the following assumptions:

- If the value of Shapiro–Wilk (sig.) > 0.05, this means that the data are normally distributed;
- If the value of Shapiro–Wilk (sig.) < 0.05, this means that the data are not normally distributed.

Table 4 illustrates the (sig.) values of the Shapiro–Wilk test for both variables, are > 0.05. This means that the data can be said to be normally distributed even though the data are at risk of a lower bound of true significance because of its small size.

Table 4. Tests of normality resulted from the LR model on the studied data.

	Kolmogorov–Smirnov [a]			Shapiro–Wilk		
	Statistic	Df	Sig.	Statistic	Df	Sig.
FG	0.135	19	0.200 *	0.916	19	0.094
Wcon	0.131	19	0.200 *	0.940	19	0.269

* This is a lower bound of the true significance; [a] Lilliefors significance correction; Df signifies degree of freedom.

Later on, the autocorrelation test (the Durbin–Watson (DW)) was used to determine the correlation of variables in the regression model along the time.

Based on the Table 5, we can see that the DW value is equal to 1.915, which means there are no autocorrelation symptoms in the current model.

Table 5. Durbin–Watson test for autocorrelation of the residuals resulted from the LR model on the studied data.

Model	R	R Square	Adjusted R Square	Std. Error of the Estimate	Durbin-Watson
1	0.875 [a]	0.766	0.753	9380.63939	1.915

[a] Predictors: (Constant), SQRT(k-Wcon).

Finally, the hypothesis test is intended to determine the effect of Wcon as the independent variable to the FG as the dependent variable. The T_{test} is used to determine the partial effect of the independent variable on the dependent variable. This test is carried out by comparing the T_{count} with the T_{table} with the level of significance of 95% ($\alpha = 0.05$). The assumptions of this test are as follows:

- If the $T_{count} > T_{table}$ or sig. (p-value) < 0.05, then the variable Wcon influences FG;
- and if the $T_{count} < T_{table}$ or sig. (p-value) > 0.05, then the variable Wcon does not influence FG.

Table 6 displays that the T_{count} is 7.468. The T_{table} is 1.729 ($n = 19$, $p = 0.05$); the result is $T_{count} > T_{table}$, and the sig. (p-value) is also < 0.05. It means that the variable Wcon is significantly influencing the FG variation.

Table 6. T_{test} applied on the studied data.

Model	T	Sig.
Wcon	7.468	0.000

3.5. Average Demand and Production of the Crops during the Period 2000–2018

The calculation of the crop demand allowed us to re-estimate the productions with the aim to achieve the best model fit between crops in terms of minimizing the FG in Egypt. This analysis was performed by taking into consideration the annual average of the variables used during the 19-year period (2000–2018). Our model recommended the best solution in comparison to the real situation of the crops production. Figure 8 illustrates an analogy between the calculated average demand and the recommended average production of the studied crops during the 19-years period. For many crops such as maize, potatoes, sugar beet, legumes, vegetables, fruit, nuts, and aromatic plants, we notice that the production is equal to the demand, which reflects the saturation of the constraint (the maximum production should not exceed the demand for each crop for the reason mentioned above). For the wheat, barley, rice and cotton crops, their demands are higher than their productions. This can be elucidated by the fact that the above-mentioned crops (especially cotton and wheat) are more valuable in Egypt, thus their price is higher as well. Consequently, the government revenues benefit from purchasing these valuable crops from the farmers at reasonable prices and then exporting them at high prices besides importing other varieties with lower prices and in larger quantities, in order to achieve the economic balance in the country [54].

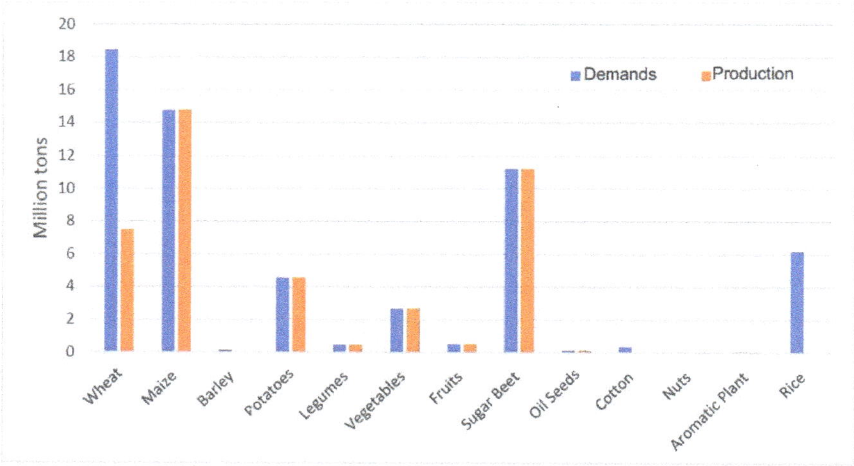

Figure 8. The average demand and the average recommended production of the crops. Resource: own calculating.

3.6. *The Deficit between the Crop Demands and Production of the Crops from 2000–2018*

With regard to the mathematical model used, the main goal is to reduce the FG in the country as much as possible for thirteen important crops. After calculating the annual productions and demands redistribution for minimizing the FG, the new resulted annual production amounts were subtracted from the new annual demands, to assess the deficit produced from the suggested redistribution carried out by our linear model on the basis of the crops' priority. As the production for each crop should not exceed the demand, we have determined the term deficit as the difference between the demand and the production. In the best case, the deficit is equal to zero, which means a self-sufficiency was reached for that crop. With regard to the four crops mentioned above, the results were not equal to zero. This result can be related to several factors, among which are the preferences of the farmers in terms of choosing these crops over the years, and the political strategy issues related to the economic balance and the laws applied on exportations and importations of the country (mentioned above). To achieve the self-sufficiency in the agricultural sector at the long term, considering only some strategic crops in this study, we can say that the suggestions provided by our mathematical model can be considered as ideal solutions if applied appropriately, regardless the political conditions of the country. Indeed, Figure 9 illustrates the deficit between the crops demand and the production from the year 2000 to 2018. For all the crops, the deficit in the suggested model resulted in zero, except for the wheat, barley, cotton, and rice crops.

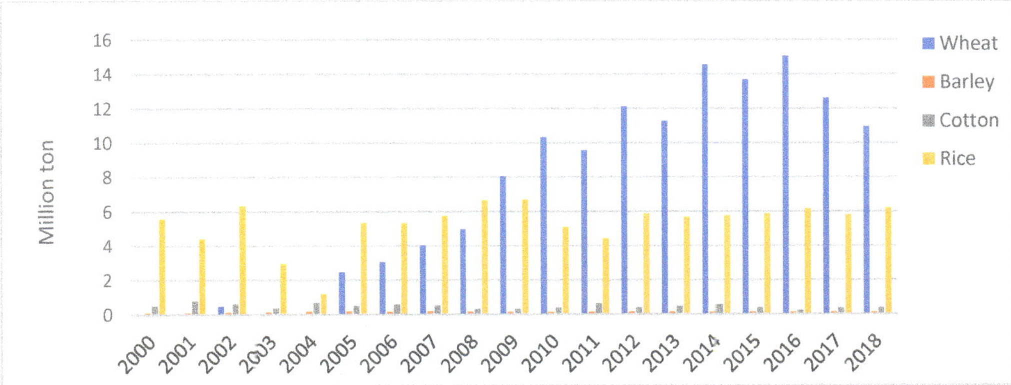

Figure 9. The deficit between the crops' demand and production of the crops from 2000–2018 for the wheat, barley, cotton, and rice crops. Resource: Own calculating

4. Conclusions and Recommendations

In this study, we have concluded important results that may help the government and/or the decision-makers to improve the reality of agricultural production and farmers.

By calculating the FG from 2000–2018, we noticed an increased with the years, starting from 2005 to 2017. Maybe the essential reason was the increase in the population growth rate that has led to an increase in food demands.

Through the estimation of water consumption and the comparison with renewable water resources for the agricultural sector in the country, an immense difference, reaching around 25 billion m^3, between the water consumed for the studied crops and the total amount of renewable water was detected. Perhaps the reasons for this may be in the traditional irrigation methods used to irrigate crops, where water losses are large and also have environmental causes.

We also noticed that the FG and water consumption are positively correlated over the years and that was proved statistically. The model was quite good in predicting the FG variation, nevertheless, it needs to be improved by incorporating other significant covariates adding fair contribution to recompense the error term. Likewise, the available data were restricted to a 19-year duration, which gives a quite small number of observations needed for a better statistical analysis.

The main objective of the study was to identify the size of the FG in Egypt to help in reducing it. So, by calculating the crop demands and re-estimating the productions, it will help in achieving the best model fit between crops in terms of minimizing the FG in Egypt. For many crops such as maize, potatoes, sugar beet, legumes, vegetables, fruit, nuts, and aromatic plants, we noticed that the production is equal to the demand, which reflects the saturation of the constraint of maximum production which should not exceed the demand for each crop. For the wheat, barley, rice and cotton crops, their demands are higher than their productions.

Regarding the crops land reallocation and by comparing with the real allocation, it was found that all the crops except rice, wheat, potatoes, sugar beet, and maize, are reallocated relatively close to the real situation.

Among the research recommendations are the following:

- Study the effect of annual international crop prices on the crops needed and the FG to find the most appropriate way of importing crops, especially strategic ones.
- The possibility of developing a mathematical model to redistribute the optimum yield through fixing precise constraints and assumptions so that the number of possibilities that give better results increases.

- Vertical expansion (increase in hectare productivity) through new varieties that are resistant and/or tolerant to environmental conditions, for example.
- Horizontal expansion (increasing the cultivated area outside the Nile Valley). This comes about by confronting the problem of water scarcity in these lands by introducing strains that tolerate drought and water stress.
- The government should give more importance to solving FG issues and to increasing the production by supporting the farmers as well as using efficient irrigation techniques to reduce water use.

Author Contributions: Conceptualization, M.A. and B.D.; methodology, M.A. and B.D.; software, B.D. and M.A.; validation, M.A.; formal analysis, M.A. and B.D.; investigation, I.S. and B.D.; resources, M.A.; data curation, M.A.; writing—original draft preparation, M.A.; writing—review and editing, M.A. and B.D.; visualization, I.S. and B.D.; supervision, I.S. All authors have read and agreed to the published version of the manuscript.

Funding: This publication was supported by the construction EFOP-3.6.3-VEKOP-16–2017-00007 ("Young researchers from talented students—Supporting scientific career in research activities in higher education"). The project was supported by the European Union, co-financed by the European Social Fund.

Institutional Review Board Statement: Not applicable.

Informed Consent Statement: Not applicable.

Data Availability Statement: Data sharing not applicable.

Conflicts of Interest: The authors declare no conflict of interest.

References

1. Godfray, H.C.J.; Beddington, J.R.; Crute, I.R.; Haddad, L.; Lawrence, D.; Muir, J.F.; Pretty, J.; Robinson, S.; Thomas, S.M.; Toulmin, C. Food security: The challenge of feeding 9 billion people. *Science* **2010**, *327*, 812–818. [CrossRef] [PubMed]
2. Godfray, H.C.J.; Garnett, T. Food security and sustainable intensification. *Philos. Trans. R. Soc. B Biol. Sci.* **2014**, *369*, 20120273. [CrossRef]
3. Mc Carthy, U.; Uysal, I.; Badia-Melis, R.; Mercier, S.; O'Donnell, C.; Ktenioudaki, A. Global food security–Issues, challenges and technological solutions. *Trends Food Sci. Technol.* **2018**, *77*, 11–200. [CrossRef]
4. Timmer, C.P. *Food Security and Scarcity: Why Ending Hunger Is so Hard*; University of Pennsylvania Press: Philadelphia, PA, USA, 2015.
5. Clover, J. Food security in sub-Saharan Africa. *Afr. Secur. Stud.* **2003**, *12*, 5–15. [CrossRef]
6. Lang, T.; Barling, D. Food security and food sustainability: Reformulating the debate. *Geogr. J.* **2012**, *178*, 313–326. [CrossRef]
7. Costanza, R.; Fisher, B.; Ali, S.; Beer, C.; Bond, L.; Boumans, R.; Danigelis, N.L.; Dickinson, J.; Elliott, C.; Farley, J. Quality of life: An approach integrating opportunities, human needs, and subjective well-being. *Ecol. Econ.* **2007**, *61*, 267–276. [CrossRef]
8. Bircher, J.; Kuruvilla, S. Defining health by addressing individual, social, and environmental determinants: New opportunities for health care and public health. *J. Public Health Policy* **2014**, *35*, 363–386. [CrossRef]
9. Grainger, M. World Summit on Food Security (UN FAO, Rome, 16–18 November 2009). *Dev. Pract.* **2010**, *20*, 740–742. [CrossRef]
10. Babu, S.; Gajanan, S.N.; Sanyal, P. *Food Security, Poverty and Nutrition Policy Analysis: Statistical Methods and Applications*; Academic Press: Cambridge, MA, USA, 2014.
11. Maxwell, S.; Smith, M. Household food security: A conceptual review. *Househ. Food Secur. ConceptsIndic. Meas.* **1992**, *1*, 1–72.
12. Eide, A.; Oshaug, A.; Eide, W.B. The Food Security and the Right to Food in International Law and Development. *Transnat'l L. Contemp. Probs.* **1991**, *1*, 415.
13. Ehrlich, P.R.; Ehrlich, A.H.; Daily, G.C. Food security, population and environment. *Popul. Dev. Rev.* **1993**, *19*, 1–32. [CrossRef]
14. Ringler, C.; Biswas, A.K.; Cline, S. *Global Change: Impacts on Water and Food Security*; Springer: Berlin/Heidelberg, Germany, 2010.
15. Capone, R.; Bilali, H.E.; Debs, P.; Cardone, G.; Driouech, N. Food system sustainability and food security: Connecting the dots. *J. Food Secur.* **2014**, *2*, 13–22.
16. Lacirignola, C.; Adinolfi, F.; Capitanio, F. Food security in the Mediterranean countries. *New Medit* **2015**, *14*, 2–10.
17. Woertz, E. Agriculture and Development in the Wake of the Arab Spring. In *Combining Economic and Political Development*; Brill Nijhoff: Boston, MA, USA, 2017; pp. 144–169.
18. Asseng, S.; Kheir, A.M.; Kassie, B.T.; Hoogenboom, G.; Abdelaal, A.I.; Haman, D.Z.; Ruane, A.C. Can Egypt become self-sufficient in wheat? *Environ. Res. Lett.* **2018**, *13*, 094012. [CrossRef]
19. Conway, G. *The Doubly Green Revolution: Food for all in the Twenty-First Century*; Cornell University Press: Ithaca, NY, USA, 1998.

20. Ouda, S.A.; Zohry, A.E.-H.; Alkitkat, H.; Morsy, M.; Sayad, T.; Kamel, A. *Future of Food Gaps in Egypt: Obstacles and Opportunities*; Springer: Berlin/Heidelberg, Germany, 2017.
21. Tuttle, J.N. *Private-Sector Engagement in Food Security and Agricultural Development*; Center for Strategic and International Studies: Washington, DC, USA, 2012.
22. Abdelkader, A.; Elshorbagy, A.; Tuninetti, M.; Laio, F.; Ridolfi, L.; Fahmy, H.; Hoekstra, A. National water, food, and trade modeling framework: The case of Egypt. *Sci. Total Environ.* **2018**, *639*, 485–496. [CrossRef]
23. ElMassah, S. Would climate change affect the imports of cereals? The case of Egypt. *Handb. Clim. Chang. Adapt. DOI* **2013**, *10*, 978–999.
24. Patel, J.N.; Bhavsar, P.N. Optimal distribution of water resources for long-term agricultural sustainability and maximization recompense from agriculture. *ISH J. Hydraul. Eng.* **2021**, *27*, 110–116. [CrossRef]
25. Shenava, N.; Shourian, M. Optimal Reservoir Operation with Water Supply Enhancement and Flood Mitigation Objectives Using an Optimization-Simulation Approach. *Water Resour. Manag.* **2018**, *32*, 4393–4407. [CrossRef]
26. Dantzig, G.B. *Linear Programming and Extensions. Princeton Landmarks in Mathematics and Physics*; Princeton University Press Princeton: Princeton, NJ, USA, 1963.
27. Smith, D.V. Systems analysis and irrigation planning. *J. Irrig. Drain. Div.* **1973**, *99*, 89–107. [CrossRef]
28. Afshar, A.; Mariño, M.A. Optimization models for wastewater reuse in irrigation. *J. Irrig. Drain. Eng.* **1989**, *115*, 185–202. [CrossRef]
29. Boles, J.N. Linear programming and farm management analysis. *J. Farm Econ.* **1955**, *37*, 1–24. [CrossRef]
30. Heady, E.O. Simplified presentation and logical aspects of linear programming technique. *J. Farm Econ.* **1954**, *36*, 1035–1048. [CrossRef]
31. McCorkle, C.O. Linear programming as a tool in farm management analysis. *J. Farm Econ.* **1955**, *37*, 1222–1235. [CrossRef]
32. Swanson, E.R. Programmed solutions to practical farm problems. *J. Farm Econ.* **1961**, *43*, 386–392. [CrossRef]
33. Zgajnar, J.; Erjavec, E.; Kavcic, S. Optimisation of production activities on individual agricultural holdings in the frame of different direct payments options. *Acta Agric. Slov.* **2007**, *90*, 45–56.
34. Chapagain, A.; Hoekstra, A. Water Footprints of Nations.Volume 2: Appendices in Value of Water Research Report Series. 2004. Available online: https://waterfootprint.org/media/downloads/Report16Vol2_1.pdf (accessed on 8 March 2021).
35. Allen, R.G.; Pereira, L.S.; Raes, D.; Smith, M. *Crop Evapotranspiration-Guidelines for Computing Crop Water Requirements-FAO Irrigation and Drainage Paper 56*; FAO: Rome, Italy, 1998.
36. Droogers, P.; Allen, R.G. Estimating reference evapotranspiration under inaccurate data conditions. *Irrig. Drain. Syst.* **2002**, *16*, 33–45. [CrossRef]
37. Hofwegan, P.V. Virtual Water- Conscious Choices. In Proceedings of the 4th World Water Forum, E-conference Synthesis, 22 April 2004. Available online: https://www.worldwatercouncil.org/sites/default/files/Thematics/virtual_water_final_synthesis.pdf (accessed on 8 March 2021).
38. Hargreaves, G.H.; Merkley, G.P. *Irrigation Fundamentals: An Applied Technology Text for Teaching Irrigation at the Intermediate Level*; Water Resources Publication: CO, USA, 1998.
39. Valiantzas, J.D. Simplified forms for the standardized FAO-56 Penman–Monteith reference evapotranspiration using limited weather data. *J. Hydrol.* **2013**, *505*, 13–23. [CrossRef]
40. Van der Gulik, T.; Nyvall, J. *Crop Coefficients for Use in Irrigation Scheduling*; Ministry of Agriculture, Food and Fisheries of British Columbia: Victoria, BC, Canada, 2001.
41. Chapagain, A.K.; Hoekstra, A.Y. *Water Footprints of Nations*; UNESCO-IFE: Delft, The Netherlands, 2004.
42. Boussard, J.-M.; Daudin, J.-J. *La Programmation Linéaire Dans les Modèles de Production*; Elsevier Mason SAS: Amsterdam, The Netherlands, 1988.
43. Moghazy, N.H.; Kaluarachchi, J.J. Sustainable Agriculture Development in the Western Desert of Egypt: A Case Study on Crop Production, Profit, and Uncertainty in the Siwa Region. *Sustainability* **2020**, *12*, 6568. [CrossRef]
44. Yao, J.; Zhang, X.; Murray, A.T. Spatial optimization for land-use allocation: Accounting for sustainability concerns. *Int. Reg. Sci. Rev.* **2018**, *41*, 579–600. [CrossRef]
45. Ghonem, M. *Egypt: Review of the Agrifood Cooperative Sector*; Country highlights FAO Investment Centre: Rome, Italy, 2019.
46. Tellioglu, I.; Konandreas, P. *Agricultural Policies, Trade and Sustainable Development in Egypt*; ICTSD and FAO: Geneva, Switzerland, 2017.
47. Soliman, I.; Fabiosa, J.F.; Bassiony, H. *A Review of Agricultural Policy Evolution, Agricultural Data Sources, and Food Supply and Demand Studies in Egypt*; Center for Agricultural and Rural Development (CARD) at Iowa State University: Iowa, IA, USA, 2010.
48. Goueli, A.; El Miniawy, A. Food and agricultural policies in Egypt. *Options Méditerranéennes Série Cah.* **1994**, *7*, 7–68.
49. Nin-Pratt, A.; El-Enbaby, H.; Figueroa, J.L.; ElDidi, H.; Breisinger, C. *Agriculture and Economic Transformation in the Middle East and North Africa: A Review of the Past with Lessons for the Future*; FAO and IFPRI: Rome, Italy, 2018.
50. Amer, M.H.; Abd El Hafez, S.A.; Abd El Ghany, M.B. *Water Saving In Irrigated Agriculture in Egypt*; LAP LAMBERT Academic Publishing: Saarbrücken, Germany, 2017.
51. Omran, E.-S.E.; Negm, A.M. *Technological and Modern Irrigation Environment in Egypt: Best Management Practices & Evaluation*; Springer Nature: Berlin/Heidelberg, Germany, 2020.
52. Garson, G.D. *Testing Statistical Assumptions*; Statistical Associates Publishing: Asheboro, NC, USA, 2012.

53. Yap, B.W.; Sim, C.H. Comparisons of various types of normality tests. *J. Stat. Comput. Simul.* **2011**, *81*, 2141–2155. [CrossRef]
54. Perrihan, A.-R. *How to Feed Egypt*; The Cairo Review of Global Affairs: Cairo, Epypt, 2013.

sustainability

MDPI

Article

Effect of Drying Methods on Chemical Profile of Chamomile (*Matricaria chamomilla* L.) Flowers

Teuta Benković-Lačić [1,*], Iva Orehovec [2], Krunoslav Mirosavljević [1], Robert Benković [1], Sanja Ćavar Zeljković [3,4], Nikola Štefelová [3], Petr Tarkowski [3,4] and Branka Salopek-Sondi [2]

[1] Biotechnical Department, University of Slavonski Brod, 35000 Slavonski Brod, Croatia; kmirosavljevic@unisb.hr (K.M.); rbenkovic@unisb.hr (R.B.)
[2] Department for Molecular Biology, Ruđer Bošković Institute, 10000 Zagreb, Croatia; iva.orehovec@irb.hr (I.O.); branka.salopek.sondi@irb.hr (B.S.-S.)
[3] Czech Advanced Technology and Research Institute, Palacky University, 78371 Olomouc, Czech Republic; sanja.cavar@upol.cz (S.Ć.Z.); nikola.stefelova@upol.cz (N.Š.); petr.tarkowski@upol.cz (P.T.)
[4] Centre of the Region Haná for Biotechnological and Agricultural Research, Department of Genetic Resources for Vegetables, Medicinal and Special Plants, Crop Research Institute, 78371 Olomouc, Czech Republic
* Correspondence: tblacic@unisb.hr

Abstract: Chamomile (*Matricaria chamomilla* L.) is used in the food industry, stomatology, pharmacy, and medicine due to the beneficial properties of chamomile flowers, which are due to the content of terpenoids, but also flavonoids and phenolic acids. This study aims to determine and compare the effects of the drying method on the metabolic profile of chamomile flowers from sustainable, organic practice. The flowers were dried using four different methods: in the sun at a temperature of around 30 °C for 4 days, in the shade at an average temperature of 20–25 °C for 7 days, in a dryer at a temperature of 105 °C for 24 h, and in a climate chamber at a temperature of 60 °C for 48 h. The drying method affects the color, aroma, dry biomass, and chemical profile of chamomile flowers. The biggest color change was between fresh chamomile flowers and chamomile flowers dried in a climate chamber at 105 °C for 24 h, and the smallest change was observed in flowers dried in the sun. The highest contents of polyphenolic compounds and antioxidant activity were measured in flower samples dried in the sun. Drying the flowers at 105 °C caused a significant decrease in total phenols and total flavonoids compared to the drying methods in the sun and shade. Drying at 60 °C for two days had the most significant negative effect on polyphenolic compounds. GC-MS analysis of chamomile essential oil revealed a total of 49 compounds. The most abundant compounds in all samples were α-bisabolol oxide A (19.6 to 24.3%), bisabolol oxide B (19.3 to 23.2%), and β-farnesene E (15.9 to 25.5%). β-Farnesene was identified in significantly lower amounts in sun-dried flowers compared to others, indicating its sensitivity to high light intensity. Volatile compounds spiroether Z, spiroether E, and matricarin were significantly reduced in samples dried at a temperature of 105 °C compared to others, which agrees with the aroma of dried flowers. Discrimination between samples based on chemical profiles showed similarity between samples dried in the sun and in the shade compared to samples dried at higher temperatures.

Keywords: antioxidant activity; chamomile; drying; essential oils; phenolic compounds; volatile compounds

Citation: Benković-Lačić, T.; Orehovec, I.; Mirosavljević, K.; Benković, R.; Ćavar Zeljković, S.; Štefelová, N.; Tarkowski, P.; Salopek-Sondi, B. Effect of Drying Methods on Chemical Profile of Chamomile (*Matricaria chamomilla* L.) Flowers. *Sustainability* **2023**, *15*, 15373. https://doi.org/10.3390/su152115373

Academic Editor: Surendra Singh Bargali

Received: 12 September 2023
Revised: 11 October 2023
Accepted: 18 October 2023
Published: 27 October 2023

1. Introduction

Chamomile (*Matricaria chamomilla* L.) is a plant from the Asteraceae family, genus Matricaria, characterized by five species widespread throughout Europe, central and southwestern Asia, northern Africa, western North America, and Macaronesia [1]. The stem of the plant is erect, hollow, and grows up to 60 cm in height. The leaves are located alternately, pinnately divided, and the flower heads grow individually on the stems. They are made up of bisexual yellow tubular flowers in the center and white petals on the edge. They bloom from May until the beginning of autumn. They are aromatic and have a pleasant smell. The fruit is a light-brown achene.

Chamomile (*Matricaria chamomilla* L.) is a well-known annual plant that has been used worldwide for thousands of years [2] in the food industry, stomatology, medicine—including dermatology, gastroenterology, otolaryngology, pulmonology, internal medicine, radiotherapy, and pediatrics [3]—pharmaceutical, and cosmetic industries due to its proven beneficial properties, such as antimicrobial [4], anti-inflammatory [5], analgesic, sedative, and antiseptic properties [6], as well as anti-diarrheal, antioxidant [7], anticancer [8], and more. Chamomile possess benefits for human health and is traditionally used to treat ulcers, wounds, eczema, skin irritation, rash, gout, bruises, cracked nipples, mastitis, burns, ear and eye infections, neuralgia, conjunctivitis, rheumatic pain, hemorrhoids, smooth muscle relaxation, and other conditions [9–12]. The beneficial effects of dried chamomile flowers are also reflected in the relief of cramps and other inflammatory conditions of the digestive tract and menstrual cramps, and they can also be used as a mild sleep aid [13]. Most of the mentioned beneficial properties of chamomile flowers are due to the content of terpenoids and phenols [14], of which flavonoids and phenolic acids are the most important [2]. More than 120 secondary metabolites have been identified in chamomile flowers, including 36 flavonoids, 28 terpenoids, and others. Previously, it has been reported that chemical components, such as α-bisabolol and cyclic ethers, have significant antimicrobial properties; umbelliferone is fungistatic, while chamazulene and α-bisabolol are potent antiseptics [15]. Most of these compounds are available to the human body after the extraction process from plant tissues, e.g., extraction with solvents, hot water, alkalis, etc., but this may lead to degradation of the bioactive constituents. Methods for extraction with organic solvents, such as ethanol, acetone, and methanol, with different volume fractions of water are much more suitable for obtaining the higher yields of these chemical constituents. Apart from the type of farming system, the main characteristics of dried flowers are largely influenced by the drying conditions and methods [16]. Drying is one of the oldest methods of preserving food and plants, which aims to reduce the moisture content without affecting the quality of the raw material and can be performed by different methods, such as drying in the sun, in the shade, in conventional ovens, and in climate chambers; spray drying; and freeze drying, and each of them has its advantages and disadvantages [17]. The most widely used method for drying aromatic and medicinal herbs is shade or sun drying, but in this case, it is difficult to control factors such as temperature and humidity, and the possibility of contamination of the dried material is greater [18]. There are scientific papers in the literature dealing with the methods of drying chamomile flowers and dry extracts, and they continue to be researched to reliably investigate the efficacy of the different drying and extraction methods since the bioactive compounds contained in chamomile flowers are often sensitive and present in low concentrations [19–21]. Borsato et al. [22] found that chamomile flowers dried at 95 °C changed from an attractive yellow to an unattractive brown caramel color. Many authors have noted that the chemical composition of flowers and essential oil of aromatic plants varies depending on the drying method, thus recognizing the need for uniform results [23,24].

Agricultural systems that avoid the use of synthetic chemical pesticides, fertilizers, growth regulators, and other harmful chemicals are the basis of organic agriculture [25]. Organic agriculture preserves biological diversity, promotes soil fertility, and adapts production methods to tradition and locality, while growing diverse crops that tolerate growing conditions without chemicals. Sustainable agriculture, which can be described in many ways, includes ecological, biological, biodynamic, bioecological, regenerative, organic, conservation agriculture [26]. It can also be defined as an agricultural practice that does not affect the chemical pollution of agricultural resources, such as soil and water, increasing and maintaining biodiversity and ensuring food safety. Sustainable cultivation has an impact on environmental protection and promotes rural development [27]. Chamomile (*Matricaria chamomilla* L.) is an extremely grateful plant for cultivation according to ecological principles and sustainable agricultural practices, whose principles we adhered to during research. Modern technological methods, which include climate chambers and various ovens, and traditional drying methods, such as sun and shade drying, can be counted among sustainable, green drying methods because they do not use non-renewable energy sources, but continue to have different effects on the extraction of secondary plant compounds, especially polyphenolic compounds. In addition to the drying method, the extraction of chemical compounds in essential oils is also affected by the type of solvent. Therefore, this study aims to determine and compare the effects of the drying method on the metabolic profile of chamomile (*Matricaria chamomilla* L.) flowers from sustainable, organic seeds and cultivation in the region of Slavonia, the eastern part of Croatia.

2. Materials and Methods

2.1 Plant Growth and Drying Methodology

The study was conducted in Brod-Posavina County, Slavonia region, in the eastern part of Croatia, about 8 km west of Slavonski Brod, on the experimental field "Slobodnica" of the University of Slavonski Brod (45°09′58.5″ N and 17°56′52.9″ E) at 87 m altitude. The climate of Brod-Posavina County is a temperate continental climate with warm summers and moderately cold winters. The experimental fields are located in an area where the monthly temperature is above 10 °C for more than four months, the average temperature in the warmest month is below 22 °C, and the average annual precipitation is 700–800 mm. The area is characterized by winter temperatures that can fall below 0 °C and summer temperatures that can rise above 30 °C [28] and average annual humidity of 72%. The area along the Sava River is predominantly characterized by alluvial amphigley soils that are moistened by soil and surface water. The experimental fields are located in an area where excessive wetting by surface water occasionally occurs—pseudogley [29].

Organically grown seeds of *Matricaria chamomilla* L. were purchased from domestic organic production by the company Espresso d.o.o. from Lužani, Brod-Posavina County. The seeds were sown by hand on 28 October 2021, on a 0.6-hectare experimental field, and the sowing rate was 8 kg per ha. Seeds started to germinate after 7 days, and the plant density was 34 plant/m^2. An organic cultivation system was used, and the flowers (Figure 1) were harvested by hand on 15 May 2022 and immediately taken to the laboratory of the Biotechnical Department of the University of Slavonski Brod, where the samples were dried using various drying methods.

Figure 1. Chamomile plants in the flowering stage in the experimental field of the University of Slavonski Brod, Croatia.

The flower samples consisted of 250 g in four replicates and were randomly selected for each drying method. Four types of drying of the plant material were performed:

- drying in the sun (temperature around 30 °C for 4 days)—SUN;
- drying in the shade (average temperature of 20–25 °C for 7 days)—SH;
- drying in the dryer (Memmert, model UNE 200) (temperature of 105 °C for 24 h)—D;
- drying in the climatic chamber (Memmert, model ICH L eco) (temperature of 60 °C for 48 h)—KK60.

The average relative humidity during the drying of chamomile flowers was 64%, while the average temperature was 18.7 °C, according to the state hydrometeorological institute. Dried samples were used for further analysis.

2.2. Polyphenols and Antioxidant Activity Measurements

For the measurement of polyphenolic compounds and antioxidant activities, 60 mg of dried tissue was extracted in 2 mL of 80% methanol. Tissue homogenization was performed by a Mixer Mill MM 400 (Retsch, Haan, Germany) for 2 min at 30 Hz, after which the extracts were mixed in a tube rotator for 30 min at 15 rpm. The extracts were then centrifuged for 10 min at 13,000 rpm (Eppendorf centrifuge), and the supernatants were used for spectroscopy measurements on a spectrofluorometer microplate reader (Tecan Infinite M200, Männedorf, Switzerland). Antioxidative activities were measured by using a UV/VIS spectrophotometer (BioSpec-1601 E, Shimadzu, Kyoto, Japan).

Total phenolics (TP) were determined following the Folin–Ciocalteu method, as reported by Singleton and Rossi [30]. TP data were expressed as equivalents of gallic acid per dry weight (mg GAE g^{-1} dw). Total phenolic acids (TPA) were measured using Arnow's reagent, according to the European Pharmacopoeia (2004) [31], and the results were presented as equivalents of caffeic acid per dry weight (mg CAE g^{-1} dw). Total flavonoids (TFL) were measured using the $AlCl_3$ method [32], and results were presented as equivalents of catechin per dry weight (mg CE g^{-1} dw). Antioxidant activity was measured by the DPPH radical scavenging capacity assay, according to Brand-Williams et al. [33]. The obtained results of the DPPH assay were expressed as µmol Trolox equivalents per gram dry weight (µmol TE g^{-1} dw).

2.3. Gas Chromatography–Mass Spectrometry (GC/MS) Analysis of Essential Oils and n-Hexane Chamomile Extracts

Essential oils were obtained by 1 h hydrodistillation in a Clevenger apparatus of 8–10 g of homogenized dried plant material. In detail, finely powdered plant material was placed in a round-bottom flask, and 500 mL of water was added. Heat was applied to the flask, causing the water to boil and produce steam, which carried the volatile essential oils along with it. After 1 h, the essential oil was then separated from the aqueous layer and transferred into the vial and kept at 4 °C until analysis [34].

For the preparation of chamomile extracts in *n*-hexane, homogenized dried plant material (100 mg) was extracted with 1800 µL *n*-hexane, containing 0.001% *n*-tridecane as internal standard, and then sonicated for 10 min in an ultrasonic bath. After centrifugation at 14,500× *g*, 1000 µL of supernatant was transferred into a new vial for analysis. Each extract was isolated in triplicate.

Volatile constituents were determined by GC/MS using an Agilent 7890A gas chromatograph fitted with a fused silica HP-5MS UI capillary column (30 m × 0.25 mm, 0.25 µm film thickness), coupled to an Agilent 5975C mass selective detector.

GC/MS analysis and the identification of the volatile constituents were conducted following the method reported by Cavar Zeljkovic et al. [35], but with some modifications in the operating conditions as follows: inlet pressure 9.35 psi, injector temperature 250 °C, detector temperature 280 °C, and split ratio for essential oils 1:9.

2.4. Statistics

Statistical analysis was performed in RStudio (R Software version 4.1.0) using packages compositions, agricolae, ggplot2, pls.

Multiple one-way ANOVAs followed by Duncan's tests for multiple comparisons (on data in logit-scale) were performed to evaluate the impact of drying on the essential oils and the hexane extracts. The effect of drying was further examined and visualized via compositional PLS-DA biplots (taking the type of drying as the response and expressing the essential oils and hexane extracts compositions in clr coefficients).

3. Results and Discussion

3.1. Dried Sample Characteristics

The dried samples differed visually in color, depending on the drying methods used. As can be seen in Figure 2, the samples dried in the sun (SUN) and shade (SH) were brightly colored and resembled the fresh material. Their aromas were easily recognizable and intense. The samples dried at 105 °C (D) and 60 °C (KK60) lost their natural color and became brownish, especially the D samples. D samples also lost their characteristic chamomile aroma, while KK60 samples were still recognizable, although the intensity of the aroma was lower compared to air-dried samples. Borsato et al. (2005) [22] previously reported that chamomile flowers dried at 95 °C changed from an attractive yellow to an unattractive brown caramel color. We noticed that when drying at higher temperatures, the chamomile flower changed color from a pleasant yellow (SUN) to an unattractive brown (D) (Figure 2).

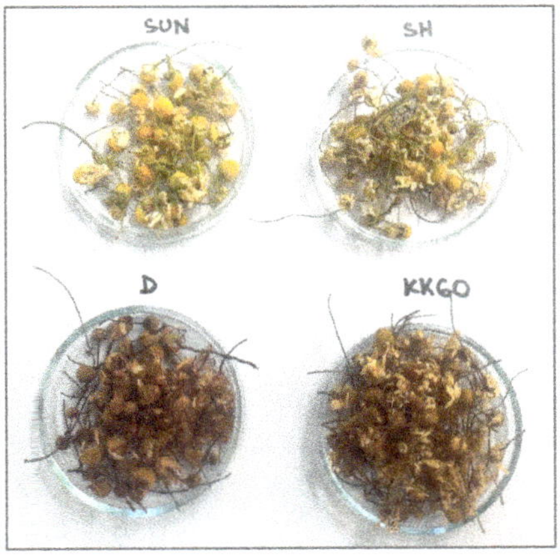

Figure 2. Samples of chamomile flowers dried in the sun for 4 days (SUN), in shade for 7 days (SH), in drier at 105 °C for 24 h (D), and in climatic chamber at 60 °C for 2 days (KK60).

Chamomile flowers were dried to their constant weight by various drying methods. The size of the sample was 250 g. After drying, the weight of chamomile flowers ranged from 42.03 g (D) to 56.97 g (SUN). The average weight of four repetitions in the SUN treatment was 54.95 g, the SH treatment was 52.09 g, the D treatment was 44.27 g, and the KK60 treatment was 52.42 g (Figure 3).

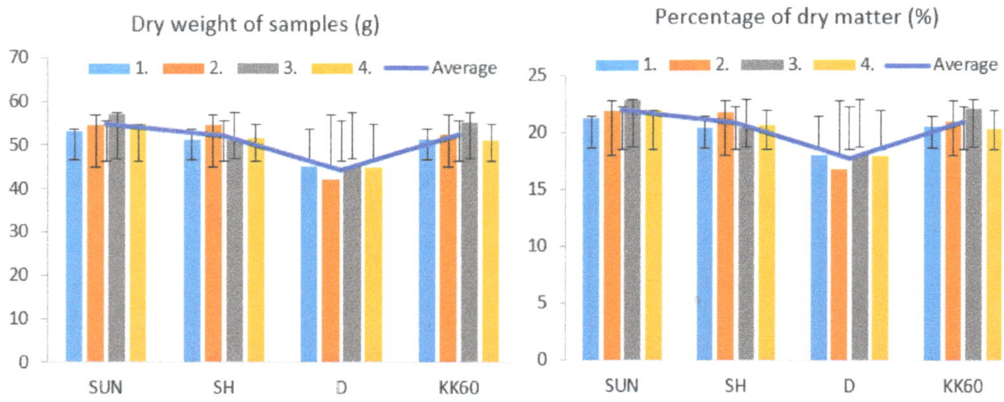

Figure 3. Dry weight and percentage of dry matter of flower samples (g) after sun drying for 4 days (SUN), shade drying for 7 days (SH), drying in the dryer at 105 °C for 24 h (D), and drying in the climatic chamber at 60 °C for 48 h (KK60). Values are an average of 4 replicates ± SD.

Dried samples had an average moisture content between 16.81 and 22.79%. The lowest moisture content of 16.81% was observed in the D treatment and the highest moisture content of 22.79% in the SUN treatment. The average moisture content of four repetitions in the SUN treatment was 21.98%, in the SH treatment was 20.84%, in the D treatment was 17.71% g, and in the KK60 treatment was 20.96% (Figure 3). Similar results were reported by Abbas et al. (2021) [19]. It is important to note that this moisture content in the samples is not suitable for storing and selling the product.

3.2. Polyphenols and Antioxidant Activity

Total polyphenols (TP), total flavonoids (TFL), total phenolic acids (TPA), and antioxidant activity measured by the DPPH assay were evaluated using spectroscopic methods and are presented in Figure 4. As can be seen, the highest contents of polyphenolic compounds and antioxidant activity were measured in flower samples dried in the air exposed to the sun (SUN); TP content was 32.02 mg GAE g^{-1} dw, TFL content was 12.36 mg CE g^{-1} dw, TPA content was 15.14 mg CAE g^{-1} dw, and antioxidant activity was 0.66 mg TE g^{-1} dw. Drying the flowers in the shade (SH) did not influence significantly TP, TFL, TPA, or antioxidant activity, compared to SUN samples. Short drying of flowers at 105 °C caused a significant decrease of TP and TFL (23.62 mg GAE g^{-1} dw and 7.51 mg CE g^{-1} dw, respectively) compared to the SUN and SH drying methods. There was a trend of a slight decrease of TPA and antioxidant activity in flowers dried at 105 °C compared to air-dried samples (SUN and SH), although it was not statistically significant. The most prominent negative effect on polyphenolic compounds in chamomile flowers was caused by drying at 60 °C for two days (KK60); TP, TFL, and TPA contents were decreased by 41.3%, 65.2%, and 44.2%, respectively, and antioxidant activity was reduced 39.4%, compared to SUN samples. Different methods of drying chamomile flowers and their influence on the content of chemical components have been investigated by other authors. Harbourne et al. (2009) [36] determined a statistically significant decrease in the phenol concentration in samples dried at higher temperatures (80 °C). Based on our results, short drying at 105 °C caused fewer negative changes in polyphenol content and antioxidative activity compared to the prolonged drying at 60 °C.

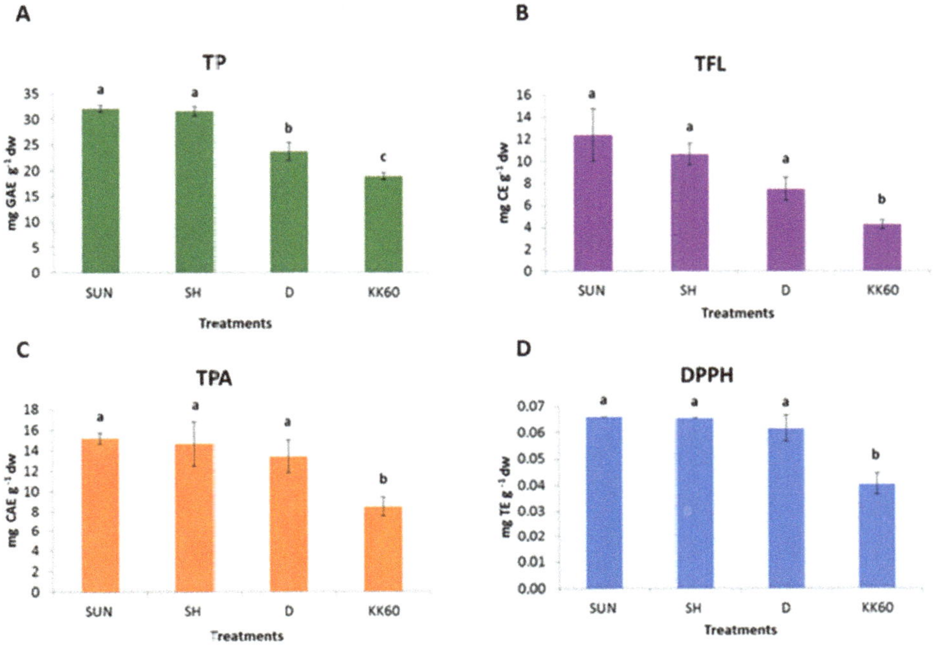

Figure 4. Polyphenolic compounds: (**A**) total polyphenols (TP), (**B**) total flavonoids (TFL), (**C**) total phenolic acids (TPA), and (**D**) antioxidant activity measured by DPPH assay in chamomile flowers dried in air exposed to direct sun for 4 days (SUN), dried in air exposed to shade (SH), dried in the drier at 105 °C for 24 h (D), and dried in the climate chamber at 60 °C for 2 days (KK60). Data are average, n = 4 ± SD. Different letters denote significant differences in compound level between treatments ($p < 0.05$, one-way ANOVA, Duncan's test).

3.3. Essential Oils

GC-MS analysis of chamomile essential oil identified 49 compounds in total (Figure 5). Salamon et al. (2023) [37] recorded between 23 and 43 chemical components in the essential oil extracted from chamomile flowers in different sites in Albania, and Aćimović et al. (2021) [38] between 47 and 57 chemical components in essential oil. The most abundant compounds in all samples are α-bisabolol oxide A (ABOLA) (19.6 to 24.3%), bisabolol oxide B (BIOB) (19.3 to 23.2%), and β-farnesene E (BFAR) (15.9 to 25.5%). These several essential oils chemotypes are known as antimicrobial, spasmolytic, antiepileptic, mitogenic, and insecticidal mediums [39]. The highest level of ABOLA was found in KK-60, then SUN, D, and SH, while the highest level of BIOB was found in the SUN sample, then in D, SH, and KK-60. Hajaji et al. (2018) [40] obtained results that showed that α-bisabolol (ABOL) can activate the programmed process of cell death in the promastigote stage of the parasite. BFAR was the highest in flowers dried at high temperatures for 24 h, while it was the lowest in samples exposed to the sun. Chamazulene (CHAM) was the highest in sun-dried samples (SUN sample, 4.3%), and it was significantly decreased upon high-temperature drying (D sample, 1.5%). Chamazulene is a well-known constituent of chamomile extracts and may be in part responsible for the chamomile aroma, which was well preserved in sun- and shade-dried samples compared to other drying treatments. Ghasemi et al. (2016) [41] determined that the content of chamazulene and α-bisabolol in chamomile oil is influenced by environmental conditions and the genetic background. We demonstrated that their content is also influenced by drying methods, which is in agreement with data obtained by Abbas et al. (2021) [19]. On the other hand, GMUU was low in air-dried samples, SUN and SH (2.2 and 2%, respectively), and significantly more abundant in the D sample (4.1%).

Low-abundant compounds were grouped, and they contributed 5.1 to 7.1% of the mixture. The terpenes composition of the essential oil is dependent on the stage of maturity and the cultivation system [42], as well as the chamomile cultivars [43].

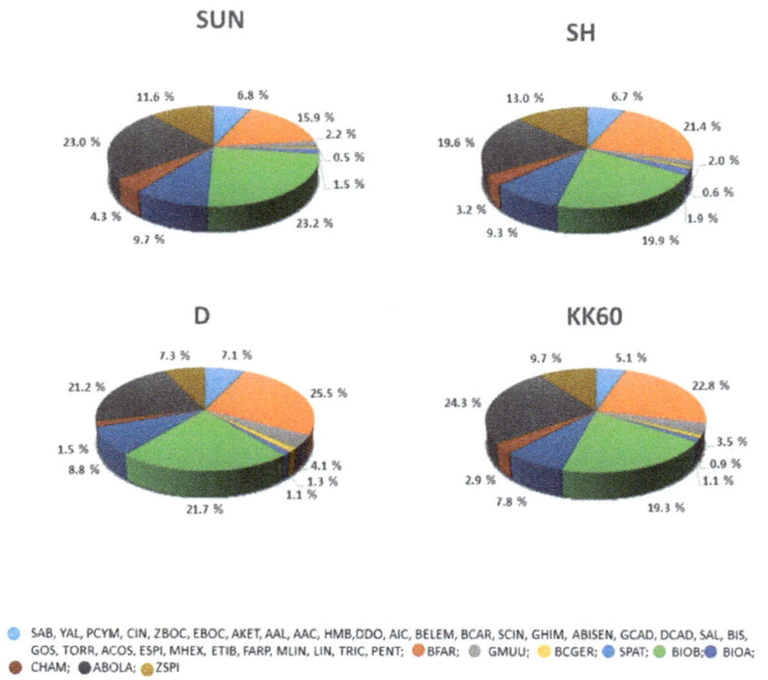

Figure 5. Abundance of essential oil compounds in chamomile flowers dried in air exposed to direct sun for 4 days (SUN), in air exposed to shade (SH), dried in a drier at 105 °C for 24 h (D), and dried in a climate chamber at 60 °C for 2 days (KK60). Bicyclogermacrene (BCGER), Bisaboladien-4-ol-2,7Z (BIS), α-Bisabolene Z (ABISEN), α-Bisabolol (ABOL), α-Bisabolol oxide A (ABOLA), epi-α-Bisabolol (EPIAB), Bisabolone oxide A (BIOA), Bisabolol oxide B (BIOB), γ-Cadinene (GCAD), δ-Cadinene (DCAD), Chamazulene (CHAM), α-Costol (ACOS), Dauca-5,8-diene (DAU), β-Farnesene E (BFAR), (2E,6E)-Farnesyl pentanoate (FARP), Farnesol 2Z,6Z (FAR), Gossonorol (GOS), γ-Himachalene (GHIM), Linoleic acid (LIN), Methyl linoleate (MLIN), α-Muurolene (AMUU), γ-Muurolene (GMUU), Methyl hexadecanoate (MHEX), Pentacosane (PENT), Salvial-4(14)-en-1-one (SAL), dehidro-Sesquicineole (SCIN), Spathulenol (SPAT), Spiroether Z (ZSPI), Spiroether E (ESPI), (E)-Tibetin spiroether (ETIB), n-Tricosane (TRIC), (E)-Tibetin spiroether (ETIB), and Torreyol (TORR) (See Table S1).

3.4. *Volatile Compounds Extracted by n-Hexane*

GC-MS analysis identified 15 volatile compounds extracted by n-hexane (Figure 6).

Some of the compounds have been already identified in the essential oil reported above, like ABOLA, BFAR, ZSPI, etc. The drying methodology affected the level of certain compounds. The most abundant volatile compound found in chamomile flowers was Spiroether Z (ZSPI). Similar results were obtained by Aćimović et al. (2021) [38]. ZSPI did not change significantly in SUN, SH, and KK60 samples (34–36%), but decreased significantly after flower drying at high temperature (D sample, to 28%). Spiroether E (ESPI) and Matricarin (MATR) were also significantly decreased in D samples compared to SUN, SH, and KK60. The amounts of volatile compounds agree with the aroma of dried samples. Spiroether E and Spiroether Z from the essential oil of chamomile showed specific repression in the production of alphatoxin AFG (1) by Aspergillus parasiticus [44]. On the other hand, Tridecane (TRID) and Pentacosane (PENT) were significantly increased in flowers dried at high temperatures compared to other samples. β-Farnesene E (BFAR) was identified in significantly lower amounts in SUN-dried flowers compared to others, suggesting its sensitivity to high light intensity. BFAR compound is a pheromone for some insects and may have importance in some forms of ecological pest control [37]. α-Bisabolol oxide A (ABOLA) and α-Bisabolol oxide B (ABOLB) were highest in samples dried at higher temperatures: SUN samples and SUN and D samples, respectively. Bicyclogermacrene (BCGER), Spathulenol (SPAT), α-Bisabolol (ABOL), (E)-Tibetin spiroether (ETIB), and *n*-Tricosane (TRIC) were identified in low amounts in all samples (below 1% in mixture). α-Bisabolol was the highest in the SUN treatment but without statistically significant differences compared to others.

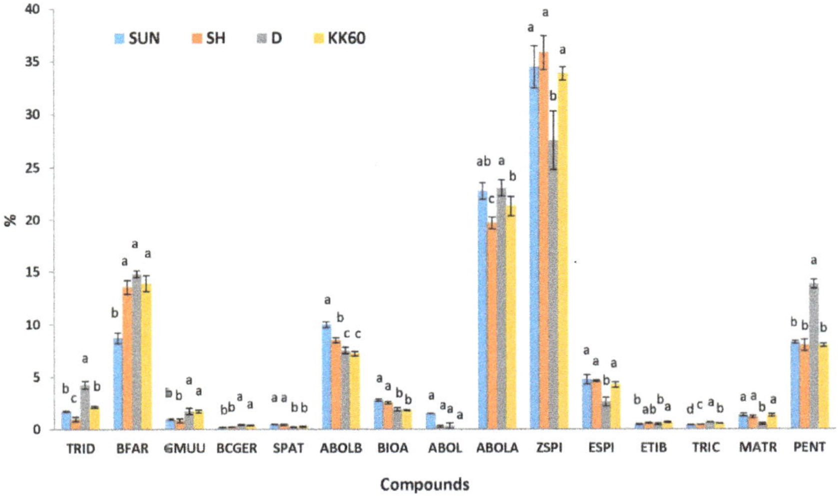

Figure 6. Volatile compounds extracted by *n*-hexane and identified by GC-MS and presented as % in mixture: Tridecane (TRID), β-Farnesene E (BFAR), γ-Muurolene (GMUU), Bicyclogermacrene (BCGER), Spathulenol (SPAT), α-Bisabolol (ABOL), α-Bisabolol oxide A (ABOLA), α-Bisabolol oxide B (ABOLB), α-Bisabolone oxide A (BIOA), Spiroether Z (ZSPI), Spiroether E (ESPI), (E)-Tibetin spiroether (ETIB), *n*-Tricosane (TRIC), Matricarin (MATR), Pentacosane (PENT). Data are average, n = 4 ± SD. Different letters denote significant differences in compound level between treatments ($p < 0.05$, one-way ANOVA, Duncan's test).

3.5 Relationship between Drying Methods and Metabolomics Profile

Our findings show that the drying methodology influences the metabolomics profile of chamomile flowers. PLS-DA biplots (Figures 7 and 8) provide a closer look at these relationships. The effect of drying technology on polyphenolic compounds and antioxidant activity is visualized in the PLS-DA biplot shown in Figure 7. The first two PLS components capture 97.92% of the variability in the analyzed traits (the vast majority, 93.76%, being explained by the first PLS component). It can be seen that higher content of polyphenolic and antioxidant activity was measured in the SUN and SH samples, while lower content was observed in the KK60 samples. Azizi et al. (2009) [45] determined that the maximum content of chamomile essential oil was obtained when drying at lower temperatures and in shaded areas.

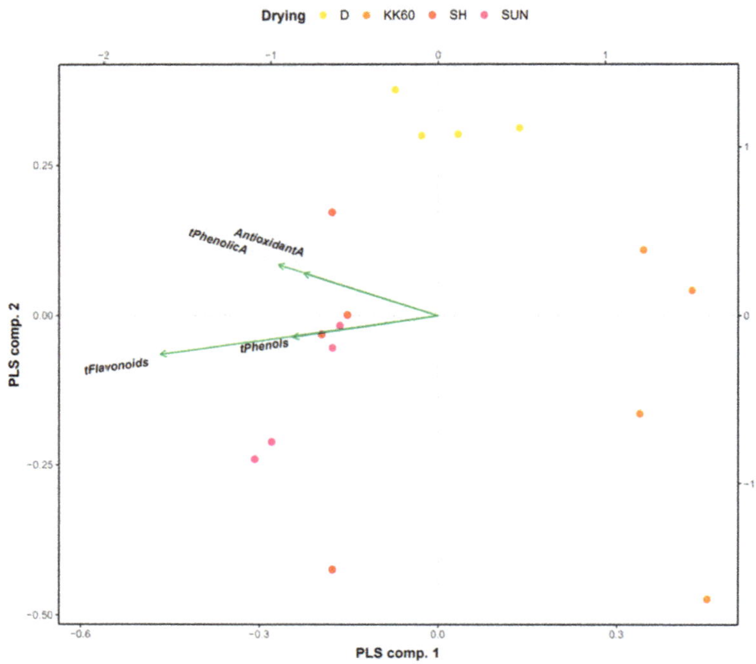

Figure 7. PLS-DA biplot showing the relationship between drying methodology and polyphenolic compounds, as well as antioxidant activity.

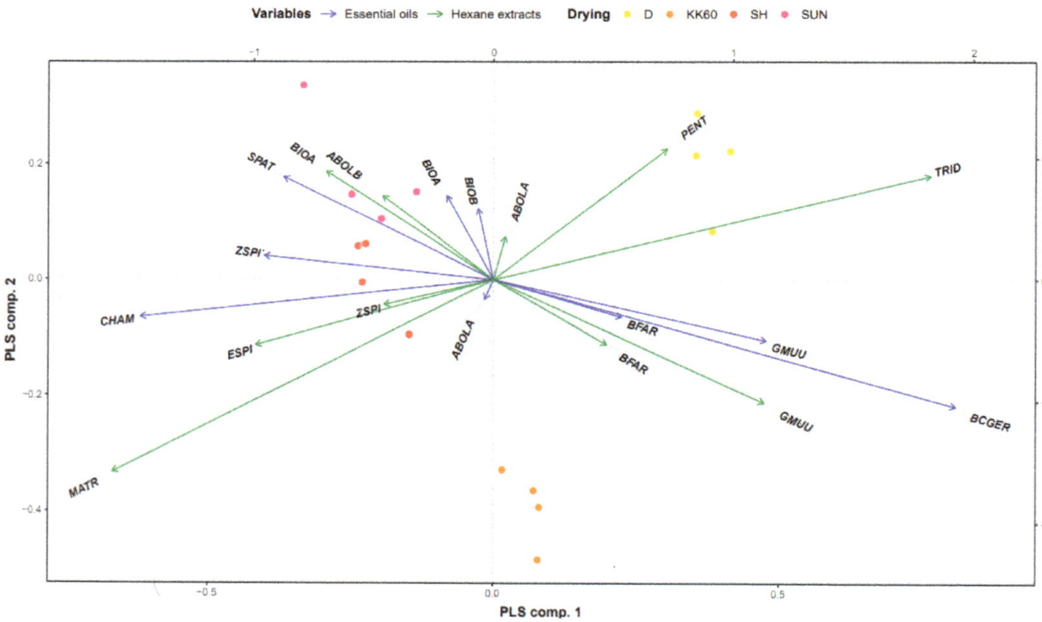

Figure 8. PLS-DA biplot showing the relationship between drying methodology and polyphenolic compounds and essential oils and *n*-hexane compounds.

The impact of the drying method on GC-MS-analyzed compounds (essential oils and *n*-hexane compounds) is visualized in the PLS-DA biplot shown in Figure 8. The first two PLS components capture 81.36% of the variability in the considered compounds (the majority, 71.41%, being explained by the first PLS component). The SUN and SH samples showed similar metabolomic profiles, while the D samples differed from them more than the KK60 samples. Compounds with relatively high (resp. low) levels in the SUN and SH samples are located on the left (resp. on the right). The D samples were characterized especially by a relatively high abundance of TRID and PENT and contrarily by a relatively low abundance of MATR, CHAM, ESPI, and ZSPI.

4. Conclusions

Based on the obtained results, we can conclude that the drying method affects the color, aroma, and weight of dry flowers, as well as the quality and composition of the essential chamomile oil. The biggest color change was between fresh chamomile flowers and chamomile flowers dried in a climate chamber at 105 °C for 24 h, followed by samples dried in a dryer at 60 °C for 48 h, and the smallest change was observed in flowers dried in the sun and flowers dried in the shade. The findings suggested that the contents of total phenols, total flavonoids, total phenolic acids, and antioxidant activity were higher in flower samples dried in the sun and in the shade compare to at higher temperatures.

Sun- and shade-dried samples were more similar in essential oils and volatile compound metabolic profiles than samples dried at higher temperatures. The volatile compounds spiroether Z (ZSPI), spiroether E (ESPI), and matricarin (MATR) were similarly abundant in the sun, in the shade, and drying in the climatic chamber at 60 °C for 48 h, but significantly decreased in drying in the dryer at 105 °C for 24 h.

Different methods of drying can cause a decrease and/or increase in some compounds, which may be desirable in certain phytomedical situations, so it is important to know and apply different methods after harvesting when drying chamomile flowers.

Supplementary Materials: The following supporting information can be downloaded at: https://www.mdpi.com/article/10.3390/su152115373/s1, Table S1: Essential oil compounds in chamomile flowers dried on air exposed to direct sun for 4 days (SUN), on air exposed to shadow (SH), dried in a drier at 105 °C for 24 h (D), and drying in the climate chamber.

Author Contributions: Conceptualization, T.B.-L.; methodology, T.B.-L., R.B., K.M., I.O., B.S.-S., S.Ć.Z., N.Š. and P.T.; validation, S.Ć.Z. and B.S.-S.; formal analysis, I.O., S.Ć.Z., N.Š. and R.B.; investigation, T.B.-L., R.B., K.M., I.O. and B.S.-S.; resources, T.B.-L., K.M., B.S.-S. and P.T.; data curation, T.B.-L., S.Ć.Z., P.T. and B.S.-S.; writing—original draft preparation, T.B.-L. and B.S.-S.; writing—review and editing, K.M., R.B., I.O., S.Ć.Z., N.Š. and P.T.; visualization, R.B., I.O. and N.Š.; supervision, K.M and S.Ć.Z.; project administration, T.B.-L. All authors have read and agreed to the published version of the manuscript.

Funding: University of Slavonski Brod and Brod-Posavina County (project: The influence of different methods of soil tillage and fertilization on the yield of field crops). This research was also funded by the project no. MZE-RO0423 by the Ministry of Agriculture, Czech Republic.

Institutional Review Board Statement: Not applicable.

Informed Consent Statement: Not applicable.

Data Availability Statement: Not applicable.

Conflicts of Interest: The authors declare no conflict of interest.

References

1. Chauhan, R.; Singh, S.; Kumar, V.; Kumar, A.; Kumari, A.; Rathore, S.; Kumar, R.; Singh, S. A Comprehensive Review on Biology, Genetic Improvement Agro and Process Technology of German Chamomile (*Matricaria chamomilla* L.). *Plants* **2022**, *11*, 29. [CrossRef] [PubMed]
2. Kolanos, R.; Stice, S.A.; Gupta, R.C.; Lall, R.; Srivastava, A. *German Chamomile in Nutraceuticals*, 2nd ed.; Academic Press: Cambridge, MA, USA, 2021; pp. 757–772.

3. Schilcher, H.; Imming, P.; Goeters, S. Pharmacology and toxicology. In *Chamomile Industrial Profiles*; Franke, R., Schilcher, H., Eds.; CRC Press: Boca Raton, FL, USA, 2005; pp. 251–265.
4. Silva, N.; Barbosa, L.; Seito, L.; Fernandes, A., Jr. Antimicrobial activity and phytochemical analysis of crude extracts and essential oils from medicinal plants. *Nat. Prod. Res.* **2012**, *26*, 1510–1514. [CrossRef] [PubMed]
5. Carnat, A.; Carnat, A.P.; Fraisse, D.; Ricoux, L.; Lamaison, J.L. The aromatic and polyphenolic composition of Roman camomile tea. *Fitoterapia* **2004**, *75*, 32–38. [CrossRef] [PubMed]
6. Gardiner, P. Complementary, holistic, and integrative medicine: Chamomile. *Pediatr. Rev.* **2007**, *28*, 16–18. [CrossRef] [PubMed]
7. Sebai, H.; Jabri, M.A.; Souli, A.; Rtibi, K.; Selmi, S.; Tebourbi, O. Antidiarrheal and antioxidant activities of chamomile (*Matricaria recutita* L.) decoction extract in rats. *J. Ethnopharmacol.* **2014**, *152*, 327–332. [CrossRef]
8. Patel, D.; Shukla, S.; Gupta, S. Apigenin and cancer chemoprevention: Progress, potential and promise. *Int. J. Oncol.* **2007**, *30*, 233–245. [CrossRef]
9. Martens, D. Chamomile: The herb and the remedy. *J. Chiropr. Acad. Homeopathy* **1995**, *6*, 15–18.
10. Awang, D.V.C. *Tyler's Herbs of Choice: The Therapeutic Use of Phytomedicinals*; Taylor and Francis Group: Abingdon, UK; CRC Press: Boca Raton, FL, USA, 2006; p. 292.
11. Sándor, Z.; Mottaghipisheh, J.; Veres, K.; Hohmann, J.; Bencsik, T.; Horváth, A. Evidence supports tradition: The in vitro effects of roman chamomile on smooth muscles. *Front. Pharmacol.* **2018**, *9*, 323. [CrossRef]
12. Dai, Y.-L.; Li, Y.; Wang, Q.; Niu, F.-J.; Li, K.-W.; Wang, Y.-Y.; Wang, J.; Zhou, C.-Z.; Gao, L.-N. Chamomile: A Review of Its Traditional Uses, Chemical Constituents, Pharmacological Activities and Quality Control Studies. *Molecules* **2023**, *28*, 133. [CrossRef]
13. Raal, A.; Orav, A.; Püssa, T.; Valner, C.; Malmiste, B.; Arak, E. Content of essential oil, terpenoids and polyphenols in commercial chamomile (*Chamomilla recutita* L. Rauschert) teas from different countries. *Food Chem.* **2012**, *131*, 632–638. [CrossRef]
14. Maschi, O.; Dal Cro, E.; Galli, G.V.; Caruso, D.; Bosisio, E.; Dell Agli, M. Inhibition of human cAMP-Phosphodiesterase as a mechanism of the spasmolytic effect of *Matricaria recutita* L. *J. Agric. Food Chem.* **2008**, *56*, 5015–5020. [CrossRef] [PubMed]
15. Srivastava, J.K.; Gupta, S. Extraction characterization stability and biological activity of flavonoids isolated from chamomile flowers. *Mol. Cell. Pharmacol.* **2009**, *1*, 138–147. [CrossRef] [PubMed]
16. Sundaram, J.; Timothy, D.; Durance, T.D.; Wang, R. Porous scaffold of gelatin–starch with nanohydroxyapatite composite processed via novel microwave vacuum drying. *Acta Biomater.* **2008**, *4*, 932–942. [CrossRef] [PubMed]
17. Lee, S.-Y.; Ferdinand, V.; Siow, L.-S. Effect of drying methods on yield, physicochemical properties, and total polyphenol content of chamomile extract powder. *Front. Pharmacol. Sec. Ethnopharmacol.* **2022**, *13*, 1–8. [CrossRef]
18. Žlabur, J.Š.; Žutić, I.; Radman, S.; Pleša, M.; Brnčić, M.; Barba, F.J. Effect of different green extraction methods and solvents on bioactive components of chamomile (*Matricaria chamomilla* L.) flowers. *Molecules* **2020**, *25*, 810. [CrossRef]
19. Abbas, A.M.; Seddik, M.A.; Gahory, A.-A.; Salaheldin, S.; Soliman, W.S. Differences in the aroma profile of chamomile (*Matricaria chamomilla* L.) after different drying conditions. *Sustainability* **2021**, *13*, 5083. [CrossRef]
20. Abutaleb, M.; Ragab, T.I.; Abdeldaim, Y.; Mohamed, A. Effect of new solar-drying designs for chamomile essential oil yield and its chemical constituents in Egypt. *Arab Univ. J. Agric. Sci.* **2021**, *29*, 375–385. [CrossRef]
21. Molnar, M.; Mendešević, N.; Šubarić, D.; Banjari, I.; Jokić, S. Comparison of various techniques for the extraction of umbelliferone and herniarin in *Matricaria chamomilla* processing fractions. *Chem. Cent. J.* **2017**, *11*, 78. [CrossRef]
22. Borsato, A.V.; Doni-Filho, L.; Ahrens, D.C. Drying of chamomile [*Chamomilla recutita* (L.) Raeuchert] under five air temperatures. *Rev. Bras. De Plantas Med.* **2005**, *7*, 77–85.
23. Hamrouni-Sellami, I.; Wannes, W.A.; Bettaieb, I.; Berrima, S.; Chahed, T.; Marzouk, B. Qualitative and quantitative changes in the essential oil of *Laurus nobilis* L. leaves as affected by different drying methods. *Food Chem.* **2011**, *126*, 691–697. [CrossRef]
24. Chen, Q.; Bi, J.; Wu, X.; Yi, J.; Zhou, L.; Zhou, Y. Drying kinetics and quality attributes of jujube (*Zizyphus jujuba* Miller) slices dried by hot-air and short- and medium-wave infrared. Radiation. *LWT Food Sci. Technol.* **2015**, *64*, 759–766. [CrossRef]
25. Stockdale, E.A.; Lampkin, N.H.; Hovi, M. Agronomic and environmental implications of organic farming systems. *Adv. Agron.* **2001**, *70*, 261–327.
26. Lee, L. Sustainability: Living within one's own ecological means. *Sustainability* **2009**, *1*, 1412–1430.
27. Beckford, C.; Campbell, D.; Barker, D. Sustainable Food Production Systems and Food Security: Economic and Environmental Imperatives in Yam Cultivation in Trelawny, Jamaica. *Sustainability* **2011**, *3*, 541–561. [CrossRef]
28. Japundžić-Palenkić, B.; Benković, R.; Benković-Lačić, T.; Antunović, S.; Japundžić, M.; Romanjek Fajdetić, N.; Mirosavljević, K. Pepper Growing Modified by Plasma Activated Water and Growth Conditions. *Sustainability* **2022**, *14*, 15967. [CrossRef]
29. Benković-Lačić, T.; Japundžić Palenkić, B., Mirosavljević, K.; Rakić, M.; Obradović, V.; Japundžić, M.; Benković, R. Morphological, pomological, and nutritional value of wild and cultivated rosehip (*Rosa canina* L.) genotypes in Slavonia, Croatia. *Acta Agrobot.* **2022**, *75*, 1–19. [CrossRef]
30. Singleton, V.L.; Rossi, J.A. Colorimetry of total phenolics witphosphomolybdic–phosphotungstic acid reagents. *Am. J. Enol. Vitic.* **1965**, *16*, 144–158. [CrossRef]
31. Council of Europe. *European Pharmacopoeia*, 4th ed.; Council of Europe: Strasbourg, France, 2004; pp. 2377–2378.
32. Zhishen, J.; Mengcheng, T.; Jianming, W. The determination of flavonoid contents in mulberry and their scavenging effects on superoxide radicals. *Food Chem.* **1999**, *64*, 555–559. [CrossRef]

33. Brand-Williams, W.; Cuvelier, M.E.; Berset, C. Use of a free radical method to evaluate antioxidant activity. *LWT Food Sci. Technol.* **1995**, *28*, 25–30. [CrossRef]
34. Adams, R.P. *Identification of Essential Oil Components by Gas Chromatography/Mass Spectrometry*, 4th ed.; Allured Publishing Corporation: Carol Stream, IL, USA, 2007.
35. Cavar Zeljkovic, S.; Komzakova, K.; Siskova, J.; Karalija, E.; Smekalova, K.; Tarkowski, P. Phytochemical variability of selected basil genotypes. *Ind. Crops Prod.* **2020**, *157*, 112910. [CrossRef]
36. Harbourne, N.; Jacquier, J.C.; O'Riordan, D. Optimisation of the extraction and processing conditions of chamomile (*Matricaria chamomilla* L.) for incorporation into a beverage. *Food Chem.* **2009**, *115*, 15–19. [CrossRef]
37. Salamon, I.; Ibraliu, A.; Kryvtsova, M. Essential Oil Content and Composition of the Chamomile Inflorescences (*Matricaria recutita* L.) Belonging to Central Albania. *Horticulturae* **2023**, *9*, 47. [CrossRef]
38. Aćimović, M.; Lončar, B.; Kiprovski, B.; Stanković Jeremić, J.; Todosijević, M.; Pezo, L.; Jeremić, J. Chamomile essential oil quality after postharvest separation treatments. *Rataratvo I Povrtlarstvo* **2021**, *58*, 72–78. [CrossRef]
39. Satyal, P.; Shrestha, S.; Setzer, W.N. Composition and Bioactivities of an (E)-β-Farnesene Chemotype of Chamomile (*Matricaria chamomilla*) Essential Oil from Nepal. *Nat. Prod. Commun.* **2015**, *10*, 1453–1457. [CrossRef]
40. Hajaji, S.; Sifaoui, I.; López-Arencibia, A. Leishmanicidal activity of α-bisabolol from Tunisian chamomile essential oil. *Parasitol. Res.* **2018**, *117*, 2855–2867. [CrossRef] [PubMed]
41. Ghasemi, M.; Babaeian Jelodar, N.; Modarresi, M.; Bagheri, N.; Jamali, A. Increase of Chamazulene and α-Bisabolol Contents of the Essential Oil of German Chamomile (*Matricaria chamomilla* L.) Using Salicylic Acid Treatments under Normal and Heat Stress Conditions. *Foods* **2016**, *5*, 56. [CrossRef]
42. Franz, C.; Voemel, A.; Hoelzl, J. Variation in the essential oil of *Matricaria chamomilla* depending on plant age and stage of development. *Acta Hortic.* **1978**, *73*, 229–238. [CrossRef]
43. Kumar, S.; Das, M.; Singh, A.; Ram, G.; Mallavarapu, G.R.; Ramesh, S. Composition of the essential oils of the flowers, shoots and roots of two cultivars of *Chamomilla recutita*. *J. Med. Aromat. Plant Sci.* **2001**, *23*, 617–623.
44. Yoshinari, T.; Yaguchi, A.; Takahashi-Ando, N.; Kimura, M.; Takahashi, H.; Nakajima, T.; Sugita-Konishi, Y.; Nagasawa, H.; Sakuda, S. Spiroethers of German chamomile inhibit production of aflatoxin G and trichothecene mycotoxin by inhibiting cytochrome P450 monooxygenases involved in their biosynthesis. *FEMS Microbiol. Lett.* **2008**, *284*, 184–190. [CrossRef]
45. Azizi, M.; Rahmati, M.; Ebadi, T.T.; Hasanzadeh, M. The effects of different drying methods on weight loss rate, essential oil and chamazolene contents of chamomile (*Matricaria recutita*) flowers. *Iran. J. Med. Aromat. Plants* **2009**, *25*, 192–201.

sustainability

Article

A Circular Economy Model to Improve Phosphate Rock Fertiliser Using Agro-Food By-Products

Lea Piscitelli [1,*], Zineb Bennani [1], Daniel El Chami [2] and Donato Mondelli [1]

1 CIHEAM Bari, Via Ceglie 9, Valenzano, 70010 Bari, BA, Italy
2 TIMAC AGRO Italia S.p.A., S.P.13, Località Ca' Nova, 26010 Ripalta Arpina, CR, Italy
* Correspondence: piscitelli@iamb.it

Abstract: Phosphorus (P) is an essential nutrient for the plant life cycle. The agricultural management of phosphorus is complicated by the inefficient use of phosphorus by plants, consequent environmental losses, and the rapid consumption of slowly renewed phosphate rock (PR). These issues represent a huge environmental burden and jeopardise food production. In this study, we proposed the combination of this fertiliser with food-processing by-products such as olive pomace, barley spent grain, and citrus pomace to increase phosphate rock solubility and the efficient use of P. Phosphate rock, by-products, and mixtures of phosphate rock and by-products were placed into litterbags and buried in sand. Periodically, one replicate per treatment was collected for the destructive measurement of total and water-soluble phosphorus. In parallel, pH, organic matter, and ash content were measured to investigate the mechanisms behind changes in P content. The mixtures' P-release values ranged between 80% and 88%, whereas phosphate rock lost 23% of its P over 30 days. Phosphate rock showed a constant water-soluble P fraction at the four sampling times, whereas the mixtures exhibited a highly water-soluble P fraction that tended to decrease over time. Specifically, citrus pomace led to the significant and rapid release of phosphorus, barley spent grain maintained the highest water-soluble fraction over 30 days, and olive pomace was not the best-performing product but still performed better than pure phosphate rock. Moreover, the increased solubility of phosphate rock in mixtures was significantly ($p < 0.001$) ascribed to the reduction in pH. The results of this experiment are promising for in vivo trials and suggest the possibility of simple and easily achievable solutions for more sustainable production systems and effective P-fertilisation strategies. Proposing such easily applicable and inexpensive solutions can reduce the distance between research achievements and field applications.

Keywords: crop nutrition; organic farming; sustainable agriculture; circular economy; agro-food by-products

Citation: Piscitelli, L.; Bennani, Z.; El Chami, D.; Mondelli, D. A Circular Economy Model to Improve Phosphate Rock Fertiliser Using Agro-Food By-Products. *Sustainability* **2022**, *14*, 16228. https://doi.org/10.3390/su142316228

Academic Editor: Antonis A. Zorpas

Received: 20 October 2022
Accepted: 1 December 2022
Published: 5 December 2022

1. Introduction

Although soil may have a high total P concentration, it is barely available because of the ease of formation of insoluble complexes with cations [1]. This makes P the least accessible macronutrient and one of the most deficient nutrients in agricultural soil [2].

Following Liebig's law of the minimum, which has been validated by various experiments [3–6], insufficiency of P in soils becomes a limiting factor for crops in terms of their ability to exploit other nutrients efficiently and to attain optimised growth. To increase yields, overcome P deficiency, and compensate for the retrogradation phenomenon, farmers in the past century have tended to overuse fertilisers [7], causing damage to natural ecosystems [8–10]. Only a small portion of the P used in agriculture is efficiently used for food production, while the rest contributes to eutrophication. Most of the current efforts towards reversing this trend focus on the recovery of P contained in plant residues and the reduction of run-off [11]. A new line of research has successfully explored the potential of chemical strategies to increase phosphorus use efficiency, for example through the use of humic–metal–phosphate acid complexes [12,13]. However, these strategies are not yet available for organic agriculture. In this context, there is a shortage of work and research on increasing the efficient use of available P resources.

For decades now, the most widely used P fertiliser has been phosphate rock, due to its relatively low cost [2,14,15]. Although this resource is renewable, the speed with which it is being consumed is leading to its depletion, and the dominance of a one-way model of use is leading to P accumulation in the environment and an additional burden for ecosystems [16].

To counter the existential challenges that humans are facing, e.g., climate change, resource depletion, and population increases [17], these agricultural trends need to change [18,19]. Sustainable agriculture is an effective alternative to intensification for adaptation to climate change and the improvement of farms' ecosystem services [20]. In particular, sustainable practices in crop nutrition require the integration of traditional and scientific knowledge to innovate [21] and minimise pressures on natural resources without compromising yields and food security.

In this context, numerous co-application techniques have been proposed in the literature, such as the addition of biomasses [22–25], microorganisms [26–28], and inorganic substances to agricultural soils [29–32]. However, many of the proposed solutions are complex and expensive, limiting their practical application at the field level.

In addition to the development of specific mechanisms of action, all these proposed techniques have resulted in higher phosphate rock solubility due to pH reduction [33,34]. Basically, in this experiment, we hypothesise that the acidifying action is due to the waste biomasses' specific action.

On the other hand, following a circular economy model, acidification can be triggered by using waste products and by-products from the agro-food industry, which can lead to the solubilisation of phosphate rock through the direct release of organic acids, the loss of protons, or the production of CO_2 during the decomposition process.

The concept of the circular economy appeared first in Kneese [35] and soon after in Pearce and Turner [36] to describe an economy which turns production waste into inputs. A decade later, after several market events that occurred between 2000 and 2010, the notion began appearing in the industrial and environmental policies of China, Europe, and the United States of America, consecutively [37–39], associated with the aim of minimising dependence on natural resources, decreasing waste, and reducing the life-cycle emissions of economies.

In agriculture, the circular economy model was first promoted in the European action plans and strategies proposed by the European Commission under the EU Green Deal initiative (https://ec.europa.eu/info/strategy/priorities-2019-2024/european-green-deal_en (accessed on 15 October 2022)). Within this framework, the innovative use of cheap and accessible agro-food waste with no valuable alternative reuse pathway can contribute to the valorisation and sustainable transformation of agriculture with low carbon costs [40].

Therefore, in this study, in contrast to the available literature exploring more expensive co-application methods, we aimed to carry out the co-application of agro-food by-products in an incubation experiment to evaluate their performance in increasing phosphate rock solubility, and to identify the best by-product and the best mixture rate. To this end, we selected three food-processing by-products based on their availability and their chemical characteristics [41]. Agro-food by-products (more generally, waste biomasses) are locally produced, easily available, and often represent a cost due to the need for their disposal. Their valorisation for the fertiliser sector would increase the latter's circularity and contribute to more sustainable agricultural practices.

2. Materials and Methods

This experiment was developed in a field belonging to CIHEAM Bari (Apulia region), in Southern Italy. The incubation of pots was carried out in 2019 during the month of June.

The climate was typically Mediterranean, with a monthly average temperature above 25 °C and the absence of precipitation [42].

All the by-products were locally collected and are representative of the waste biomasses produced during food transformation in the Mediterranean basin. The selected by-products were citrus pomace (C), olive pomace (O), and barley spent grain (B) collected from local transformation sectors. The by-products were placed in a greenhouse until they were completely air-dried and they were ground before their chemical characterisation. They were then brought to the laboratory for the determination of their pH and organic matter and ash contents, as well as their total and water-soluble phosphorus contents. The by-products' chemical characteristics are presented in Table 1.

Table 1. Chemical characteristics of air-dried and ground brewer's spent grain (B), citrus pomace (C), and olive pomace (O).

		B	C	O
pH	H_2O	5.3 ± 0.1	3.4 ± 0.1	5.5 ± 0.1
Organic Matter	%	96.0 ± 0.1	89.0 ± 0.2	94.0 ± 0.2
Ash	%	4.2 ± 0.1	11.2 ± 0.2	6.2 ± 0.1
Total P	g/kg	12.8 ± 0.4	64.6 ± 5.7	6.3 ± 0.6

The phosphate rock used in this study was provided by TIMAC AGRO Italia (Pheosol line–Fosfonature 26) and was a soft natural rock containing 26% P_2O_5. This commercial product was analysed in terms of its pH, exhibiting a pH of 6.2 ± 0.1.

After chemical characterisation, sand was used to incubate the litterbags containing different treatments. The sand had a slightly alkaline pH (7.9) and low available phosphorus (<6 mg/kg).

The treatments in the litterbags were designated as PR, containing 15 g of phosphate rock; B, C, and O, containing 45 g of each by-product separately; and their mixtures, BPR, CPR, and OPR, with 15 g of phosphate rock and 45 g of each by-product. Three litterbags per treatment were then buried in pots containing only sand, and four pots per treatment were arranged and left in an open field for a total of 30 days. Every 10 days (T1, T2, T3) one litterbag per pot was collected for the destructive analysis of dry weight, pH, organic matter, total P, and water-soluble P. The first chemical analysis (T0) was performed on treatments that had not been incubated. The above-cited parameters were analysed, while the total P was calculated based on the sum of total P contained in the elements included in the treatment.

As for the characterisation of by-products and PR, the pH in the water was measured with a 3:50 w:v ratio and that of organic matter was determined through dry combustion at 550 °C [43]. Total phosphorus was measured colourimetrically after the acid digestion of samples, whereas water-soluble P was quantified colourimetrically from water sample extracts at 1:100 w:v [44].

Data processing was carried out through analysis of variance, and the significance of differences was identified using Fisher's least significant difference (LSD) test at a 5% probability level among treatments. In graphs, the means with significant differences ($p \leq 0.05$) are labelled with different letters, whereas values with no significant differences are not labelled.

3. Results

The pH of the PR treatment did not change across the four observations (from T0 to T3). On the other hand, the other treatments underwent some modifications over time, with few differences between by-products and by-products with phosphate rock (Figure 1). B and BPR exhibited pH values of approximately 5.3 at T0 and 6.7 at T3, C and CPR had values of approximately 3.4 at T0 and 6.1 at T3, and O and OPR had values of approximately 5.5 at T0 and 6.0 at T3. Treatments containing barley spent grain and olive pomace had already reached a pH of 6.0 at the first collection of litterbags (T1), whereas treatments containing citrus pomace only reached a pH value of 6.0 after twenty days (T2).

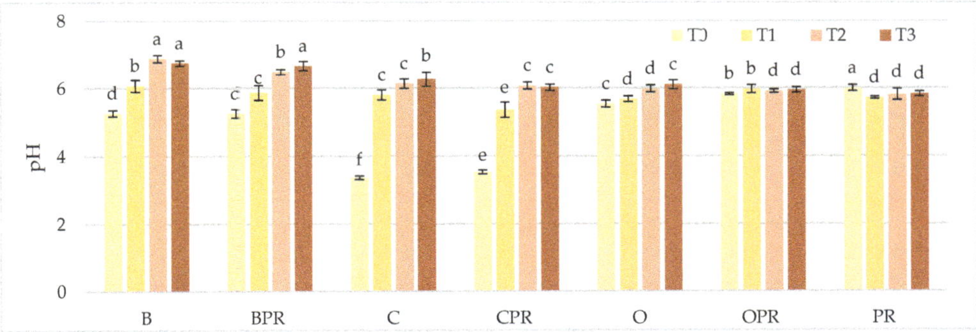

Figure 1. Changes in pH values of different treatments over 30 days. pH values of treatments were compared statistically across the four sampling times, and data with different letters indicate significant differences ($p \leq 0.05$).

Figure 2 shows through a bar graph the organic matter (OM, g) and ash content (Ash, g) per litterbag at each sampling time and, additionally, the trends of weight loss (WL, %). In all treatments, organic matter decreased significantly, whereas ash content did not differ over time. Moreover, the ash content was clearly different between by-products and by-products with phosphate rock. Comparing each by-product and the same by-product with the addition of phosphate rock, the reduction in organic matter content was similar. The only difference was observed in T1 and T2 in the case of the barley spent grain treatments, with BPR having a slightly higher organic matter content than B, and in T2 and T3 in the case of the citrus pomace treatments, which revealed a higher organic matter content in C than in CPR.

In contrast to the reduction in organic matter, the weight loss values followed upward trends. Even in this case, the trends of by-products and by-products plus phosphate rock were similar, with the only significant differences observed in the citrus pomace treatments. Indeed, the weight losses of CPR were constantly lower than those of C and reached 45% at the last sampling time (T3), approximately six percentage points less than C (51%).

Treatments containing only by-products exhibited lower total phosphorus values compared to their PR-added analogues (Figure 3). The PR treatment group exhibited a total phosphorus reduction of about 27 g/kg over a period of 30 days and displayed the highest total phosphorus content. Considering the total phosphorus losses, PR had the lowest loss (23.5%), whereas C had the highest (98.4%); treatments O and B had losses of approximately 80–85%; CPR had a loss of about 70%; and BPR and OPR exhibited losses of approximately 27–30%. Among the by-products, the greatest losses occurred during the first 10 days, with the highest reduction observed for C (96%). In contrast, in the PR treatment group, the total phosphorus reduction was about 9% every 10 days.

Water-soluble phosphorus is a fraction of total phosphorus, and the treatments' trends are alike. Indeed, in treatments containing only by-products, the highest water-soluble phosphorus amount was detected in T0, and this remained steady across the other sampling times. With respect to PR, the highest amount of water-soluble phosphorus was measured in T0, and the solubility in water slowly and constantly decreased over time. Regarding treatments with by-products with added phosphate rock, T0 exhibited the highest water-soluble phosphorus amount, with CPR being more abundant than the others, and unvarying fractions were observed for BPR from T1 to T3. In contrast, a constant reduction in water-soluble phosphorus aliquots was visible for CPR and OPR.

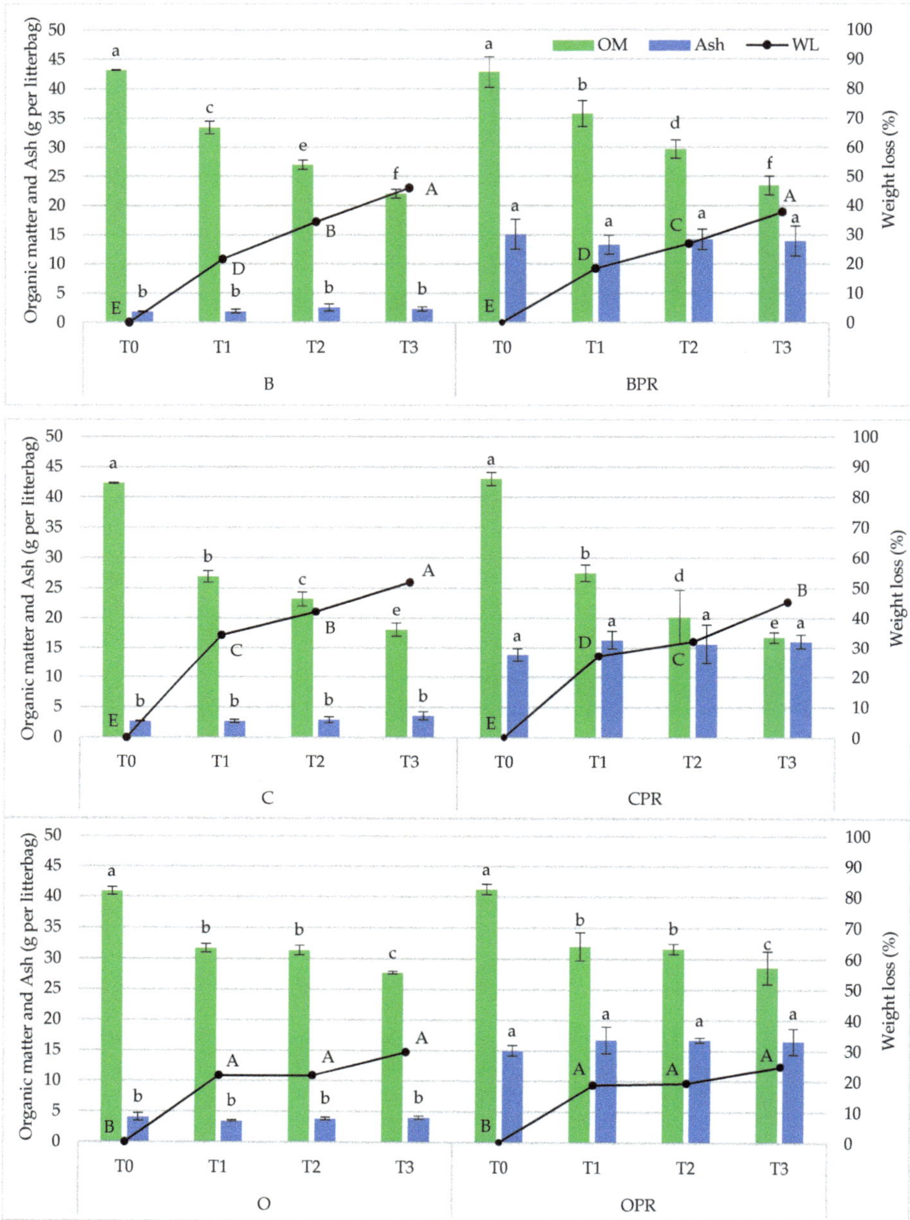

Figure 2. Litterbags' organic matter and ash content (g per litterbag) and weight losses (%) for treatments coupled with common by-products. By-products alone and those with phosphate rock were compared statistically across the four sampling times. Data with different letters indicate significant differences ($p \leq 0.05$).

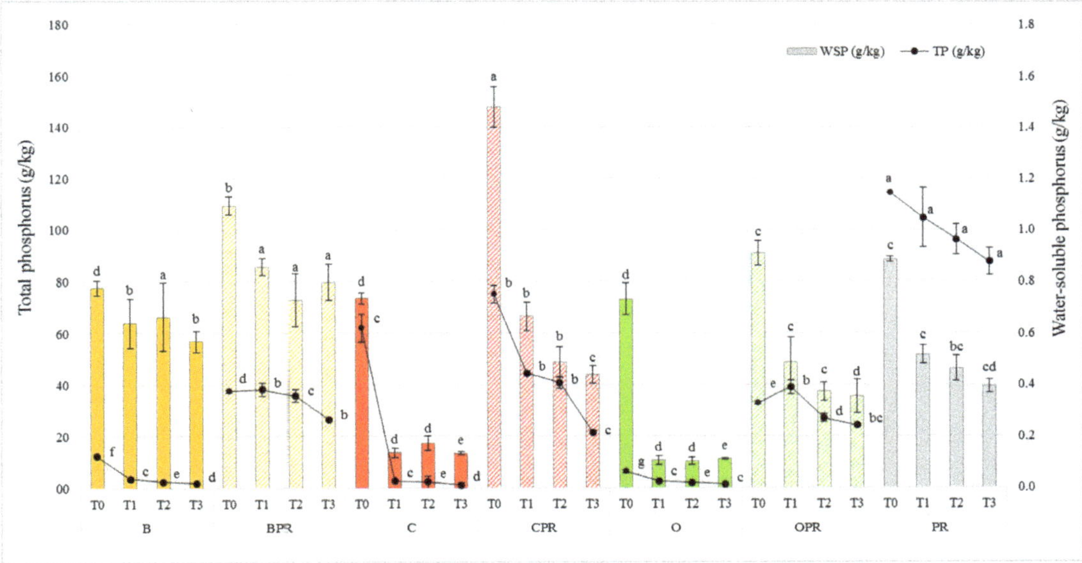

Figure 3. Total phosphorus (trends, g/kg) and water-soluble phosphorus (bars, g/kg) measured in litterbags with various treatments. Statistical comparisons among values for each sampling time were carried out, and data with different letters indicate significant differences ($p \leq 0.05$).

The high correlation coefficients of barley spent grain- and citrus pomace-based treatments indicate the significant ($p < 0.001$) impact of organic matter on pH (Figure 4). On the other hand, the coefficient of determination—which underlined a strong linear relationship in the case of barley spent grain and citrus pomace—was weaker for olive pomace ($R^2 \approx 0.2$).

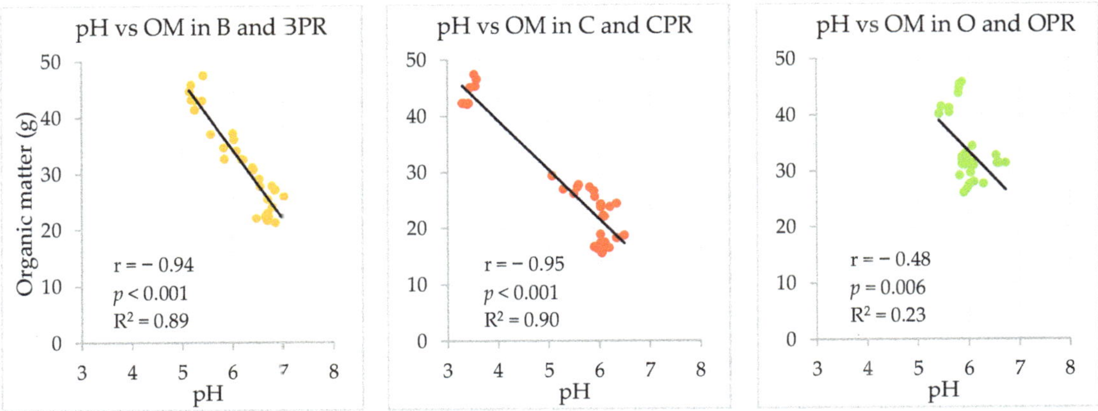

Figure 4. Correlations, significance level, and linear relationships among pH and organic matter (OM) in by-products and by-products plus phosphate rock.

Figure 5 shows that a high correlation was observed between organic matter and weight losses and a low and nonsignificant ($p > 0.001$) correlation was observed between weight losses and ash content. Moreover, the coefficient of determination for organic matter (g per litterbag) and weight loss (%) was approximately 0.8, and a weak relationship was observed for ash content (g per litterbag) and weight loss (%). On the other hand, the diagram on the right of Figure 5 (ASH vs. WL) clearly displays the separation between treatments containing only by-products (light grey circles) and by-products plus phosphate rock (dark grey circles).

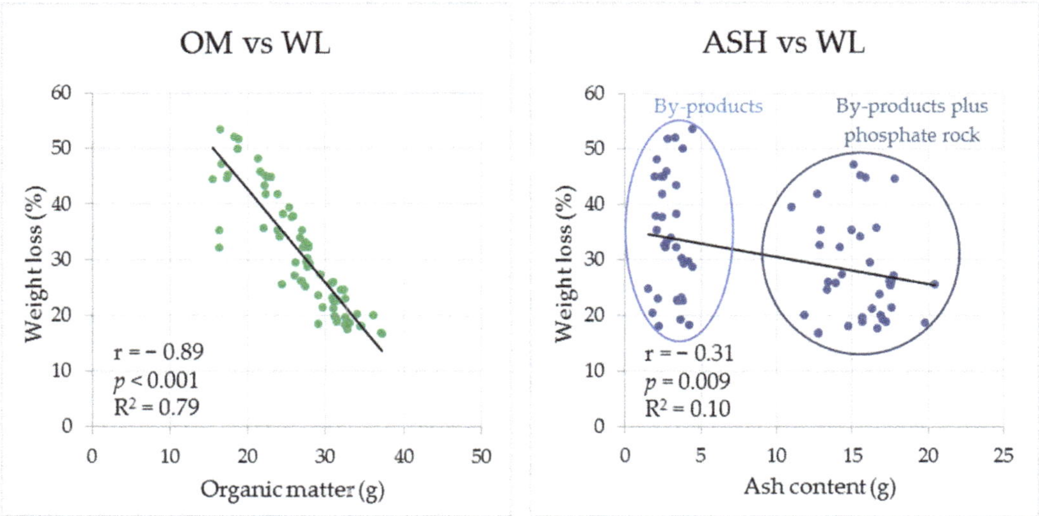

Figure 5. Correlations, significance level, and linear relationships among organic matter (OM) or ash content (ASH) and weight loss (WL).

Table 2 presents the correlation coefficient values, significance levels, and coefficients of determination related to the pH, water-soluble phosphorus (WSP), and total phosphorus (TP) values for each by-product and by-product plus phosphate rock. The pH values of olive pomace plus phosphate rock had the weakest correlation with water-soluble and total phosphorus, whereas BPR and CPR showed a high and significant correlation. The same situation was visible even in the case of coefficient of determination values.

Table 2. Correlations, significance level, and linear relationships among pH, water-soluble phosphorus (WSP), and total phosphorus (TP) for by-products plus phosphate rock.

	pH vs. WSP			pH vs. TP			WSP vs. TP		
	BPR	CPR	OPR	BPR	CPR	OPR	BPR	CPR	OPR
r	−0.85	−0.99	−0.58	−0.86	−0.97	−0.58	0.88	0.98	0.97
p	<0.001	<0.001	0.019	<0.001	<0.001	0.019	<0.001	<0.001	<0.001
R^2	0.72	0.97	0.33	0.74	0.93	0.33	0.77	0.97	0.94

4. Discussion

Considering the evolution of the pH values, the treatments that showed the most significant changes were C and CPR. Indeed, the pH varied from acidic to slightly acidic (Figure 1). This specific modification, as well as the more general reduction in acidity observed for all the by-product-based treatments, can be explained in terms of the loss of organic acids naturally present in these by-products [45,46]. The hypothesis of a reduction in acidity due to the loss of organic compounds was corroborated by the linear negative relationships among pH and organic matter shown in Figure 4 and is supported by the findings of Tumbure et al. [47].

The mutual exchange between the contents of litterbags and sand led to a modification of the pH, which reached 6 for all treatments after 30 days. This change was guided by the loss of mass, mostly organic matter (Figure 2) and subsequent weight losses. According to Zukswert and Prescott [48] the biggest reduction in mass occurred in the earliest days, and this happened even in our case, in which the angular coefficients were higher in the interval between T0 and T1 for all the considered treatments (Figure 2). Moreover, the mass loss was mostly ascribable to the organic matter, whereas ash content did not vary over time in any treatments. The treatments that lost the least weight were O and OPR, and this result was consistent with the slight modifications of pH observed for these groups, and this could be due to the high lignin content in olive pomace. Lignin is present in olive pomace [49] and, together with cellulose and hemicellulose, is one of the main compounds of this food-processing by-product. Additionally, this compound is recalcitrant to thermal and physicochemical degradation [50,51], and only a few specific microbial strains can decompose it [52]. In contrast, barley spent grain and citrus pomace treatments exhibited organic matter losses and thus weight losses of about 40% and 50%, respectively. These mass losses could be due to the permeability of the litterbags' tissue [53] and to the interaction of the by-products with external surrounding factors. Indeed, barley spent grain and citrus pomace both exhibit steady interaction with microbial communities—a direct link in the case of barley spent grain because of naturalised microbial charge [54], as well as an indirect connection through the richness in molecules that are attractive for microorganisms [55]. Nevertheless, even in the case of barley spent grain and citrus pomace treatments, there was a consistency in terms of the wide modification in pH values and organic matter reductions, thus supporting the hypothesis of the loss of organic acids.

The modification in terms of weight loss in the PR treatments was negligible and thus is not reported in Figure 2; for all the other treatments, most of the mass losses were organic, and there was a good fit between weight loss and organic matter, with an r-value of approximately 0.9 (Figure 5, left). On the other hand, the correlation of ash content and weight loss was small, and treatments were well distributed over the Y-axis and separated on the X-axis according to the addition of phosphate rock to by-products. These diagrams together support what was already underlined in Figure 2 regarding the limited losses of ash in comparison with the notable decreases in organic matter content. According to Prescott and Vesterdal [56], the decomposition of plant biomass and its transformation can follow several pathways that depend on external and site-specific conditions. In environments with low amounts of natural organic matter and poor biological activities, such as sand, the decomposition of labile organic matter is fostered by an emphasised priming effect [57]. On the other hand, summer temperatures can play a crucial role in organic matter decomposition. Indeed, Pérez et al. [58] found that after four months, the biomass weight of litterbags decreased by about 2% and that leaf litter decomposition was slower in winter than in the hotter seasons.

Total phosphorus was lower in treatments containing a single by-product; only the C treatment had a high total phosphorus content, but this was mostly labile and was lost during the first 10 days (Figure 3). It is important to underline that the higher level of total phosphorus in OPR at T1 compared to T0 can be justified by the relative loss in organic matter during the first 10 days (already shown in Figure 2). In turn, this led to a reduction in the total mass of the litterbags that did not correspond to a related and coherent loss of phosphorus, with a consequently slowed release in the absence of additives such as microorganisms [59]. On the other hand, by-products plus phosphate rock had high initial total phosphorus contents, and, at T3, BPR had lost 27% of its total phosphorus, and OPR and CPR had lost 30% and 70%, respectively. On the other hand, PR lost about 24% of its initial total phosphorus over 30 days, thus suggesting a good performance of all the mixtures in increasing the solubility of phosphate rock. Relatedly, several studies have documented the adoption of different strategies or practices for increasing phosphate rock solubility, such as the addition of zeolite and pillared clay [31], nanoparticles [32], a combination with acid mine waste [29], co-composting [22], a combination with green waste [23], co-application with manure [24] or other amendments [25], and enrichment with microorganisms [27,28]. The effects of all these strategies were directly or indirectly ascribable to acidifying effects. Indeed, phosphate rock efficacy is higher under acidic conditions, and even artificial combination with an organic acid increases phosphorus solubility [60].

The crucial role of pH in increasing the solubility of phosphorus from phosphate rock was corroborated by the high correlation between pH and TP (Table 2). Moreover, pH significantly ($p < 0.001$) affected WSP, especially in the case of BPR and CPR (Table 2), in which a progressive reduction in pH corresponded to a higher water-soluble fraction. Water-soluble phosphorus dynamics over 30 days mirrored the total phosphorus trends, as underlined by the high positive correlation between TP and WSP shown in Table 2. Despite TP and WSP's high correlation, it is important to highlight that the water-soluble fraction indicates phosphorus that is rapidly available. Therefore, its evaluation, as well as the maintenance of an adequate WSP supply, is crucial for the use of these mixtures in soil–plant systems [61]. In this sense, PR maintained a constant and adequate water-soluble phosphorus fraction. On the other hand, CPR showed the highest phosphorus release from the preparation of the mixture. Moreover, BPR showed a constantly higher WSP over the 30 days. The knowledge regarding these BPR and CPR behaviours can be exploited for the programming of a fertilisation plan in order to tailor phosphorus release to the needs of crops. In contrast, the WSP of OPR was close to that of PR. This result, together with the TP values, suggests that most of the solubilising action occurred in the first 10 days, in which there was a high release of phosphorus.

The results of this experiment confirm the role of pH in increasing the solubility of phosphate rock and underline the potential uses of by-products in combination with this fertiliser and as a substitute for more complex technologies, strategies, and products. The findings contribute to the achievement of more sustainable management of phosphate rock [62] and phosphorus flow [63], and the valorisation of waste biomasses through soil incorporation [64].

At a global level, there is an urgent need to reduce the distance between scientific research achievements and applications [11], but this transfer of knowledge can only be fostered if the proposed solutions are easily applicable and inexpensive; the present work meets these requirements.

5. Conclusions

In this study, we evaluated the combination of phosphate rock with three widely available food-processing by-products at the benchmark level with the aim of increasing the solubility of phosphate rock and the consequent phosphorus release for crop nutrition.

In a simplified environment consisting of litterbags incubated in sand, all the selected by-products showed a high capacity for solubilisation of phosphate rock. The weight losses in litterbags containing only by-products and by-products combined with phosphate rock were mainly caused by losses of organic matter. Two out of three by-products showed a high correlation between pH and organic matter content. Moreover, pH was positively correlated with total phosphorus and water-soluble phosphorus in the litterbags containing mixtures. The combination of phosphate rock with the considered by-products resulted in more solubilised phosphorus than phosphate rock alone after 30 days, and barley spent grain and citrus pomace seemed to be the most promising by-products for field use due to the high correlations observed between pH and total phosphorus.

Indeed, citrus pomace had a high natural phosphorus content and, combined with phosphate rock, produced a great and rapid release of phosphorus (within 10 days). In contrast, barley spent grain maintained the highest water-soluble phosphorus content over the entire experiment. Nevertheless, olive pomace combined with phosphate rock still exhibited better performance than phosphate rock alone, even though the mechanisms of phosphorus solubilisation were not strictly ascribable to pH. Beyond the performances of the single by-products, this study demonstrated the in vitro efficacy of products that must be valorised and that can contribute to strengthening soil organic matter. In addition, the more efficient use of phosphate rock will reduce the squandering of this fertiliser and the related problem of eutrophication. The advantage of this combination is its simplicity and affordability, which can make its practical application more easily adaptable for farm use. In turn, agricultural practices such as this one can be spread widely in order to foster agricultural sustainability.

The limitations of this experimental set-up include the use of a single ratio of partitioning, the limited types of biomasses used, and lack of evidence regarding the mixtures' performances in soil–plant systems. Indeed, this study is the first step towards more complex work which could lead to the development of innovative products for agriculture. The results support the fulfilment of the EU and UN policies aiming towards a more circular agro-food sector and sustainable agriculture.

In this context, the results illustrated in this research add to the scientific literature by highlighting promising by-products for valorisation in crop nutrition, and they should also motivate the scientific community to find simple, easily achievable, scalable, and cheap solutions to support farmers in transitioning to more sustainable production systems. The industrial sector could also build on these results and conduct pre-industrial research to explore the normative, technical, and environmental aspects of incorporating these by-products efficiently into the production process. The benefits extend to the agricultural sector, which needs to innovate to increase nutrient use efficiency, especially in organic agriculture. The outcomes of this study represent the first step in a broader project to explore other potential by-products from local agro-food industries and to test these mixtures in more complex systems, such as soil–plant systems.

Author Contributions: Conceptualisation, L.P.; methodology, L.P. and D.M.; software, L.P.; validation, L.P., D.E.C. and D.M.; formal analysis, L.P. and Z.B. investigation, L.P. and Z.B.; data curation, L.P.; writing—original draft preparation, L.P.; writing—review and editing, D.E.C. and D.M.; visualisation, L.P., D.E.C. and D.M.; supervision, D.E.C. and D.M. All authors have read and agreed to the published version of the manuscript.

Funding: This research received no external funding.

Institutional Review Board Statement: This study do not require ethical approval.

Informed Consent Statement: Not applicable.

Conflicts of Interest: The authors declare no conflict of interest.

References

1. Cordell, D.; Rosemarin, A.; Schröder, J.J.; Smit, A.L. Towards global phosphorus security: A systems framework for phosphorus recovery and reuse options. *Chemosphere* **2011**, *84*, 747–758. [CrossRef] [PubMed]
2. Hellal, F.; El-Sayed, S.; Zewainy, R.; Amer, A. Importance of phosphate pock application for sustaining agricultural production in Egypt. *Bull. Natl. Res. Cent.* **2019**, *43*, 11. [CrossRef]
3. Hiddink, J.G.; Kaiser, M.J. Implications of Liebig's Law of the Minimum for the Use of Ecological Indicators Based on Abundance. *Ecography* **2005**, *28*, 264–271. [CrossRef]
4. Wang, J.; Baerenklau, K.A. Crop response functions integrating water, nitrogen, and salinity. *Agric. Water Manag.* **2014**, *139*, 17–30. [CrossRef]
5. Ferreira, I.E.P.; Zocchi, S.S.; Baron, D. Reconciling the Mitscherlich's law of diminishing returns with Liebig's law of the minimum. Some results on crop modeling. *Math. Biosci.* **2017**, *293*, 29–37. [CrossRef]
6. Rogers, D.; Weaver, D.; Summers, R.; Dobbe, E.; Master, R.; McFerran, R.; Mussell, G.; Dawson, L.; Mercy, J.; Richards, P.; et al. Critical phosphorus values from the Better Fertiliser Decisions for Pastures project Early insights from validation trials. *Crop Pasture Sci.* **2021**, *72*, 731–741. [CrossRef]
7. Withers, P.J.; Sylvester-Bradley, R.; Jones, D.L.; Healey, J.R.; Talboys, P.J. Feed the crop not the soil: Rethinking phosphorus management in the food chain. *Environ. Sci. Technol.* **2014**, *48*, 6523–6530. [CrossRef]

8. Wurtsbaugh, W.A.; Paerl, H.W.; Dodds, W.K. Nutrients, eutrophication and harmful algal blooms along the freshwater to marine continuum. *WIREs Water* **2019**, *6*, e1373. [CrossRef]
9. Lin, S.S.; Shen, S.L.; Zhou, A.; Lyu, H.A. Assessment and management of lake eutrophication: A case study in Lake Erhai, China. *Sci. Total Environ.* **2021**, *751*, 141618. [CrossRef]
10. Liu, L.; Zheng, X.; Wei, X.; Kai, Z.; Xu, Y. Excessive application of chemical fertilizer and organophosphorus pesticides induced total phosphorus loss from planting causing surface water eutrophication. *Sci. Rep.* **2021**, *11*, 23015. [CrossRef]
11. Approaching peak phosphorus. *Nat. Plants* **2022**, *8*, 979. [CrossRef]
12. Urrutia, O.; Javier Erro, J.; Guardado, I.; San Francisco, S.; Mandado, M.; Baigorri, R.; Yvin, J.C.; Garcia-Mina, J.M. Physico-chemical characterization of humic-metal-phosphate complexes and their potential application to the manufacture of new types of phosphate-based fertilizers. *J. Plant Nutr. Soil Sci.* **2013**, 1–9. [CrossRef]
13. Erro, J.; Urrutia, O.; Baigorri, R.; Aparicio-Tejo, P.; Irigoyen, I.; Torino, F.; Mandado, M.; Yvin, J.C.; Garcia-Mina, J.M. Organic Complexed Superphosphates (CSP): Physicochemical Characterization and Agronomical Properties. *J. Agric. Food Chem.* **2008**, *60*, 2008–2017. [CrossRef]
14. Rafael, R.B.A.; Fernández-Marcos, M.L.; Cocco, S.; Ruello, M.L.; Weindorf, D.C.; Cardelli, V.; Corti, G. Assessment of potential nutrient release from phosphate rock and dolostone for application in acid soils. *Pedosphere* **2018**, *28*, 44–58. [CrossRef]
15. Munir, A.; Adel, G.; Saud, S.A.O.; Khaled, D.A.; Mahmoud, N. Acidulated activation of phosphate rock enhances release, lateral transport and uptake of phosphorus and trace metals upon direct-soil application. *Soil Sci. Plant Nutr.* **2019**, *65*, 183–195. [CrossRef]
16. Powers, S.M.; Chowdhury, R.B.; MacDonald, G.K.; Metson, G.S.; Beusen, A.H.W.; Bouwman, A.F.; Hampton, S.E.; Mayer, B.K.; McCrackin, M.L.; Vaccari, D.A. Global opportunities to increase agricultural independence through phosphorus recycling. *Earth's Future* **2019**, *7*, 370–383. [CrossRef]
17. UN. Make the SDGs a Reality. Department of Economic and Social Affairs of the United Nations (UN). Available online: https://sdgs.un.org/goals (accessed on 15 October 2022).
18. Pretty, J.; Bharucha, Z.P. Sustainable intensification in agricultural systems. *Ann. Bot.* **2014**, *114*, 1571–1596. [CrossRef]
19. Withers, P.J.A.; Doody, D.G.; Sylvester-Bradley, R. Achieving Sustainable Phosphorus Use in Food Systems through Circularisation. *Sustainability* **2018**, *10*, 1804. [CrossRef]
20. El Chami, D.; Daccache, A.; El Moujabber, M. How can sustainable agriculture increase climate resilience? A systematic review. *Sustainability* **2020**, *12*, 3119. [CrossRef]
21. El Chami, D. Towards sustainable organic farming systems. *Sustainability* **2020**, *12*, 9832. [CrossRef]
22. Korzeniowska, J.; Stanisławska-Glubiak, E.; Hoffmann, J.; Górecka, H.; Jóźwiak, W.; Wiśniewska, G. Improvement of the solubility of rock phosphate by co-composting it with organic components. *Pol. J. Chem. Technol.* **2013**, *15*, 10–14. [CrossRef]
23. Bustamante, M.A.; Ceglie, F.G.; Aly, A.; Mihreteab, H.T.; Ciaccia, C.; Tittarelli, F. Phosphorus availability from rock phosphate: Combined effect of green waste composting and sulfur addition. *J. Environ. Manag.* **2016**, *182*, 557–563. [CrossRef] [PubMed]
24. Poblete-Grant, P.; Biron, P.; Bariac, T.; Cartes, P.; Mora, M.d.L.L.; Rumpel, C. Synergistic and Antagonistic Effects of Poultry Manure and Phosphate Rock on Soil P Availability, Ryegrass Production, and P Uptake. *Agronomy* **2019**, *9*, 191. [CrossRef]
25. Sabah, N.U.; Tahir, M.A.; Sarwar, G.; Luqman, M.; Aziz, A.; Manzoor, M.Z.; Aftab, M. Biosolubilization of phosphate rock using organic amendments: An innovative approach for sustainable maize production in Aridisols—A review. *Sarhad J. Agric.* **2022**, *38*, 617–625. [CrossRef]
26. Alori, E.T.; Glick, B.R.; Babalola, O.O. Microbial phosphorus solubilization and its potential for use in sustainable agriculture. *Front. Microbiol.* **2017**, *2*, 971. [CrossRef] [PubMed]
27. Billah, M.; Khan, M.; Bano, A.; Nisa, S.; Hussain, A.; Dawar, K.M.; Munir, A.; Khan, N. Rock Phosphate-Enriched Compost in Combination with Rhizobacteria; A Cost-Effective Source for Better Soil Health and Wheat (*Triticum aestivum*) Productivity. *Agronomy* **2020**, *10*, 1390. [CrossRef]
28. Barin, M.; Asadzadeh, F.; Hosseini, M.; Hammer, E.C.; Vetukuri, R.R.; Vahedi, R. Optimization of Biofertilizer Formulation for Phosphorus Solubilizing by *Pseudomonas fluorescens* Ur21 via Response Surface Methodology. *Processes* **2022**, *10*, 650. [CrossRef]
29. Santos, W.O.; Hesterberg, D.; Mattiello, E.M.; Vergütz, L.; Barreto, M.S.; Silva, I.R.; Souza Filho, L.F. Increasing Soluble Phosphate Species by Treatment of Phosphate Rocks with Acidic Waste. *J. Environ. Qual.* **2016**, *45*, 1988–1997. [CrossRef]
30. Zhang, X.-M.; Li, Y.; Hu, C.; He, Z.-Q.; Wen, M.-X.; Gai, G.-S.; Huang, Z.-H.; Yang, Y.-F.; Hao, X.-Y.; Li, X.-Y. Enhanced Phosphorus Release from Phosphate Rock Activated with Lignite by Mechanical Microcrystallization: Effects of Several Typical Grinding Parameters. *Sustainability* **2019**, *11*, 1068. [CrossRef]
31. Teles, A.P.B.; Rodrigues, M.; Pavinato, P.S. Solubility and Efficiency of Rock Phosphate Fertilizers Partially Acidulated with Zeolite and Pillared Clay as Additives. *Agronomy* **2020**, *10*, 918. [CrossRef]
32. Avşar, C. A novel assessment strategy for nanotechnology in agriculture: Evaluation of nanohydroxyapatite as an alternative phosphorus fertiliser. *Kem. Ind.* **2022**, *71*, 327–334. [CrossRef]
33. Sharma, S.B.; Sayyed, R.Z.; Trivedi, M.H.; Gobi, T.A. Phosphate solubilizing microbes: Sustainable approach for managing phosphorus deficiency in agricultural soils. *SpringerPlus* **2013**, *2*, 587. [CrossRef]

34. Nesme, T.; Colomb, B.; Hinsinger, P.; Watson, C.A. Soil phosphorus management in organic cropping systems: From current practices to avenues for a more efficient use of P resources. In *Organic Farming, Prototype for Sustainable Agricultures*; Bellon, S., Penvern, S., Eds.; Springer: Dordrecht, The Netherlands, 2014; pp. 23–45. [CrossRef]
35. Kneese, A.V. The Economics of Natural Resources. *Popul. Dev. Rev.* **1988**, *14*, 281–309. [CrossRef]
36. Pearce, D.W.; Turner, R.K. *Economics of Natural Resources and the Environment*; Johns Hopkins University Press: Baltimore, MD, USA, 1989; 392p.
37. EPA. The National Recycling Strategy: Part One of a Series on Building a Circular Economy. Available online: https://www.epa.gov/recyclingstrategy/national-recycling-strategy (accessed on 15 October 2022).
38. EC. Circular Economy Action Plan: For a Cleaner and More Competitive Europe. The European Commission (EC). Available online: https://environment.ec.europa.eu/strategy/circular-economy-action-plan_en (accessed on 15 October 2022).
39. Zhu, J.; Fan, C.; Shi, H.; Shi, L. Efforts for a Circular Economy in China: A Comprehensive Review of Policies. *J. Ind. Ecol.* **2019**, *23*, 110–118. [CrossRef]
40. Tripathi, N.; Hills, C.D.; Singh, R.S.; Atkinson, C.J. Biomass waste utilisation in low-carbon products: Harnessing a major potential resource. *NPJ Clim. Atmos. Sci.* **2019**, *2*, 35. [CrossRef]
41. Patsios, S.I.; Kontogiannopoulos, K.N.; Mitrouli, S.; Plakas, K.V.; Karabelas, A.J. Characterisation of agricultural waste co- and by-products. *Agrocycle-EU Horizon 2020* **2016**.
42. Piscitelli, L.; Colovic, M.; Aly, A.; Hamze, M.; Todorovic, M.; Cantore, V.; Albrizio, R. AdaptiveAgricultural Strategies for FacingWater Deficit in Sweet MaizeProduction: A Case Study of a Semi-Arid Mediterranean Region. *Water* **2021**, *13*, 3285. [CrossRef]
43. *UNI CEN/TS 15370-1:2006*; Solid Biofuels—Method for the Determination of Ash Melting Behaviour—Part 1: Characteristic Temperatures Method. CEN: Brussels, Belgium, 2006.
44. *ISO—15958*; Fertilizers—Extraction of Water-Soluble Phosphorus. ISO: Geneva, Switzerland, 2019. Available online: https://www.iso.org/obp/ui/#iso:std:iso:5316:ed-1:v1:en (accessed on 3 March 2019).
45. Chanalia, P.; Gandhi, D.; Anjana, B.S.; Singh, J.; Dhanda, S. Antioxidant activity and nutritional value of Citrus limetta and Ananas comosus pomace. *J. Food Sci. Nutr. Ther.* **2018**, *4*, 004–007. [CrossRef]
46. Solé, M.M.; Pons, L.; Conde, M.; Gaidau, C.; Bacardit, A. Characterization of Wet Olive Pomace Waste as Bio-Based Resource for Leather Tanning. *Materials* **2021**, *14*, 5790. [CrossRef]
47. Tumbure, A.; Bretherton, M.; Bishop, P.; Hedley, M. Phosphorus recovery from an igneous phosphate rock using organic acids and pyrolysis condensate. *Sci. Afr.* **2022**, *15*, e01098. [CrossRef]
48. Zukswert, J.; Prescott, C. Relationships among leaf functional traits, litter traits, and mass loss during early phases of leaf litter decomposition in 12 woody plant species. *Oecologia* **2017**, *185*, 305–316. [CrossRef] [PubMed]
49. Cequier, E.; Aguilera, J.; Balcells, M.; Canela-Garayoa, R. Extraction and characterization of lignin from olive pomace: A comparison study among ionic liquid, sulfuric acid, and alkaline treatments. *Biomass Convers. Biorefinery* **2019**, *9*, 241–252. [CrossRef]
50. Ghouma, I.; Jeguirim, M.; Guizani, C.; Ouederni, A.; Limousy, L. Pyrolysis of Olive Pomace: Degradation Kinetics, Gaseous Analysis and Char Characterization. *Waste Biomass Valoris.* **2017**, *8*, 1689–1697. [CrossRef]
51. Wang, Z.; Zhu, X.; Deuss, P.J. The effect of ball milling on birch, pine, reed, walnut shell enzymatic hydrolysis recalcitrance and the structure of the isolated residual enzyme lignin. *Ind. Crops Prod.* **2021**, *167*, 113493. [CrossRef]
52. Singh, A.K.; Bilal, M.; Iqbal, H.M.N.; Meyer, A.S.; Raj, A. Bioremediation of lignin derivatives and phenolics in wastewater with lignin modifying enzymes: Status, opportunities and challenges. *Sci. Total Environ.* **2021**, *777*, 145988. [CrossRef]
53. Krishna, M.P.; Mohan, M. Litter decomposition in forest ecosystems: A review. *Energy Ecol. Environ.* **2017**, *2*, 236–249. [CrossRef]
54. Bianco, A.; Budroni, M.; Zara, S.; Mannazzu, I.; Fancello, F.; Zara, G. The role of microorganisms on biotransformation of brewers' spent grain. *Appl. Microbiol. Biotechnol.* **2020**, *104*, 8661–8678. [CrossRef]
55. Zannini, D.; Dal Poggetto, G.; Malinconico, M.; Santagata, G.; Immirzi, B. Citrus Pomace Biomass as a Source of Pectin and Lignocellulose Fibers: From Waste to Upgraded Biocomposites for Mulching Applications. *Polymers* **2021**, *13*, 1280. [CrossRef]
56. Prescott, C.E.; Vesterdal, L. Decomposition and transformations along the continuum from litter to soil organic matter in forest soils. *For. Ecol. Manag.* **2021**, *498*, 119522. [CrossRef]
57. Chen, L.; Liu, L.; Qin, S.; Yang, G.; Fang, K.; Zhu, B.; Kuzyakov, Y.; Chen, P.; Xu, Y.; Yang, Y. Regulation of priming effect by soil organic matter stability over a broad geographic scale. *Nat. Commun.* **2019**, *10*, 5112. [CrossRef]
58. Pérez, P.; Barro, R.; Pérez, J.; Fernández, M.J.; Moyano, A.; Ciria, P. Nutrient Release through Litterfall in Short Rotation Poplar Crops in Mediterranean Marginal Land. *Forests* **2021**, *12*, 1185. [CrossRef]
59. Ghorbanzadeh, N.; Mahsefat, M.; Farhangi, M.B.; Khalili Rad, M.; Proietti, P. Short-term impacts of pomace application and *Pseudomonas* bacteria on soil available phosphorus. *Biocatal. Agric. Biotechnol.* **2020**, *23*, 101742. [CrossRef]
60. Jamal, A.; Khan, A.; Sharif, M.; Jamal, H. Application of Different Organic Acids on Phosphorus Solubility from Rock Phosphate. *J. Hortic. Plant Res.* **2018**, *2*, 43–48. [CrossRef]
61. Huang, L.; Mao, X.; Wang, J.; Chen, X.Y.; Wang, G.; Liao, Z. The effect and mechanism of improved efficiency of physicochemical pro-release treatment for low-grade phosphate rock. *J. Soil Sci. Plant Nutr.* **2014**, *14*, 316–331. [CrossRef]

62. Idrissi, H.; Taha, Y.; Elghali, A.; El Khessaimi, Y.; Aboulayt, A.; Amalik, J.; Hakkou, R.; Benzaazoua, M. Sustainable use of phosphate waste rocks: From characterization to potential applications. *Mater. Chem. Phys.* **2021**, *260*, 124119. [CrossRef]
63. Wu, J.; Hartmann, T.H.; Chen, W.S. Toward sustainable management of phosphorus flows in a changing rural–urban environment: Recent advances, challenges, and opportunities. *Curr. Opin. Environ. Sustain.* **2019**, *40*, 81–87. [CrossRef]
64. Okolie, J.A.; Epelle, E.I.; Tabat, M.E.; Orivri, U.; Amenaghawon, A.N.; Okoye, P.O.; Gunes, B. Waste biomass valorization for the production of biofuels and value-added products: A comprehensive review of thermochemical, biological and integrated processes. *Process Saf. Environ. Prot.* **2022**, *159*, 323–344. [CrossRef]

sustainability

MDPI

Article

Biostimulants for Resilient Agriculture: A Preliminary Assessment in Italy

Rita Leogrande [1], Daniel El Chami [2,*], Giulio Fumarola [3], Michele Di Carolo [3], Giuseppe Piegari [2], Mario Elefante [2], Donato Perrelli [3] and Crescenza Dongiovanni [3]

1 Council for Agricultural Research and Economics, Research Centre for Agriculture and Environment (CREA-AA), Via C. Ulpiani 5, I-70125 Bari, Italy; rita.leogrande@crea.gov.it
2 TIMAC AGRO Italia S.p.A., S.P.13, Località Ca' Nova, I-26010 Ripalta Arpina, Italy; giuseppe.piegari@roullier.com (G.P.); mario.elefante@roullier.com (M.E.)
3 Centre for Research, Experimentation and Training in Agriculture Basile Caramia (CRSFA), I-70010 Locorotondo, Italy; giuliofumarola@crsfa.it (G.F.); micheledicarolo@crsfa.it (M.D.C.); donatoperrelli@crsfa.it (D.P.); enzadongiovanni@crsfa.it (C.D.)
* Correspondence: daniel.elchami@roullier.com; Tel.: +39-0373-669-233; Fax: +39-0373-669-291

Abstract: In agriculture, plant biostimulants have become necessary to meet the United Nations sustainable development goals (UN-SDGs) and advance the European Green Deal. In particular, seaweed-based biostimulants have received a greater acceptance for their several benefits in crop growth and yield. In this study, we evaluated the effects of foliar applications of a vegetable- and brown-algae-based extract (*Ascophyllum nodosum* (L.) Le Jol. on grapes (*Vitis vinifera* L. cv. Montepulciano) and olives (*Olea europaea* L. cv. Coratina) and its agronomic performance in two field experiments in the Apulia region, which is known for its modern agricultural sector. The results highlight that the crop responses differ in grape and olive orchards. The biostimulant application determined significant increases in bunch development (+9.5%) and bunch weight (+10%) compared to the untreated control. In the olive orchard, the yield was not significantly influenced by biostimulant application, whereas we observed quality improvement in the olive oil of the treated plants compared to the control. To better understand the mechanisms behind this difference, the research concludes by suggesting that further research pursues in-depth studies and high scientific and technical proficiency to determine and optimise the rates and timing of applications.

Keywords: biostimulants; vegetable extract; seaweed extract; agrosystems; resilience; Apulia (Italy)

Citation: Leogrande, R.; El Chami, D.; Fumarola, G.; Di Carolo, M.; Piegari, G.; Elefante, M.; Perrelli, D.; Dongiovanni, C. Biostimulants for Resilient Agriculture: A Preliminary Assessment in Italy. *Sustainability* **2022**, *14*, 6816. https://doi.org/10.3390/su14116816

Academic Editors: Imre J. Holb and Anastasios Michailidis

Received: 8 March 2022
Accepted: 23 May 2022
Published: 2 June 2022

1. Introduction

The two-way relationship between agriculture and climate change poses a serious constraint on reaching food security and availability [1]. The action plan for sustainable development, which began in 1992 with Agenda 21 and continued in 2015 with Agenda 2030, set specific objectives for this transformative development [2]. As a response, the European Union launched the "European Green Deal" in 2020 to sustainably develop the member states' economies without increasing resource deterioration (https://ec.europa.eu/info/strategy/priorities-2019-2024/european-green-deal_en, 22 February 2022). The agricultural strategy of the deal, the Farm to Fork strategy (F2F), established different targets to adapt European agriculture to climate change and increase its resilience [3]. Furthermore, the main targets of the F2F strategy include the reduction of fertilisers by 20% and nutrient losses by at least 50% [3] by adopting different sustainable practices for crop production [4].

Within this context, the available scientific evidence has documented the importance of organic and mineral nutrition for overall sustainable development [5], in particular, for crop growth and health [6], soil fertility [7,8], and productivity [9,10]. However, reaching the peak of crop nutrient uptake, the new paradigm for sustainable agriculture, based on the integration of scientific and technological advances in nutrition, puts soil, health, and the environment at the centre to efficiently manage highly productive agrosystems [4,11].

For this purpose, biostimulants represent a promising field in crop nutrition, are fundamental to reaching sustainable agrosystems, and are resilient against climate change [4,12,13]. Biostimulants were first defined by Zhang and Schmidt [14]. More recently, Du Jardin [15] described them as "a substance or micro-organism that, when applied to seeds, plants, or the rhizosphere, stimulates natural processes to enhance or benefit nutrient uptake, nutrient efficiency, tolerance to abiotic stress, or crop quality and yield".

The literature has identified different categories of biostimulants, which still need thorough assessment in agriculture to answer all aspects of their use under different conditions and crops [15]. Seaweed extracts and botanicals represent an important biostimulant category, possessing plant-growth-promoting activities that still need a thorough assessment to transform this traditional knowledge into scientific evidence [4,16–18]. The extracted substances act on plants and soils, and their action is intensively studied to understand all the physical, chemical, and biological aspects that control it and assess its effects.

Indeed, a handful of studies investigated the effects of seaweed and botanical extracts on plant metabolism and physiological health [19,20] and their phytohormone-like activity [21]. More recently, the research involved gene expression analysis, highlighting the effects of seaweed and botanical extracts on crops' metabolic regulatory pathways [22]. Others assessed the changes in soil properties (biological, physical, and biochemical) and their relation to nutrients' uptake efficiency [23–25]. Furthermore, seaweed and botanical extracts enhance the performance of plants under abiotic stresses, such as tolerance to freezing temperatures and high temperatures [26,27], drought, water, and salinity [28,29]. Agronomic efficiency has also been subject to different assessments to evaluate seaweed and botanical extracts' effects on improving productivity, product quality, and shelf life [13,30,31].

However, conflicting statements were reported on the effects of biostimulant applications in different climatic conditions occurring year after year in open-field cultivation systems. Indeed, Soppelsa et al. [32] observed positive effects of biostimulant application on yield performance in an apple orchard for two consecutive years, highlighting that the results obtained were independent of the impact of the seasonal climatic conditions. Similarly, Frioni et al. [33] showed beneficial effects of the seaweed extract foliar applications on fruit quality for several red grapevine cultivars over various climatic conditions (from cold to warm viticultural regions). In contrast, other studies noted different behaviours of the crops treated with biostimulants due to different climatic conditions [34,35]. In particular, Gutiérrez-Gamboa et al. [34] observed that seaweed applications on grapevines had a differential effect on must amino acids content, which depended on the climate conditions of the season. The authors found that during the driest season, the low dosage of the seaweed applications increased the concentration of several amino acids in musts, while in the rainy season, the concentration of certain amino acids in musts increased only with a high biostimulant dosage.

Furthermore, Chanda et al. [36] stated that not all microalgae species have significant biostimulant effects on plant growth, suggesting that the biostimulant properties are due to "species-specific" metabolites produced by particular microalgae species.

However, acknowledging that climate change effects vary in time, space, and intensity [37], adaptation in agriculture is related to sustainable agriculture and should be case-specific, and needs to be assessed and evaluated, considering specific field conditions [38]. Therefore, unlike previous studies, this research aimed to evaluate the agronomic and organoleptic effects of seaweed and botanical extract produced at the CMI Roullier laboratories (Centre Mondial de l'Innovation Roullier). Thus, we assessed the qualitative and quantitative performance of different management practices of two strategic crops in the Apulia region's (Italy) agriculture, using field experiments and satellite imaging.

2. Material & Methods

The experimental fields are located in Brindisi province in the Italian Apulia region, known for its modern agricultural sector. In particular, Apulia is placed first among the grape-growing areas of Italy and has 35.5% of the total olive area (Table 1). The climate is typically Mediterranean, with mild winters and hot summers. Figure 1 shows the average weather data collected between 1999 and 2019, with monthly average variations.

Table 1. Agricultural area of grapes and olives.

Indicator	Italy	Apulia
Average Farm Size (ha)	11.0	6.6
Total Agricultural Area (ha)	12,598,161	1,285,274
Grape Agricultural Area (ha) †	614,958	92,038
Olive Agricultural Area (ha) ‡	1,032,858	366,897

† Grape agricultural area includes grapevine and table grape varieties. ‡ Olive agricultural area has table and oil olive varieties. Source: Istat [39].

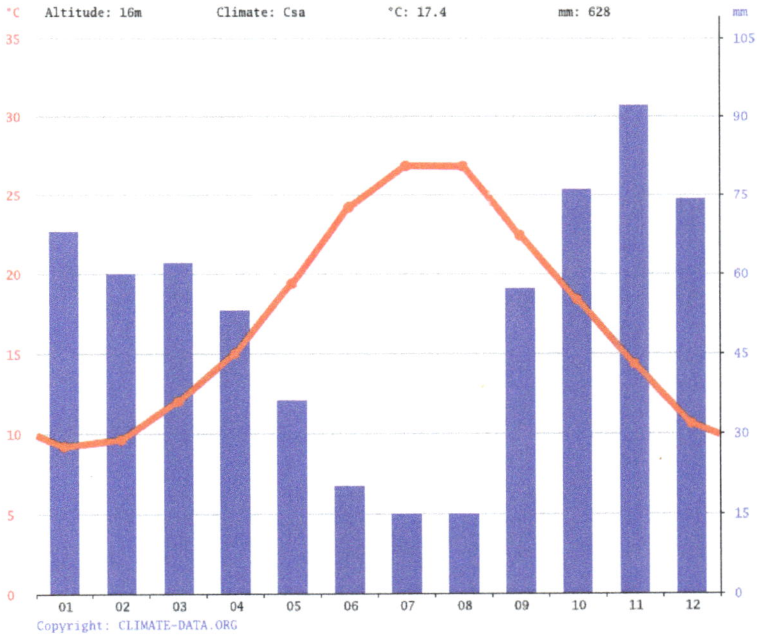

Figure 1. Average annual and monthly precipitation (mm) and temperature (°C) between 1999 and 2019 in Brindisi. License: Attribution-NonCommercial 4.0 International (CC BY-NC 4.0).

2.1. Experimental Design

The field trials were performed in *Olea europea* L. and *Vitis vinifera* L. orchards in Ostuni and Carovigno in Brindisi province (Figure 2). In both cases, the field trial was divided into four experimental plots, each of 2000 m^2, two of which were treated with vegetables and brown algae extract (*Ascophyllum nodosum* (L.) Le Jol.) in foliar spray, with the other two left untreated (Figure 2). The product was formulated in the CMI Roullier laboratories (Centre Mondial de l'Innovation Roullier) [19,22] to act on the following physiological aspects:

- Improve the photosynthetic activity of the crop and biomass development;
- Ameliorate resistance to biotic and abiotic stressors;
- Improve the overall crop nutrition.

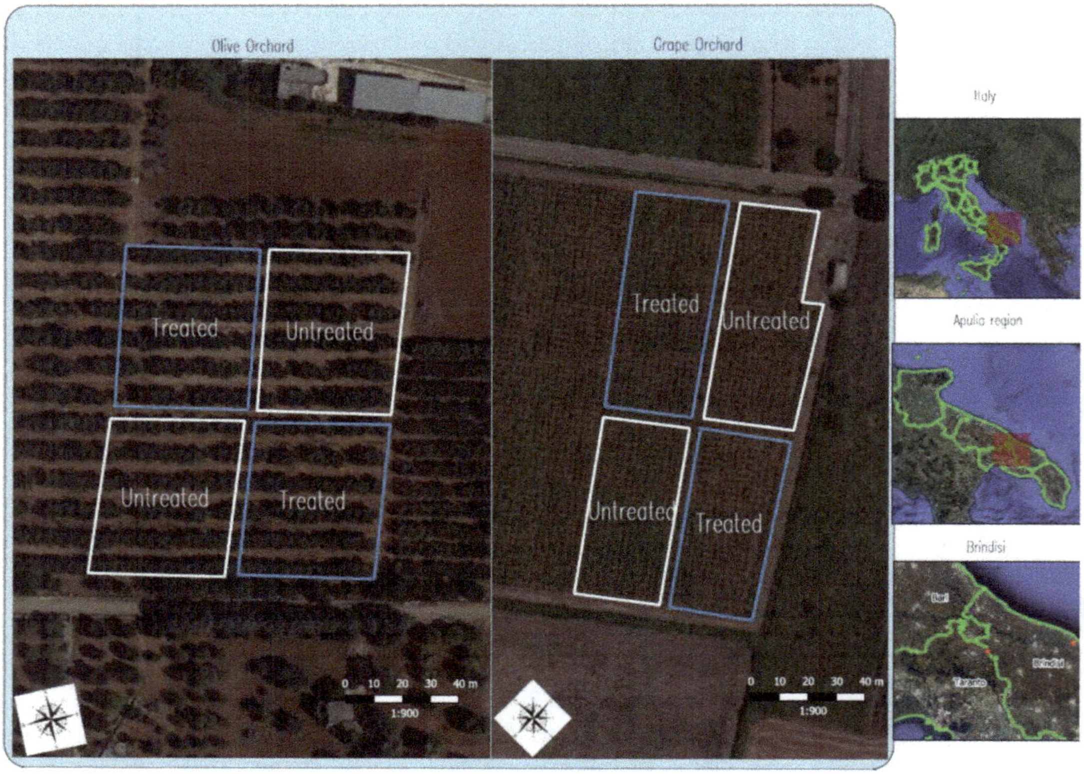

Figure 2. Maps of the case study within the regional and local context.

2.1.1. First Experiment

The trial was carried out in 50 year-old olive orchards in loamy soil at 30 m above sea level (Table 2). The orchard consisted of Coratina and Leccino cultivars, spaced at 10.0×4.0 m, and planting density was around 250 plants ha^{-1}. Two plots of 2000 m^2 were selected for each treatment within the experimental field.

Table 2. Infographics of the case study.

	Experiment 1: Olive Orchard	Experiment 2: Grape Orchard
Location	Carovigno	Ostuni
Coordinates	40°43′33.8″ N 17°43′30.5″ E	40°42′29.8″ N 17°23′27.2″ E
Variety	Leccino	Montepulciano
Age	50	31
Soil Type	Loam	Clay Loam
Soil pH	7.9	7.3
EC (μS cm^{-1})	0.63	0.14
% Sand (2–0.05 mm)	36.0	37.5
% Lime (0.05–0.002 mm)	38.3	31.7
% Clay (<0.002 mm)	25.7	30.8
Total Calcium (g/kg)	20	16

Agricultural practices were similar for all plots in fertilisation, irrigation, and phytosanitary applications. Treated plots in the olive orchard received two foliar applications (3 L ha^{-1} each), first at veraison and the second twenty days after. Applications were performed with a 1000 L atomiser equipped with six nozzles per side, each at 2.5 mm in diameter, delivering a water mixture volume of 1.500 L ha^{-1}.

2.1.2. Second Experiment

The experimental design of the grape orchard was set up in clay loam soil at 302 m above sea level on Montepulciano cultivar planted in 1991 in a tendon training system (Table 2). Plants spacing between rows and within the row was 2.5 m × 2.5 m with an east-west orientation and planting density of around 1600 plants ha^{-1}. Two plots of 2000 m^2 were selected for each treatment within the experimental field. Within each plot two subplots, consisting of 10 plants each, were identified and all subsequent investigations were carried out on these selected plants.

The agronomic practices and phytosanitary application performed in the orchard were the same for all plots. The biostimulant was applied three times at a 3 L ha^{-1} each time, using an atomiser equipped with ten nozzles each with 1.5 mm cone diameters and a distribution volume equivalent to 1000 L ha^{-1}. The first application was at 10 cm shoot length, the second at the pre-flowering stage, and the third at the post-fruit set.

2.2. Data Collection

2.2.1. First Experiment

In each plot of the olive orchard, we selected ten plants of Leccino cultivar for data collection, for a total of 20 plants per treatment. We determined the yield at harvest (first 10 days of November) and extracted oil from representative fruits of each plot (10 kg).

Free fatty acids (acidity), number of peroxides, and spectrophotometric examination in the ultraviolet (specific extinction at 232 nm (K232), specific extinction at 270 nm (K270), and the variation of specific extinction (ΔK)), were carried out in accordance with official European method of olive oil analysis (Reg. CEE 2568/1991 11/07/1991 GU CEE L248 05/09/1991 and subsequent modifications).

The total polyphenol content in the olive oil was determined by a colorimetric reaction with the Folin–Ciocalteau reagent. Briefly, 10 mL of a methanol–water (80:20) mixture was added to 10 g of olive oil. After shaking and centrifugation, two phases were obtained and the upper phase was recovered one more time to extract all residual polyphenols in the remaining oil. Then, 1 mL of extract, 1 mL of Folin–Ciocalteau reagent (Titolchimica, Pontecchio Polesine (RO), Italy), and 9 mL of 7.5% sodium carbonate (Sigma-Aldrich, Saint Louis, MO, USA) solution were introduced into a test tube. At the same time, a blank was prepared with 1 mL of Folin–Ciocalteau reagent, 9 mL of 7.5% sodium carbonate solution, and 1 mL of methanol–water mixture. After two hours, measures were carried out at 765 nm using a UV-Vis spectrophotometer Evolution 201 (Thermo Scientific, Waltham, MA, USA). The results were expressed in gallic acid by constructing a calibration curve in the range of 25–500 mg L^{-1}.

2.2.2. Second Experiment

In each plot of the grape orchard, we identified two subplots, each with 10 plants, to constitute four replicates per treatment, for a total of 40 plants per treatment for data collection.

For the selected plants, we measured the rachis length on all bunches three times during the experiment: at 10 cm shoot length (before biostimulant application), at the pre-flowering stage, and at the post-fruit set. In addition, from veraison to harvesting time, four samplings of 1 kg each were randomly performed, taking portions of bunches (3–4 grapes) from each subplot for each plant at different exposures and positions.

The grape juice obtained from each sample (4 samples per treatment) was filtered and subjected to multiparametric analysis through the FT-IR chemometric technique using the WineScan Flex instrument (FOSS, Hilleroed, Denmark). The following parameters were determined: reducing sugars (g L^{-1}); glucose (g L^{-1}); fructose (g L^{-1}); sugar degree (°Bx); pH, total acidity (g L^{-1}), volatile acidity (g L^{-1}); total dry extract (g L^{-1}); malic acid (g L^{-1}); tartaric acid (g L^{-1}); gluconic acid (g L^{-1}); citric acid (g L^{-1}); potassium (g L^{-1}); absorbance at 420 nm (A420), 520 nm (A520), 620 nm (A620); and colour tone (A420/A520).

In the first 10 days of October, at commercial harvest time, all bunches of each plant in the selected subplots (40 plants per treatment) were collected to determine bunch weight and number per plant.

2.3. Data Analysis

2.3.1. Statistical Analysis

All data are presented as means ± standard error and were statistically analysed by ANOVA according to a completely randomised design with four and two replications for grape and olive orchards, respectively. We adopted Tukey's HSD test to compare the means [40].

2.3.2. Satellite Vegetation Indices

The research compared the conventional vegetation indices typically derived from Landsat 8-Operational Land Imager and Sentinel-2-MultiSpectral Instrument data. The indices used include: (i) the Chlorophyll Index Red-Edge (CIRE) described by Gitelson and Merzlyak [41] as a good indicator of assessing production potential and understanding the nutrient status, stress due to water, and disease outbreak, etc.; (ii) the soil-adjusted vegetation index (SAVI) to assess vegetative cover where crop cover is low (e.g., grapes) or in arid regions [42]; (iii) the normalised difference vegetation index (NDVI) to estimate the density of green on an area of land [43]; (iv) the enhanced vegetation index (EVI2) to quantify vegetation greenness (compared to NDVI, EVI is more sensitive in areas with dense vegetation and it corrects for canopy background noise and atmospheric conditions) [44]; (v) the normalised difference red edge index (NDRE1) to assess N status and canopy density as indicators of crop health [41]; and (vi) the normalised difference red edge index (NDRE2) which is more accurate than the previous [45].

3. Results

The foliar biostimulant application differently affected the responses of the two crops studied.

3.1. First Experiment

Table 3 reports olive yield and olive oil chemical characteristics. The olive yield in the treated plots, although not significant, tended to be higher than in the control, with an average increase of 7.6%. Moreover, some oil chemical characteristics showed statistically significant differences. In particular, the oil acidity was significantly (p = 0.05) higher in the untreated control (0.23%) than in the treated plots (0.21%). Meanwhile, biostimulant application significantly (p = 0.01) increased total polyphenol content compared to the control. The increment in total polyphenol content was 3.6 times higher in treatment with the biostimulant than in the untreated control (Table 3).

Table 3. Yield production at harvesting time and chemical parameters of the oil §.

Treatments	Yield kg plant^{-1}	Acidity %	No. Peroxides meq O_2 kg^{-1}	K232	K270	Total Polyphenol mg kg^{-1}
Untreated control	75.27 a A	0.23 a A	5.95 a A	1.72 a A	0.09 b A	50.50 b B
Biostimulant	80.85 a A	0.21 b A	6.20 a A	1.78 a A	0.13 a A	181.00 a A

§ The values in each column followed by a different letter are significantly different according to Tukey's HSD test; with lowercase letters, significance is at $p \leq 0.05$ level and with uppercase letters, significance is at $p \leq 0.01$ level.

In Figure 3, the monitored satellite indices between 24 May 2021, before the biostimulant application, and 31 October 2021, at the end of the agricultural season, are reported. Olive plants' vegetation, chlorophyll, and N status (CIRE, SAVI, EVI2, NDVI, NDRE1, and NDRE2) showed a net increase in the treated plots compared to the untreated plots. The difference in indices' time-series means and standard deviations confirmed this increase (Figures 4 and 5), yet it was not statistically significant.

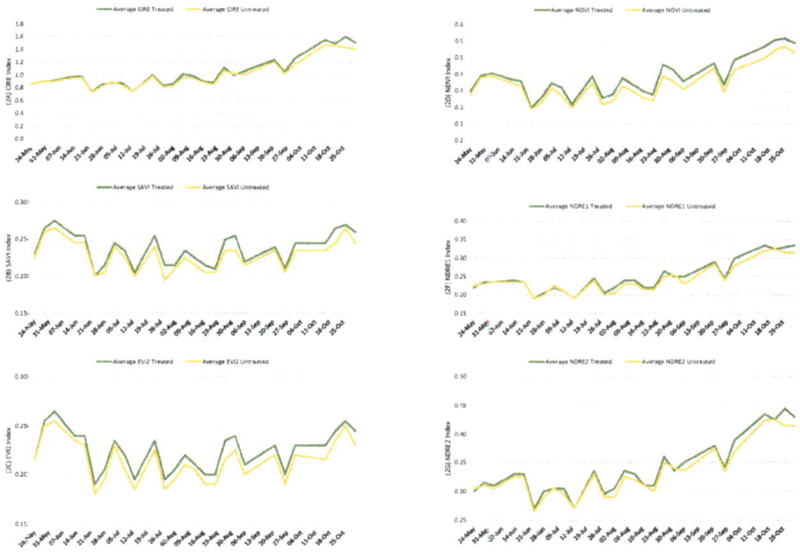

Figure 3. Comparison of (2A) CIRE, (2B) SAVI, (2C) EVI2, (2D) NDVI, (2E) NDRE1, and (2F) NDRE2 on treated and untreated olive plots in 2021.

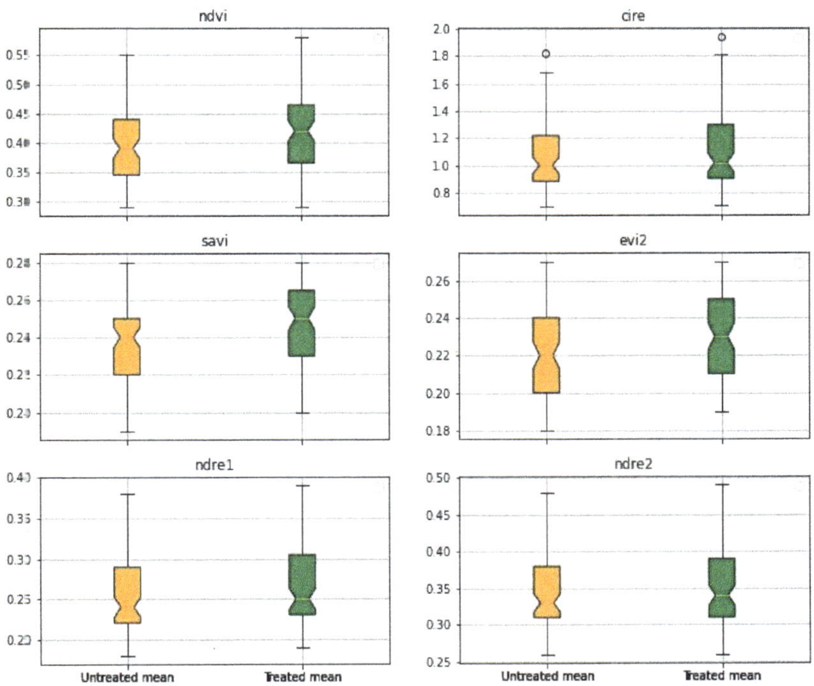

Figure 4. Satellite indices time-series mean comparison on treated and untreated olive plots.

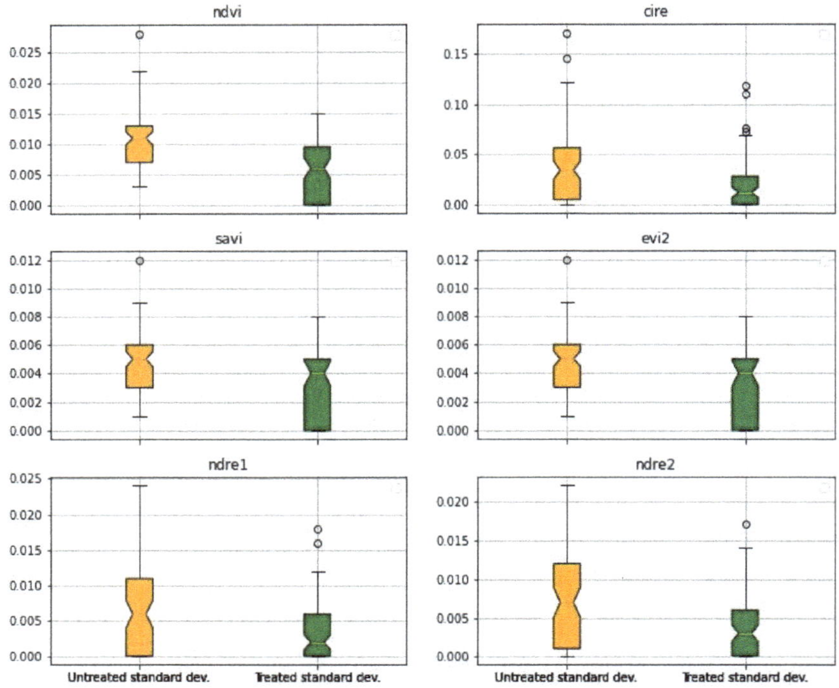

Figure 5. Satellite indices time-series standard deviation comparison on treated and untreated olive plots.

3.2. Second Experiment

After two biostimulant applications, the rachis length did not significantly differ from the untreated control (Table 4), whereas the third dose of the biostimulant, applied during the post-fruit set, induced a significant ($p = 0.01$) increase of 9.5%, on average, in bunch length compared to the control (Table 4).

Table 4. Effect of biostimulant on rachis length (cm) $^{\xi}$.

Treatments	7 June 2021 Pre-Flowering Stage	21 June 2021 Post-Fruit Set	6 July 2021 Berries Beginning to Touch
Untreated control	10.8 a A	13.3 a A	13.7 b B
Biostimulant	11.3 a A	14.5 a A	15.0 a A

$^{\xi}$ The values in each column followed by a different letter are significantly different according to Tukey's HSD test, with lowercase letters significance is at $p \leq 0.05$ level and uppercase letters significance is at $p \leq 0.01$ level.

The number of bunches was not significantly different between the treatments at harvest, whereas the bunch weight in the treatment with the biostimulant significantly increased ($p = 0.05$) compared to the control (Table 5).

Table 5. Effect of biostimulant on fruit yield at harvesting time *.

Treatments	Number of Bunches Plant^{-1}	Bunch Weight (Std) (g)
Untreated control	34.85 a A	195.1 (89.6) b A
Biostimulant	34.3 a A	214.4 (97.3) a A

* The values in each column followed by a different letter are significantly different according to Tukey's HSD test, with lowercase letters significant at $p \leq 0.05$ level and uppercase letters significant at $p \leq 0.01$ level.

All variables related to the quality of grape pulp, determined in the four investigations carried out from fruit set to berry ripening, did not show any statistically significant difference (Figures 6–9). In particular, at harvest, the content of reducing sugars, glucose, fructose, and total dry extract (Figure 6) was, on average, 217, 108, 109. and 239 g L^{-1}, respectively. In any case, fifteen days before harvesting (at 20 September 2021), the values of these variables were similar to those recorded at harvest in the treatment with the biostimulant. In contrast, in the control, they tended to increase. This behaviour was also confirmed by the °Brix values (Figure 4) that increased in the last fifteen days before harvesting by 3% in the untreated control. In contrast, it remained almost unchanged (+0.6%) in the biostimulant treatment. In addition, pH value, total acidity, malic, tartaric gluconic and citric acid, and potassium content were similar in both treatments and during the ripening of the berries (Figures 7–9).

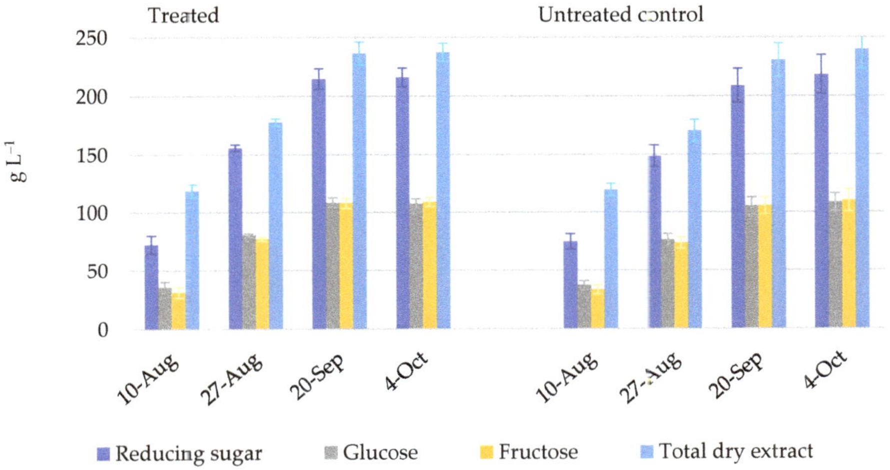

Figure 6. Evolution of reducing sugars, glucose, fructose, and total dry extract in the grape pulp during berry ripening (2021 season).

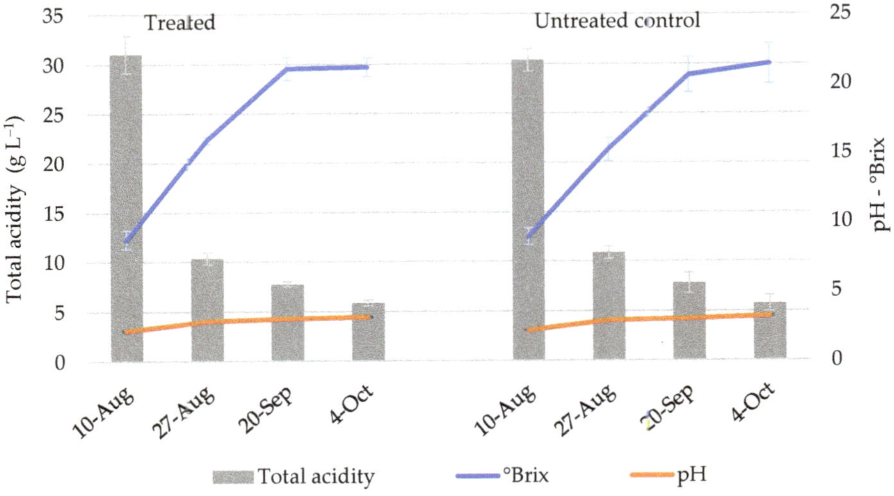

Figure 7. Evolution of total acidity, total sugar content (°Brix), and pH in the grape pulp during berry ripening (2021 season).

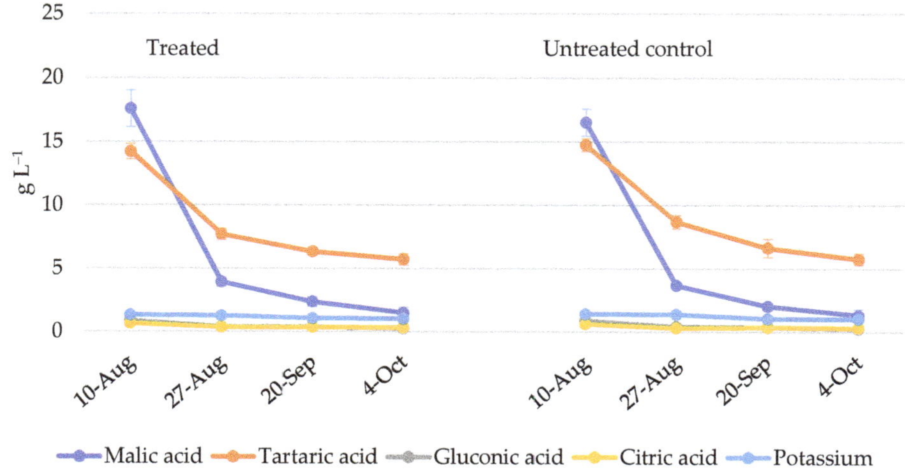

Figure 8. Evolution of acids and potassium content in the grape pulp during berry ripening (2021 season).

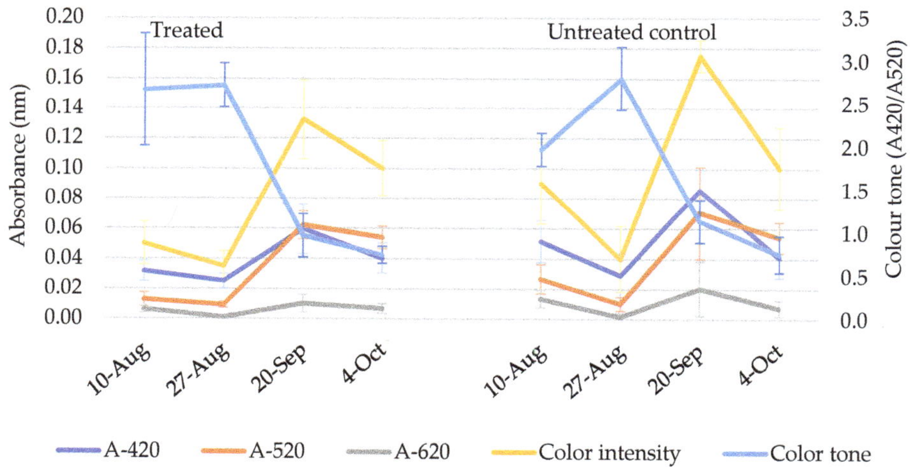

Figure 9. Absorption spectra and colour tone variation during the 2021 season.

The monitored satellite indices in the grape orchard between 24 May 2021, before the biostimulant application, and 31 October 2021, at the end of the agricultural season, are reported in Figure 10. Although not significant, the vegetation, chlorophyll, and N status (CIRE, SAVI, EVI2, NDVI, NDRE1, and NDRE2) showed a slightly higher rate in the treated plots than in the untreated. After the third biostimulant application, the differences between the two treatments were more evident in September. The indices' time-series means and standard deviations confirmed this trend (Figures 11 and 12).

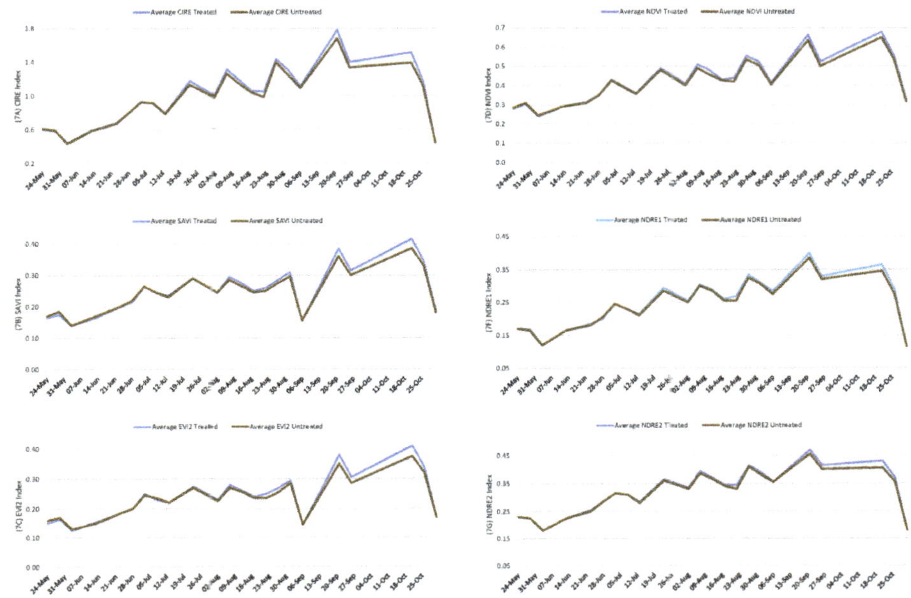

Figure 10. Comparison of (7A) CIRE, (7B) SAVI, (7C) EVI2, (7D) NDVI, (7E) NDRE1, and (7F) NDRE2 on treated and untreated grape plots in 2021.

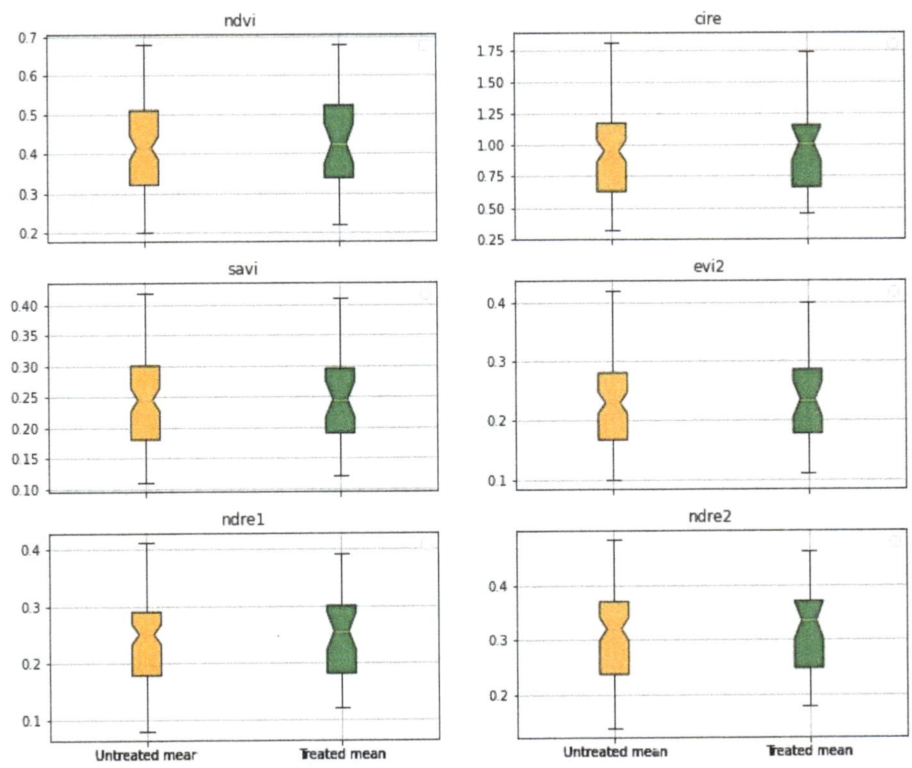

Figure 11. Satellite indices time-series mean comparison on treated and untreated grape plots.

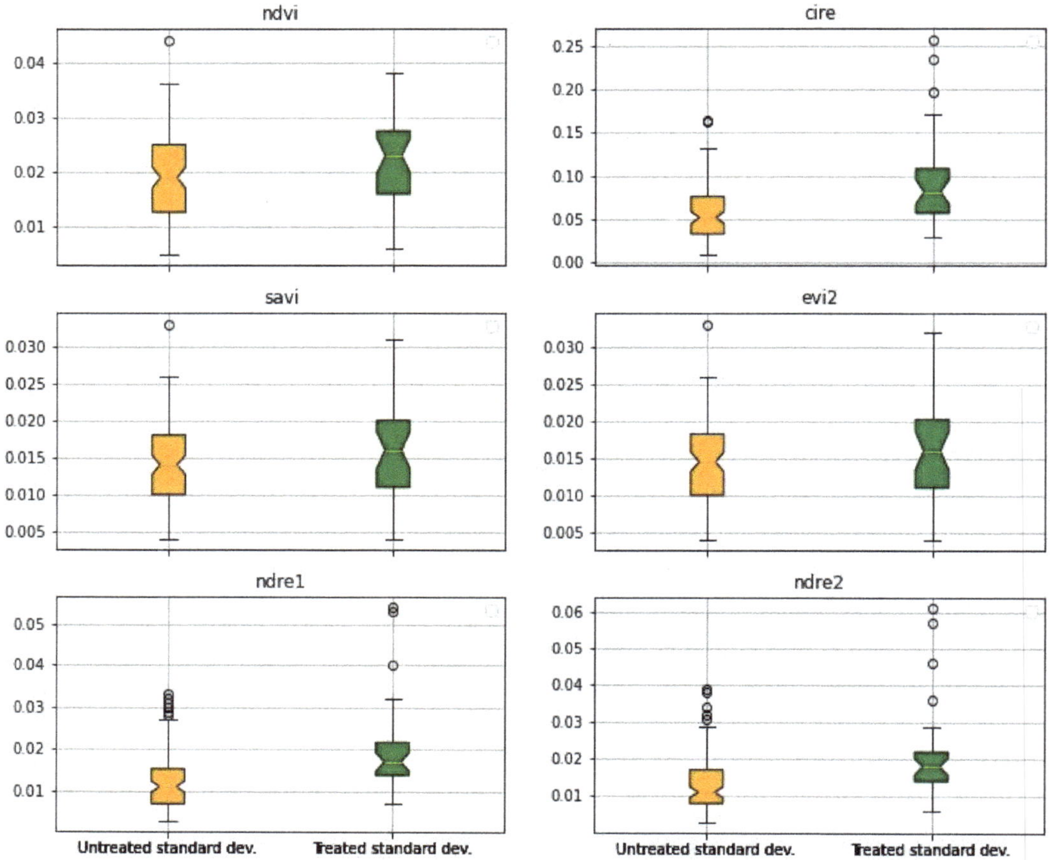

Figure 12. Satellite indices time-series standard deviation comparison on treated and untreated grape plots.

4. Discussion

Grapes (*Vitis vinifera* L. cv. Montepulciano) and olives (*Olea europaea* L. cv. Coratina) are important perennial crops that are well-adapted to the environmental conditions prevailing in the Mediterranean Basin. Since the consumption of these products is increasing, both as table fruits and industrial fruits (the viticultural and olive oil industries), sustainable practices are essential for modern agriculture to reduce environmental externalities while increasing crop yield. Therefore, biostimulant use can provide sustainable practices for farmers since their application could (i) improve nutrient use efficiency and, consequently, (ii) minimise fertiliser losses, (iii) increment total yield, and (iv) obtain a high-quality product.

This study shows that biostimulant application positively influences the agronomic performances of olive and grape plants in terms of olive oil quality improvement and grape yield quantity increase. This was also confirmed by the monitored satellite indices and is in line with the scientific literature [13].

Specifically, the lowest acidity and highest total polyphenol content observed in the treatment with biostimulant improved the olive oil shelf life. Moreover, due to its antioxidant effect, the total polyphenol content is an essential parameter in olive oil quality. Many factors may affect fruit production and physiology and, consequently, olive oil quality, like climatic conditions, cultivar, agronomic practices, fruit ripening, and harvest conditions [46–48]. Several studies reported the positive effects of biostimulant application on olive yield and oil quality [31,49–51]. In particular, Zouari et al. [50] observed that the foliar application of biostimulants could increase the oil polyphenol content over two consecutive growing seasons.

In grapevines, the three biostimulant applications mainly affected the yield and not pulp quality. Previous studies stated that the effects of biostimulants depend on several conditions, including the application rate, timing, and number of applications [51,52]. During the post-fruit set, the third biostimulant application seemed to positively affect bunch development and, consequently, the marketable yield. Indeed, biostimulant use in table grapes could change bunch morphology when applied during the development and growth of inflorescences and fruits [53–55]. Frioni et al. [33] observed that, in some cultivars of red grapevine, the biostimulant did not influence the change of pH and acidity. At the same time, the total soluble solid was slightly and positively affected during the first part of the ripening process, which partially agreed with our results.

Moreover, our findings highlight that the seaweed foliar application could be considered a viable strategy to stimulate early fruit maturation. In fact, the content of reducing total sugars, glucose, fructose, and total dry extract in grape pulp reached the plateau fifteen days before harvesting. In contrast, these variables increased in untreated control until the harvesting. Other authors also observed a faster fruit yield in the early season, after four biostimulant applications from blooming to the early development of strawberry fruits [56]. Appropriate nutrient management is crucial for optimising crop production [57]. Biostimulants should be used to enhance nutrient uptake and stimulate stress-related tolerance mechanisms [58] because they contain substances and/or micro-organisms whose function is to stimulate natural processes to enhance or benefit nutrient uptake, nutrient efficiency, tolerance to abiotic stress, and crop quality [59]. Seaweed extracts contain organic matter, mineral nutrients, and various hormones that contribute to plant growth, photosynthetic activity, and tolerance to biotic and abiotic stresses, thereby improving fruits' yield and quality [60,61]. They constitute a valuable and innovative tool to overcome nutrient limitations in different crop systems by enhancing plant resilience and improving nutrient uptake and assimilation [13,62,63].

Although it is a preliminary outcome, the results in both experimental fields confirm a general opinion that the foliar biostimulant applications could have beneficial effects on growth, yield, and quality. Still, crop responses could be unpredictable because they depend on crop characteristics, the specific phenological stage at the application time, the growing conditions, the number of applications, and timing. In fact, in our study, the biostimulant use caused a significant increase in grape yield but no effects on pulp quality. At the same time, in the olive orchard, the treatment improved oil quality without a significant yield increase. Consequently, high scientific and technical proficiency is necessary to determine and optimise the rates and timing of applications.

Therefore, our results are a first step towards understanding the numerous effects of seaweed extracts on plant responses. Consequently, further investigations should be carried out to gain a better understanding of the mode of action of the biostimulants and to assess the reliability of their application in the open field to allow accurate protocols to be established for their effective utilisation.

5. Conclusions

This paper reports the preliminary results of an investigation on using a vegetable- and seaweed-based biostimulant extract applied to grapes and olives, two strategic crops grown in the Apulia region. The results indicate that the biostimulant affects the responses of the two crops differently. In fact, the biostimulant applications in grapes increased the yield quantity but not the quality. In contrast, the olive yield did not change between treatments but the oil quality improved with the biostimulant applications.

This finding is in agreement with previous studies that found beneficial effects of biostimulants applied to several herbaceous and perennial crops. The outcome also confirms the scientific literature, which discusses the difference in response according to changes in abiotic and biotic factors. Therefore, for adaptative modern agriculture to be resilient against climate change, in-depth studies are fundamental to clarifying the mechanisms of these natural substances and identifying the optimal application techniques. Such studies will support high scientific and technical professionals in determining and optimising the rates and timing of such applications.

Author Contributions: Conceptualisation, D.E.C.; Data curation, G.P., M.E. and C.D.; Formal analysis, R.L. and C.D.; Investigation, G.F., M.D.C., G.P., M.E. and D.P.; Methodology, C.D.; Project administration, D.E.C. and C.D.; Supervision, D.E.C. and C.D.; Validation, D.E.C. and C.D.; Writing—original draft, R.L.; Writing—review & editing, D.E.C., G.F., M.D.C., D.P. and C.D. All authors have read and agreed to the published version of the manuscript.

Funding: This research received no external funding.

Acknowledgments: The authors would like to acknowledge the support of Marco Ferrari for the graphical support and Flavio Bellino's help in processing satellite images.

Conflicts of Interest: The authors declare no conflict of interest.

References

1. Wiebe, K.; Robinson, S.; Cattaneo, A. Climate Change, Agriculture and Food Security: Impacts and the Potential for Adaptation and Mitigation. In *Sustainable Food and Agriculture—An Integrated Approach*; Campanhola, C., Pandey, S., Eds.; The Food and Agricultural Organisation of the United Nations (FAO): Rome, Italy, 2019; pp. 55–74. [CrossRef]
2. UN. *Sustainable Development Goals: The 17 Goals*; Department of Economic and Social Affairs of the United Nations (UN): New York, NY, USA. Available online: https://sdgs.un.org/goals (accessed on 22 February 2022).
3. EU. *Farm to Fork Strategy: For a Fair, Healthy and Environmentally-Friendly Food System*; EU: Brussels, Belgium, 2020; p. 22. Available online: https://ec.europa.eu/food/sites/food/files/safety/docs/f2f_action-plan_2020_strategy-info_en.pdf (accessed on 22 February 2022).
4. El Chami, D.; Daccache, A.; El Moujabber, M. How Can Sustainable Agriculture Increase Climate Resilience? A Systematic Review. *Sustainability* **2020**, *12*, 3119. [CrossRef]
5. White, P.J.; Brown, P.H. Plant nutrition for sustainable development and global health. *Ann. Bot.* **2010**, *105*, 1073–1080. [CrossRef] [PubMed]
6. Barker, A.V.; Pilbeam, D.J. Introduction. In *Handbook of Plant Nutrition*, 2nd ed.; Barker, A.V., Pilbeam, D.J., Eds.; CRC Press: Boca Raton, FL, USA, 2007; pp. 3–18.
7. Jaskulska, I.; Lemanowicz, J.; Breza-Boruta, B.; Siwik-Ziomek, A.; Radziemska, M.; Dariusz, J.; Białek, M. Chemical and Biological Properties of Sandy Loam Soil in Response to Long-Term Organic–Mineral Fertilisation in a Warm-Summer Humid Continental Climate. *Agronomy* **2020**, *10*, 1610. [CrossRef]
8. Agegnehu, G.; Nelson, P.N.; Bird, M.I. Crop yield, plant nutrient uptake and soil physicochemical properties under organic soil amendments and nitrogen fertilization on Nitisols. *Soil Tillage Res.* **2016**, *160*, 239–246. [CrossRef]
9. Singh, Y.V.; Singh, K.K.; Sharma, S.K. Influence of Crop Nutrition on Grain Yield, Seed Quality and Water Productivity under Two Rice Cultivation Systems. *Rice Sci.* **2013**, *20*, 129–138. [CrossRef]
10. Fageria, N.K.; Baligar, V.C.; Li, Y.C. The Role of Nutrient Efficient Plants in Improving Crop Yields in the Twenty First Century. *J. Plant Nutr.* **2008**, *31*, 1121–1157. [CrossRef]
11. Howden, S.M.; Soussana, J.-F.; Tubiello, F.N.; Chhetri, N.; Dunlop, M.; Meinke, H. Adapting agriculture to climate change. *Proc. Natl. Acad. Sci. USA* **2007**, *104*, 19691–19696. [CrossRef]
12. El Chami, D. Towards Sustainable Organic Farming Systems. *Sustainability* **2020**, *12*, 9832. [CrossRef]
13. Chami, D.; Galli, F. An Assessment of Seaweed Extracts: Innovation for Sustainable Agriculture. *Agronomy* **2020**, *10*, 1433. [CrossRef]
14. Zhang, X.; Schmidt, R.E. The impact of growth regulators on the α-tocopherol status in water-stressed *Poa pratensis* L. *Int. Turfgrass Soc. Res. J.* **1997**, *8*, 1364–1371.
15. Du Jardin, P. Plant biostimulants: Definition, concept, main categories and regulation. *Sci. Hortic.* **2015**, *196*, 3–14. [CrossRef]
16. Craigie, J.S. Seaweed extract stimuli in plant science and agriculture. *J. Appl. Phycol.* **2011**, *23*, 371–393. [CrossRef]
17. Wells, E.; Wilkinson, M.; Wood, P.; Scanlan, C. The use of macroalgal species richness and composition on intertidal rocky seashores in the assessment of ecological quality under the European Water Framework Directive. *Mar. Pollut. Bull.* **2007**, *55*, 151–161. [CrossRef] [PubMed]
18. Henderson, J. *The Roman Book of Gardening*, 1st ed.; Routledge: Abingdon, UK, 2004; p. 164.
19. Jannin, L.; Arkoun, M.; Etienne, P.; Laîné, P.; Goux, D.; Garnica, M.; Fuentes, M.; Francisco, S.S.; Baigorri, R.; Cruz, F.; et al. Brassica napus Growth is Promoted by Ascophyllum nodosum (L.) Le Jol. Seaweed Extract: Microarray Analysis and Physiological Characterization of N, C, and S Metabolisms. *J. Plant Growth Regul.* **2013**, *32*, 31–52. [CrossRef]

20. Zhang, X.; Ervin, E.H.; Schmidt, R.E. Physiological Effects of Liquid Applications of a Seaweed Extract and a Humic Acid on Creeping Bentgrass. *J. Am. Soc. Hortic. Sci.* **2003**, *128*, 492–496. [CrossRef]
21. Khan, W.; Rayirath, U.P.; Subramanian, S.; Jithesh, M.N.; Rayorath, P.; Hodges, D.M.; Critchley, A.T.; Craigie, J.S.; Norrie, J.; Prithiviraj, B. Seaweed Extracts as Biostimulants of Plant Growth and Development. *J. Plant Growth Regul.* **2009**, *28*, 386–399. [CrossRef]
22. Billard, V.; Etienne, P.; Jannin, L.; Garnica, M.; Cruz, F.; Garcia-Mina, J.-M.; Yvin, J.-C.; Ourry, A. Two Biostimulants Derived from Algae or Humic Acid Induce Similar Responses in the Mineral Content and Gene Expression of Winter Oilseed Rape (*Brassica napus* L.). *J. Plant Growth Regul.* **2013**, *33*, 305–316. [CrossRef]
23. Stamatiadis, S.; Evangelou, E.; Yvin, J.-C.; Tsadilas, C.; Mina, J.M.G.; Cruz, F. Responses of winter wheat to *Ascophyllum nodosum* (L.) Le Jol. extract application under the effect of N fertilization and water supply. *J. Appl. Phycol.* **2014**, *27*, 589–600. [CrossRef]
24. Dobromilska, R.; Mikiciuk, M.; Gubarewicz, K. Evaluation of cherry tomato yielding and fruit mineral composition after using of Bio-algeen S-90 preparation. *J. Plant Growth Regul.* **2008**, *13*, 491–499.
25. Crouch, I.J.; Beckett, R.P.; Van Staden, J. Effect of seaweed concentrate on the growth and mineral nutrition of nutrient-stressed lettuce. *J. Appl. Phycol.* **1990**, *2*, 269–272. [CrossRef]
26. Mancuso, S.; Azzarello, E.; Mugnai, S.; Briand, X. Marine Bioactive Substances (IPA Extract) Improve Foliar Ion Uptake and Water Stress Tolerance in Potted Vitis vinifera Plants. *Adv. Hortic. Sci.* **2006**, *20*, 156–161. [CrossRef]
27. Möller, M.; Smith, M. The significance of the mineral component of seaweed suspensions on lettuce (*Lactuca sativa* L.) seedling growth. *J. Plant Physiol.* **1998**, *153*, 658–663. [CrossRef]
28. Shukla, P.S.; Shotton, K.; Norman, E.; Neily, W.; Critchley, A.; Prithiviraj, B. Seaweed extract improve drought tolerance of soybean by regulating stress-response genes. *AoB Plants* **2018**, *10*, plx051. [CrossRef] [PubMed]
29. Zhu, J.-K. Abiotic Stress Signaling and Responses in Plants. *Cell* **2016**, *167*, 313–324. [CrossRef]
30. Fan, D.; Hodges, D.M.; Critchley, A.T.; Prithiviraj, B. A Commercial Extract of Brown Macroalga (*Ascophyllum nodosum*) Affects Yield and the Nutritional Quality of Spinach In Vitro. *Commun. Soil Sci. Plant Anal.* **2013**, *44*, 1873–1884. [CrossRef]
31. Chouliaras, V.; Tasioula, M.; Chatzissavvidis, C.; Therios, I.; Tsabolatidou, E. The effects of a seaweed extract in addition to nitrogen and boron fertilization on productivity, fruit maturation, leaf nutritional status and oil quality of the olive (*Olea europaea* L.) cultivar Koroneiki. *J. Sci. Food Agric.* **2009**, *89*, 984–988. [CrossRef]
32. Soppelsa, S.; Kelderer, M.; Casera, C.; Bassi, M.; Robatscher, P.; Andreotti, C. Use of Biostimulants for Organic Apple Production: Effects on Tree Growth, Yield, and Fruit Quality at Harvest and During Storage. *Front. Plant Sci.* **2018**, *9*, 1342. [CrossRef]
33. Frioni, T.; Sabbatini, P.; Tombesi, S.; Norrie, J.; Poni, S.; Gatti, M.; Palliotti, A. Effects of a biostimulant derived from the brown seaweed Ascophyllum nodosum on ripening dynamics and fruit quality of grapevines. *Sci. Hortic.* **2018**, *232*, 97–106. [CrossRef]
34. Gutiérrez-Gamboa, G.; Garde-Cerdán, T.; Rubio-Bretón, P.; Pérez-Álvarez, E.P. Biostimulation to Tempranillo grapevines (*Vitis vinifera* L.) through a brown seaweed during two seasons: Effects on grape juice and wine nitrogen compounds. *Sci. Hortic.* **2020**, *264*, 109177. [CrossRef]
35. Zouari, I.; Mechri, B.; Tekaya, M.; Dabbaghi, O.; Cheraief, I.; Mguidiche, A.; Annabi, K.; Laabidi, F.; Attia, F.; Hammami, M.; et al. Olive oil quality influenced by biostimulant foliar fertilizers. *Braz. J. Biol. Sci.* **2020**, *7*, 3–18. [CrossRef]
36. Chanda, M.; Benhima, R.; Elmernissi, N.; Kasmi, Y.; Karim, L.; Sbabou, L.; Youssef, Z.; ElArroussi, H. Screening of microalgae liquid extracts for their bio stimulant properties on plant growth, nutrient uptake and metabolite profile of *Solanum lycopersicum* L. *Sci. Rep.* **2020**, *10*, 2820. [CrossRef]
37. IPCC. *Managing the Risks of Extreme Events and Disasters to Advance Climate Change Adaptation. A Special Report of Working Groups I and II of the Intergovernmental Panel on Climate Change*; Field, C.B., Barros, V., Stocker, T.F., Qin, D., Dokken, D.J., Ebi, K.L., Mastrandrea, M.D., Mach, K.J., Plattner, G.-K., Allen, S.K., et al., Eds.; IPCC: Geneva, Switzerland; Cambridge University Press: Cambridge, UK; New York, NY, USA, 2012; p. 582.
38. Wall, E.; Smit, B. Climate Change Adaptation in Light of Sustainable Agriculture. *J. Sustain. Agric.* **2005**, *27*, 113–123. [CrossRef]
39. Istat. *Indagine Sulla Struttura e Produzione delle Aziende Agricole: Aziende e Superfici per Coltivazione*; Istituto Nazionale di Statistica: Rome, Italy, 2016.
40. Tukey, J.W. Comparing Individual Means in the Analysis of Variance. *Biometrics* **1949**, *5*, 99–114. [CrossRef] [PubMed]
41. Gitelson, A.A.; Merzlyak, M.N. Quantitative estimation of chlorophyll-a using reflectance spectra: Experiments with autumn chestnut and maple leaves. *J. Photochem. Photobiol. B Biol.* **1994**, *22*, 247–252. [CrossRef]
42. Huete, A.R. A soil-adjusted vegetation index (SAVI). *Remote Sens. Environ.* **1988**, *25*, 295–309. [CrossRef]
43. Rouse, W., Jr.; Haas, R.H.; Schell, J.A.; Deering, D.W. Monitoring vegetation systems in the great plains with ERTS. In Proceedings of the Third Earth Resources Technology Satellite-1 Symposium, NASA SP-351, Washington, DC, USA, 10–14 December 1974; pp. 309–317.
44. Jiang, Z.; Huete, A.R.; Didan, K.; Miura, T. Development of a two-band enhanced vegetation index without a blue band. *Remote Sens. Environ.* **2008**, *112*, 3833–3845. [CrossRef]
45. Barnes, E.M.; Clarke, T.R.; Richards, S.E.; Colaizzi, P.D.; Haberland, J.; Kostrzewski, M.; Waller, P.; Choi, C.; Riley, E.; Thompson, T.L.; et al. Coincident Detection of Crop Water Stress, Nitrogen Status and Canopy Density using Ground-Based Multispectral Data. In Proceedings of the 5th International Conference on Precision Agriculture, Bloomington, IN, USA, 16–19 July 2000; Robert, P.C., Rust, R.H., Larson, W.E., Eds.; 2000.

46. Giuffrè, A.M. Evolution of Fatty Alcohols in Olive Oils produced in Calabria (Southern Italy) during Fruit Ripening. *J. Oleo Sci.* **2014**, *63*, 485–496. [CrossRef]
47. Giuffrè, A.M. Influence of cultivar and harvest year on triglyceride composition of olive oils produced in Calabria (Southern Italy). *Eur. J. Lipid Sci. Technol.* **2013**, *115*, 928–934. [CrossRef]
48. Uceda, M.; Hermoso, M.; Aguilera, M.P. The Quality of Olive Oil. In *El Cultivo del Olivo*; Barranco, D., Fernández-Escobar, R., Rallo, L., Eds.; Mundi-Prensa: Madrid, Spain, 2004; pp. 657–684.
49. Almadi, L.; Paoletti, A.; Cinosi, N.; Daher, E.; Rosati, A.; Di Vaio, C.; Famiani, F. A Biostimulant Based on Protein Hydrolysates Promotes the Growth of Young Olive Trees. *Agriculture* **2020**, *10*, 618. [CrossRef]
50. Ali, A.H.; Aboohanah, M.A.; Abdulhussein, M.A. Impact of foliar application with dry yeast suspension and amino acid on vegetative growth, yield and quality characteristics of olive. *Kufa J. Agric. Sci.* **2019**, *11*, 33–42.
51. Gamboa, G.G.; Romanazzi, G.; Garde-Cerdán, T.; Pérez-Álvarez, E.P. A review of the use of biostimulants in the vineyard for improved grape and wine quality: Effects on prevention of grapevine diseases. *J. Sci. Food Agric.* **2019**, *99*, 1001–1009. [CrossRef] [PubMed]
52. Lisek, J.; Sas-Paszt, L.; Derkowska, E.; Mrowicki, T.; Przybył, M.; Frąc, M. Growth, Yielding and Healthiness of Grapevine Cultivars 'Solaris' and 'Regent' in Response to Fertilizers and Biostimulants. *J. Hortic. Res.* **2016**, *24*, 49–60. [CrossRef]
53. Khan, A.S.; Ahmad, B.; Jaskani, M.J.; Ahmad, R.; Malik, A.U. Foliar application of mixture of amino acids and seaweed (Ascophyllum nodosum) extract improve growth and physicochemical properties of grapes. *Int. J. Agric. Biol.* **2012**, *14*, 383–388.
54. Norrie, J.; Branson, T.; Keathley, P.E. Marine plant extracts impact on grape yield and quality. *Acta Hortic.* **2002**, 315–319. [CrossRef]
55. Norrie, J.; Keathley, J.P. Benefits of ascophyllum nodosum marine-plant extract applications to 'thompson seedless′ grape production. *Acta Hortic.* **2006**, 243–248. [CrossRef]
56. Weber, N.; Schmitzer, V.; Jakopic, J.; Stampar, F. First fruit in season: Seaweed extract and silicon advance organic strawberry (Fragaria×ananassa Duch.) fruit formation and yield. *Sci. Hortic.* **2018**, *242*, 103–109. [CrossRef]
57. Tanou, G.; Ziogas, V.; Molassiotis, A. Foliar Nutrition, Biostimulants and Prime-Like Dynamics in Fruit Tree Physiology: New Insights on an Old Topic. *Front. Plant Sci.* **2017**, *8*, 75. [CrossRef]
58. Kunicki, E.; Grabowska, A.; Sękara, A.; Wojciechowska, R. The effect of cultivar type, time of cultivation, and biostimulant treatment on the yield of spinach (*Spinacia oleracea* L.). *Folia Hortic.* **2010**, *22*, 9–13. [CrossRef]
59. EBIC. Promoting the Biostimulant Industry and the Role of Plant Biostimulants in Making Agriculture More Sustainable; European Biostimulants Industry Council. Available online: http://www.biostimulants.eu/ (accessed on 30 March 2020).
60. Bulgari, R.; Franzoni, G.; Ferrante, A. Biostimulants Application in Horticultural Crops under Abiotic Stress Conditions. *Agronomy* **2019**, *9*, 306. [CrossRef]
61. Battacharyya, D.; Babgohari, M.Z.; Rathor, P.; Prithiviraj, B. Seaweed extracts as biostimulants in horticulture. *Sci. Hortic.* **2015**, *196*, 39–48. [CrossRef]
62. Mohammadipour, N.; Souri, M.K. Beneficial effects of glycine on growth and leaf nutrient concentrations of coriander (*Coriandrum sativum*) plants. *J. Plant Nutr.* **2019**, *42*, 1637–1644. [CrossRef]
63. Souri, M.K.; Bakhtiarizade, M. Biostimulation effects of rosemary essential oil on growth and nutrient uptake of tomato seedlings. *Sci. Hortic.* **2019**, *243*, 472–476. [CrossRef]

sustainability

MDPI

Article

Assessment of Ionomic, Phenolic and Flavonoid Compounds for a Sustainable Management of *Xylella fastidiosa* in Morocco

Kaoutar El Handi [1,*], Majida Hafidi [2], Khaoula Habbadi [1], Maroun El Moujabber [3], Mohamed Ouzine [1], Abdellatif Benbouazza [1], Miloud Sabri [1] and El Hassan Achbani [1]

1 Laboratory of Phyto-Bacteriology and Biocontrol, Plant Protection Unit-National Institute of Agronomic Research INRA, Meknès 50000, Morocco; khaoula405@gmail.com (K.H.); mohamed.ouzine1@usmba.ac.ma (M.O.); dacus5@hotmail.com (A.B.); miloud.sabri@uit.ac.ma (M.S.); achbaniofficiel@gmail.com (E.H.A.)

2 Laboratory of Biology, Moulay Ismail University, Meknès 50000, Morocco; hafidimaj@yahoo.fr

3 CIHEAM Bari, Istituto Agronomico Mediterraneo, Via Ceglie 9, Valenzano, 70010 Bari, Italy; elmoujabber@iamb.it

* Correspondence: kaoutar.elhandi@edu.umi.ac.ma

Abstract: Morocco belongs to the countries ranked at a high-risk level for entry, establishment, and spread of *Xylella fastidiosa*, which has recently re-emerged as a plant pathogen of global importance causing olive quick decline syndrome (OQDS). Symptomatic infection by *X. fastidiosa* leads to devastating diseases and important economic losses. To prevent such losses and damages, countries without current outbreaks like Morocco need to first understand their host plant responses to *X. fastidiosa*. The assessment of the macro and micro-elements content (ionome) in leaves can give basic and useful information along with being a powerful tool for the sustainable management of diseases caused by this devastating pathogen Herein, we compare the leaf ionome of four important autochthonous Moroccan olive cultivars ('Picholine Marocaine', 'Haouzia', 'Menara', and 'Meslalla'), and eight Mediterranean varieties introduced in Morocco ('Arbequina', 'Arbosana', 'Leccino', 'Ogliarola salentina', 'Cellina di Nardo', 'Frantoio', 'Leucocarpa', and 'Picholine de Languedoc'), to develop hypotheses related to the resistance or susceptibility of the Moroccan olive trees to *X. fastidiosa* infection. Leaf ionomes, mainly Ca, Cu, Fe, Mg, Mn, Na, Zn, and P, were determined using inductively coupled plasma optical emission spectroscopy (ICP-OES). These varieties were also screened for their total phenolics and flavonoids content Data were then involved in a comparative scheme to determine the plasticity of the pathogen. Our results showed that the varieties 'Leccino', 'Arbosana', 'Arbequina' consistently contained higher Mn, Cu, and Zn and lower Ca and Na levels compared with the higher pathogen-sensitive 'Ogliarola salentina' and 'Cellina di Nardò'. Our findings suggest that 'Arbozana', 'Arbiquina', 'Menara', and 'Haouzia' may tolerate the infection by *X. fastidiosa* to varying degrees, provides additional support for 'Leccino' having resistance to *X. fastidiosa*, and suggests that both 'Ogliarola salentina' and 'Cellina di Nardö' are likely sensitive to *X. fastidiosa* infection.

Keywords: olive quick decline syndrome; *Xylella fastidiosa*; calcium; manganese; Leccino; Leccinola salentina; olive; Moroccan olive varieties; Mediterranean olive varieties

check for updates

Citation: El Handi, K.; Hafidi, M.; Habbadi, K.; El Moujabber, M.; Ouzine, M.; Benbouazza, A.; Sabri, M.; Achbani, E.H. Assessment of Ionomic, Phenolic and Flavonoid Compounds for a Sustainable Management of *Xylella fastidiosa* in Morocco. *Sustainability* **2021**, *13*, 7818. https://doi.org/10.3390/su13147818

Academic Editor: Imre J. Holb

Received: 19 May 2021
Accepted: 6 July 2021
Published: 13 July 2021

1. Introduction

In Morocco, olive (*Olea europaea* subsp. *europaea* L.) groves have a crucial socio-economic role, representing the main source of livelihood for many local farmers. Moroccan olive groves represent the Southwesternmost part of the Mediterranean olive-growing landscape. In this country, olive cultivation and oil production are a deep-rooted tradition, both as an income for more than 450,000 farmers and a high environmental value crop, due to its role in soil protection, particularly, in mountain farms [1]. Furthermore, over the last few years, land use for olive cultivation in Morocco has increased from 946,818 ha in 2014 to 1,073,493 ha in the 2019 growing season [2] making this crop one of the most profitable and strategic horticultural crops in the country. The 'Picholine Marocaine' is the predominant variety; more than 96% of olive groves are planted with this variety [3]. Two varieties, 'Menara' and 'Haouzia', registered for cultivation in Morocco, were developed through clonal selection [4].

However, despite its importance, the olive crop presents some constraints, especially related to biotic stresses caused by plant pathogens [5]. *X. fastidiosa* is an important plant pathogen that attacks several economically important plants including the olive tree [6,7]. This pathogen has a very wide host range, including plants belonging to 595 plant species, 275 genera and 85 families [8]; however, it is well known as the causal agent of grapevine Pierce's disease (PD) and of Citrus Variegated Chlorosis (CVC) in South and North America. In Europe, *X. fastidiosa* is regulated as a quarantine pathogen; only individual outbreaks were reported until 2013 when the bacterium was detected for the first time in the Southern part of the Apulia region (Southern Italy) [8], one of the major olive-growing areas in Italy. The disease, named Olive Quick Decline Syndrome (OQDS), has a highly destructive impact on the infected trees and is characterized by leaf scorching, desiccation of leaves, twigs and branches and leads the whole tree to death within few years [9]. The olive quick decline syndrome (OQDS) caused by *X. fastidiosa* is one of the most damaging diseases threatening olive trees worldwide [9].

Studies on this disease are progressing, but quantifiable data and estimates on its spread are scarce [8]. To date, Morocco is considered a country free from this devastating bacterium [10]. However, the likelihood of *X. fastidiosa* occurrence and spread in Morocco is high due to the ineffective control measures adopted, and to the insect vectors that transmit the bacterium across the world [10]. Indeed, Morocco is 13 km from Spain and is the gateway to all African countries. Also, Morocco has significant trade with several European countries, notably France, Spain, and Italy, which may contribute to the introduction of *X. fastidiosa* indirectly via infected plant material [10].

The OQDS is characterized by leaf scorching and scattered desiccation of twigs and small branches which, in the early stages of the infection, are observed on the upper part of the canopy [11]. Leaf tips and margins turn from dark yellow to brown and eventually dry out. Over time, symptoms increase severely and extend to the rest of the canopy, which acquires a blighted appearance [12]. According to D'Attoma et al. [13], variation in leaf mineral nutrients is also associated with the olive infection by *X. fastidiosa*.

The analysis of the whole profile of trace elements and mineral nutrients can contribute to the evaluation the physiological status of the plant in inter-connection with the pathogen infection [14]. Studies on mineral elements accumulation in specific plant tissues, especially in the leaves, have been used to assess the physiological status of the plant [13]. Regarding disease development in plant hosts, the influence of specific mineral elements is well documented. However, the ionome has only been used in a few instances as a composite phenotypic character to assess the relationship between plants and *X. fastidiosa* infection. Indeed, some studies understandably indicated a correlation between the content of some ions in the leaf and the virulence of *X. fastidiosa* [13,15]. It was highly recommended that the host ionome and its variation could be considered as a potential tool for the control of diseases caused by this xylem-limited phytopathogenic bacterium [13]. Mineral transport and balance are crucial for the growth and development of plants and also microorganisms and can be a major factor in disease control and progression [16]. In host–pathogen interactions, the competition for these elements is a phenomenon known as "nutritional immunity" [17]. The ionome analysis, defined as the total profile of the mineral nutrients and trace elements found in an organism, represents a pathogen's approach to looking into the physiological status of the plant [18]. In previous studies, the ionome analyses of field-grown blueberry, pecan, grapevines, and greenhouse-grown tobacco during *X. fastidiosa* infection, highlighted significant changes between uninfected and infected plants, as well as between symptomatic and asymptomatic leaves, revealing a complex interaction between different elements in the host [14,19,20]. *X. fastidiosa* accumulates high levels of metals (Mn,

Zn and Cu) in its biofilm cells (important for the virulence of this bacterium), as compared with planktonic cells [21]. Additionally, several elements (Ca, Mg, and Fe) are known to promote the expression of virulence traits [15,22,23] whereas others (Cu and Zn) have a deleterious effect on growth and biofilm production [24]. Particular attention has been given also to the identification of secondary metabolites that are essential for plant disease resistance [25] as possible strategies for disease management. Bacterial growth, assemblage and biofilm formation could be affected by xylem sap components [15]. Phenolic acids and flavonoids have been shown to inhibit the in vitro growth of *X. fastidiosa* [25].

The aim of the present work was to find a sustainable management tool for olive trees threatened by *X. fastidiosa*. The tool consists in a thorough determination and comparison of the ionomic, phenolics and flavonoids profile of 'Picholine Marocaine', the most widespread and typical Moroccan olive variety, with those of Moroccan clonal selected varieties ('Haouzia', 'Menara', and 'Meslalla') and eight Mediterranean varieties recently introduced in Morocco ('Arbequina', 'Arbosana', 'Leccino', 'Ogliarola', 'Cellina di Nardo', 'Frantoio', 'Leucocarpa' and 'Picholine de Languedoc') to develop hypotheses related to the resistance or susceptibility of the Moroccan olive trees to *X. fastidiosa* infection. All the varieties were grown in the same experimental olive grove and under identical pedoclimatic conditions. To the best of our knowledge, we will be the first in Morocco to carry out this type of proactive research on *X. fastidiosa*. However, this research is easily replicable in all Mediterranean countries where olive trees are under an existential pathogenic threat.

2. Materials and Methods

2.1. Samples Collection

In order to assess the leaf ionomic profiles of four autochthonous Moroccan cultivars ('Picholine Marocaine', 'Haouzia', 'Menara', and 'Meslalla'), and eight Mediterranean varieties introduced in Morocco ('Arbequina', 'Arbosana', 'Leccino', 'Ogliarola', 'Cellina di Nardo', 'Frantoio', 'Leucocarpa' and 'Picholine de Languedoc'), leaf samples were collected from uninfected 16-year-old olive trees in late December 2020 from olive groves located at the experimental station of the National Agricultural Research Institute (INRA) in Ain Taoujdate, Fez-Meknes region (Morocco). For each tree, five branches were selected, and mature leaves were detached from the median part of hardwood cuttings and collected for analysis. All the olive trees were grown in the same experimental olive grove and under identical pedoclimatic conditions.

2.2. Polymerase Chain Reaction

Leaf samples were tested by PCR for the detection of *X. fastidiosa*. Total DNA was extracted from leaf petioles and midveins using a CTAB- based extraction buffer [24,26]. For PCR, the RST31/RST33 set of primers targeting the 16 S rDNA gene was used [26]. The used primers (RST31/RST33) are widely accepted for the detection of the bacterium in quarantine programs, as well as primers targeting the 16 S rDNA genomic region, which are more acceptable for the precise detection of a wider number of genetically diverse strains of *X. fastidiosa*. Reactions were conducted in a final volume of 20 µL, using 0.5 µL each of forward and reverse primer, 3 µL of total DNA template and 3 µL of 5× GoTaq polymerase (Promega). PCR was completed using a (5 PRIMEG/C Serial No 51147-2) thermocyler set to the following: 94 °C for 3 min, 35 cycles of 94 °C for 30 s, 50–55 °C for 30–45 s and 72 °C for 30 s, and a final extension of 5 min at 72 °C [27]. The resulting PCR products were visualized by electrophoresis in 1% Tris-Acetate-EDTA agarose gel stained with ethidium bromide.

2.3. Extract Preparation for Assessment of Total Phenolic and Flavonoid Content

Leaves were cut and frozen at −20 °C for later lyophilization. Then, they were ground into powder at room temperature using an IKA A11 Basic Grinder (St. Louis, MO, USA). Extraction was based on the method previously described by Sanders et al. [28] and moderately modified by Xie and Bolling [29]. First, 1 g aliquots of powder were transferred into polypropylene tubes and homogenized in 20 mL of ethanol and ultrapure water (80:20, *v/v*) at 4 °C for 15 min using an IKA T-18 Basic Ultra-Turrax homogenizer (IKAWerke GmbH & Co., Staufen, Germany). The homogenate was then centrifuged at 3000× *g* for 10 min at 4 °C, and the supernatant was removed from the residue. The residue was homogenized, and the supernatant removed, for a total of three extractions. The supernatants were then combined and filtered through Whatman No. 1 filter paper.

2.4. Total Phenolic Content

The total phenolic (TP) content of leaf extracts was determined using the Folin–Ciocalteu micro method [30]. Three Folin's reactions were made for each olive leaf sample in 1 mL microcentrifuge tubes. The reaction mixture contained 40 µL of extract, 3160 µL of ultrapure water, 200 µL of the Folin–Ciocalteu reagent and 600 µL of 20% sodium carbonate solution. After 30 min of incubation at 40 °C, absorbance was measured at λ = 765 nm (UV-1700 Shimadzu, Japan). The TP content is expressed as gallic acid equivalent per dry weight (mg GAE/gdw) for olive leaves. Three independent experiments were performed.

2.5. Total Flavonoid Content

Total flavonoid (TF) content was measured using the colorimetric method with aluminum chloride [31,32]. Absorbance was measured at 510 nm and the results were expressed as catechin equivalent per dry weight (mg CE/gdw). Three flavonoid reactions were made for each olive leaf sample.

2.6. Determination of Leaf Ionome and Soil Parameters

The soil texture is sandy-clay according to international standards, slightly calcareous, moderately rich in organic matter, phosphorus and potassium, and with a usable water reserve of 1.7 mm.cm^{-1}. After excising the petioles, the whole leaves and soil were crushed to a fine powder by a plastic mortar and pestle and sampled at 5 and 10 mg of dry weight. Samples were digested for 1 h at 100 °C in 200 µL of mineral-free concentrated nitric acid. After dilution with ultrapure, mineral-free water and centrifugation at 13,000× *g* to remove any remaining particulates, samples were analyzed by ICP-OES as described by Cobine et al. [21], with simultaneous measurement of Ca, Fe, Mg, Na, Mn, Na, P, S, and Zn. As controls, blanks of nitric acid were digested in parallel. Mineral concentrations were determined by comparing emission intensities to a standard curve created from certified mineral standards (SPEX CertiPrep). Three independent experiments were performed.

2.7. Statistical Analyses

Ionome data (individual minerals) were analyzed separately with a one-way analysis of variance (ANOVA) followed by the post-hoc Student–Newman–Keuls test. Principal component analysis was carried out using a correlation matrix. A scatter plot was created according to PC1 and PC2 using SPSS v20 software. SPSS statistical package software (SPSS for Windows, Version 20, SPSS Inc., Chicago, IL, USA) was used for the statistical analysis of data.

3. Results

3.1. Polymerase Chain Reaction

X. fastidiosa PCR detection confirmed that all plant samples analyzed were not infected. No amplified DNA was obtained from any of the tested samples using PCR, confirming the absence of the bacterium in these samples.

3.2. Determination of Leaf Ionome

The total concentrations of mineral elements of the leaves sampled are shown in Table 1. The elemental composition was compared with reference values of nutrient content in olive leaves [33]. The concentration of Mg, Mn, Na, and Zn was within these reference ranges. However, Fe and P were considered low or close to the minimal range. Comparing leaf ionomes, statistical analyses showed that 'Leccino', 'Arbozana', 'Arbiquina', 'Menara' and 'Haouzia' had higher levels of Mn, Cu, Zn, and P. The same varieties also showed lower levels of Ca, Na. Concerning the four remaining varieties, 'Picholine Marocaine', 'Picholine de Languedoc', 'Ogliarola', and 'Cellina di Nardo', data indicated lower levels of Mn, Zn, Cu. These varieties showed higher levels of Ca and Na (Table 1). Nutrient concentrations are expressed in g·kg^{-1}, except for Mn, Na, and Zn that are expressed in mg·kg^{-1}. Ca concentration was higher than the reference (1–14 mg·kg^{-1}).

Table 1. Elemental analysis of olive leaves expressed in weight per dry weight. The mean of three replicates was represented.

	Ca		Mg	P	Cu	Mn	Na	Zn	Fe
Reference *	mg/kg or g/kg	10–14	1–1.6	1–1.3	-	20–36	<200	4–9	90–124
Arbiquina	NA	13.20	2.04	0.82	17.35	31.18	33.30	7.80	67.05
Arbozana	NA	12.52	2.09	0.93	18.50	35.38	28.60	10.23	67.60
Menara	NA	13.35	2.01	0.72	17.03	31.18	35.75	7.86	69.46
Haouzia	NA	14.32	1.83	0.70	17.02	29.46	35.16	7.66	70.32
Picholine Marocaine	NA	19.50	0.98	0.49	10.40	23.40	41.35	5.13	80.36
Picholine Languedoc	NA	20.50	0.85	0.33	9.97	21.33	41.61	4.57	80.87
Frantoio	NA	14.45	1.75	0.68	15.25	28.82	35.05	7.25	72.34
Leucocarpa	NA	14.51	1.65	0.62	15.15	28.21	34.24	7.06	72.36
Leccino	NA	7.02	3.62	2.51	23.82	42.63	20.21	15.61	41.3
Meslalla	NA	15.30	1.50	0.53	14.93	26.38	40.31	7.01	73.65
Cellina di Nardò	NA	27.15	0.16	0.12	7.88	18.59	48.76	4.80	93.45
Ogliarola	NA	27.31	0.12	0.13	7.12	17.99	49.40	4.31	95.45

Element concentrations are expressed in g/kg, except for Fe, Mn, Na, and Zn that are expressed in mg/kg. * Reference concentrations were obtained from Kailis and Harris [34].

Results of soil analysis showed that the soil where olive trees are grown had a higher content in Cu, Zn, and Mn and a low level of Mg. Ca concentration in 0–35 cm soil depth was lower than the reference (3000 mg·kg^{-1}), while Mn concentration was higher than the reference (5–20 g·kg^{-1}) (Tables 2 and 3).

Table 2. Physical and chemical properties of soil.

Soil Depth	Sand	Silt	Clay	pH	EC	OM	K2O	P2O5	CaCo3
Cm	%	%	%		mS/cm	%	mg·kg^{-1}	mg·kg^{-1}	%
0–35	46.8 ± 0.4	10.20	43.00	6.50	0.10	2.50	458.80	73.30	2.70
35–70	46.10	16.10	37.60	7.80	0.10	1.60	222.50	15.10	3.10

EC: electrical conductivity, OM: organic matter, K2O: potassium, P2O5: phosphorus, CaCo3: calcium carbonates. Data represent means ± standard deviation.

Table 3. Chemical analysis of soils wherein the leaf ionome profiles were evaluated.

	Mg	Cu	Mn	Na	Zn
Soil analysis (mg·kg^{-1} Fine Fraction)	237.6	19	25.3	2109	0.8

Values represent averages of three replicate samplings. All varieties refer to soil collected from the same area.

3.3. Determination of Total Phenolic and Flavonoid Content

The total phenolic content in leaves of the studied varieties is reported in Figure 1. Regarding results, the total phenolic content of all varieties varied considerably since several varieties showed different statistically significant values (F = 157.69, p = 0.02). 'Leccino' presented the statistically significant higher phenolic content (45.8 mg·GAE/g) followed by 'Arbiquina', 'Arbozana', 'Menara', 'Haouzia', 'Frantoio', 'Leucocarpa', 'Meslalla', 'Picholine Marocaine', 'Picholine de Languedoc', 'Cellina di Nardò', and 'Ogliarola' which showed the statistically significant lower phenolic content (8.07 mg·GAE/g). Concerning the total flavonoid content, statistical analyses showed a significant difference regarding the variety's values ($F = 136.45$, $p = 0.03$). 'Leccino' showed the highest total flavonoid content (24.49 mg·GAE/g) and 'Ogliarola' the lowest total flavonoid content (4.89 mg·GAE/g) (Figure 2).

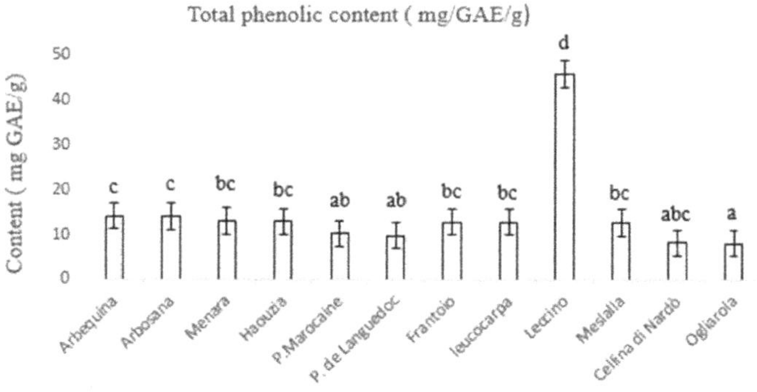

Figure 1. Total phenolic content for the studied Olive varieties (mg GAE/g).

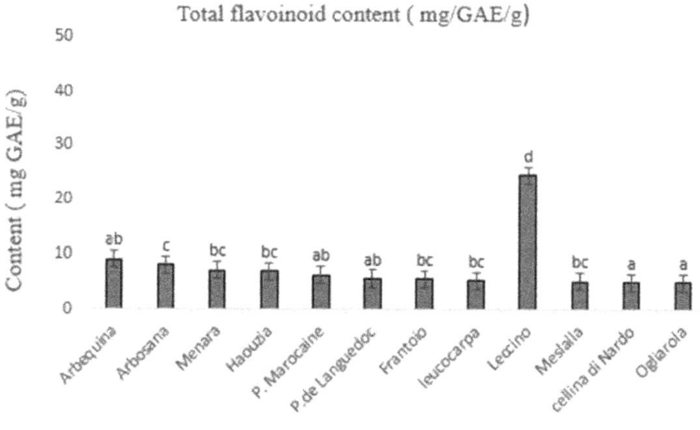

Figure 2. Total flavonoid content for the studied Olive varieties (mg GAE/g).

3.4. *Principal Component Analysis*

Principal component analysis (PCA) was used to determine the most significant descriptors in the data set. Only a principal component loading of more than 0.5 was considered as being significant for each factor. Thus, a total variance of 98.47% was explained by only two components. The first component consisted of 10 variables, which represent more than 90% of all total variables, and explained 91.4% of the total variance (Figure 3). The first component accounted for 91.4% of total variance, which is strongly correlated to Ca ($r = -0.921$), Mn ($r = 0.979$), Mg ($r = 0.987$), Na ($r = 0.970$), Zn ($r = 0.974$), Cu ($r = 0.963$), P ($r = 0.965$), Fe ($r = -0.989$), TPC ($r = 0.939$), TFC (0.865). The second function accounted for 7.06% of total inertia.

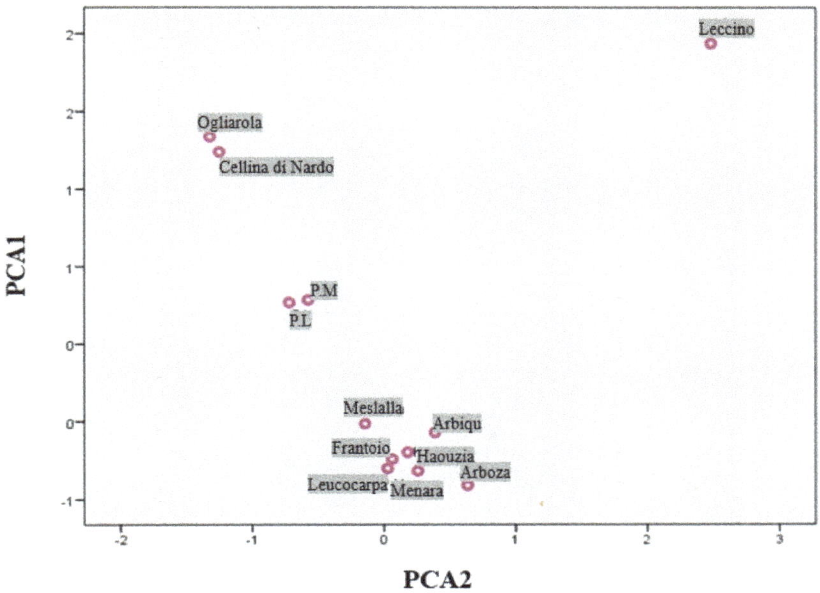

Figure 3. Scatter plot for the two principal components (PC1/PC2, 98.47% of total variance) for the studied Olive varieties based on all descriptors.

4. Discussion

Xylella fastidiosa continues to emerge as a major and devastating bacterial pathogen for innumerable crops, and no cure has been identified so far. Current management strategies are based on the use of cultivars showing resistance to the pathogen in the field such as 'Leccino' [13] and the control of vectors to limit the disease spread. The first requires including tolerant genotypes in breeding programs or replacing susceptible varieties in the current fields [35–37]. Many characteristics make *X. fastidiosa* attractive for studying the effects of nutrients in plant-pathogen interactions. Specifically, since it is xylem-limited, this bacterium is found in the vessels where mineral elements are translocated throughout the plant. Consequently, we spotlight the hypothesis that differences within the ionome of non-infected Moroccan olive trees, the total phenolic and flavonoid content can give an overview of the plasticity and immunity of the Moroccan olive sector to *X. fastidiosa* before the bacterium enters Moroccan groves [36,38].

The outcome of the present investigation shows that 'Leccino' olive variety has a higher Mn, Cu, Zn, total phenolic, and flavonoid content, and lower Ca and Na content which is in line with other previous work [13,14,39]. Manganese, Cu, and Zn are essential micronutrients for plant growth; Mn is involved in the photosynthetic machinery and in the detoxification of reactive oxygen species (ROS) [40], Cu is essential for the formation of chlorophyll [41], and Zn is involved as a cofactor in many enzymes such as alcohol dehydrogenase, carbonic anydrase, and RNA polymerase [42]. It is worth noting that these ions are also strongly involved in the plant defense machinery and in the *X. fastidiosa* virulence. Furthermore, these ions are strongly involved in the plant defense artillery against infections, including *X. fastidiosa* [43]. Specific attention is attributed to Zn ability to reduce the pathogenicity of pathogens [44]. A Zn-finger protein gene, CAZFP1, encodes a zinc-finger transcription factor that builds up in the preliminary phase of the infection of *Xanthomonas campetsris pv. vesicatoria* to pepper fruits [45]. In addition, Zn-fingers binding domains are related to the effector-triggered immune response [46]. High Zn concentrations can preserve plants by direct toxicity and by Zn-triggered organic defenses [43,47]. This confirms the importance of crop nutrient management for a sustainable agriculture [37]. *X. fastidiosa* biofilm formation is prohibited by Zn and Cu concentrations higher than 0.25 mM, and 200 μM respectively [21], and *in planta*, Zn detoxification is needed to trigger the full virulence of the pathogen [19]. Within this context, previous studies have shown that the supply to the olive canopy of a zinc-copper–citric acid biocomplex, namely Dentamet®, reduces both the field symptoms and *X. fastidiosa* subsp. *pauca* cell densities in the foliage allowing the trees to survive the infection [48]. Recently a high Mn leaf content would appear to match up with a relative level of tolerance in Leccino cultivar to *X. fastidiosa* subsp. *pauca* [13]; the present study would corroborate this feature since both 'Ogliarola salentina' and 'Cellina di Nardò' cultivars are characterized by a lower Mn content than Leccino. The Mn ion is involved in flavonoid and lignin production thereby preserving the cultivar from infection by *X. fastidiosa* subsp. *pauca* [49]. Another important ion, Ca, seems to influence biofilm formation by both extracellular ionic bridging and intracellular stimulation that relies on protein [13]. Ca increases cell attachment probably via type I pili, twitching motility and cell-to-cell attachment responsible for cell aggregation [50]. Consequently, it is a limiting factor in the initial stages of biofilm formation characterized by cell attachment, while it has a less prominent role in late stages of biofilm maturation [15].

Based on those facts, it is logical that olive varieties exhibiting high content of Mn, Cu, and Zn and low content of Ca and Na ('Arbequina', 'Arbosana', 'Menara' and 'Haouzia') would likely be more effective in resisting the development of the infection after the formation of *X. fastidiosa* biofilm. On the other hand, the varieties showing a high content of Ca and Na and reduced content of Mn, Cu, and Zn ('Frantoio', 'leucocap' 'Meslala') would be more adapted to fight the formation of *X. fastidiosa* biofilm in the first place. We also believe that olive varieties with deficiency of the mentioned ions would likely be extremely sensitive and prone to a fast and strong *X. fastidiosa* infection ('Picholine marocaine', and 'Picholine de Languedoc') based on the current data.

Besides ions, phenolic and flavonoid compounds are known for their strong antibacterial effect and their importance in the protection of plants against infections [48,49,51]. The mechanisms of antibacterial action of phenolic compounds are not yet fully deciphered but these compounds are known to involve many sites of action at the cellular level. While phenolic acids have been shown to disrupt membrane integrity, as they cause consequent leakage of essential intracellular constituents [52]. Flavonoids may link to soluble proteins located outside the cells and with bacteria cell walls thus promoting the formation of complexes [53]. Flavonoids also may act through inhibiting both energy metabolism and DNA synthesis thus affecting protein and RNA syntheses. In this study, 'Leccino' had the highest phenolic and flavonoid content, followed by 'Arbequina' and 'Arbozana', and the lowest values were observed in 'Ogliarola' and 'Cellina di Nardò', which is in concordance with a previous study [54]. The high phenolic and flavonoid content may be an indicative parameter on the ability of olive trees to fight *X. fastidiosa* infection based on other studies [25]. Previous studies on grapevine reported an increase of phenolic compounds following *X. fastidiosa* infection, since this plant possesses the ability to change its metabolism towards an excessive formation of phenolic compounds as a defense mechanism against pathogens [25,53,55]. In fact, and following *X. fastidiosa* infection, phenolic compounds (e.g., catechin and digalloylquinic acid), glycosides (e.g., astringin) and flavonoids (e.g., catechins, pyrocyanidins) were found in higher quantities around xylem tissues, where they help the xylem sap to rise and reach different parts of the plant [25]. Thus, those secondary metabolites are determinant in the defense against *X. fastidiosa* infection, and olive varieties with the highest content would, therefore, be more prominent to resist this pathogen.

5. Conclusions

The plant pathogen *X. fastidiosa* responsible for the olive quick decline syndrome is considered a quarantine pathogen, and its introduction is highly prohibited in Morocco. This study provided insights on olive varieties that could resist, or are entirely sensitive to *X. fastidiosa* infection, based on the analysis of ions, phenolic and flavonoid content.

Therefore, and as a prevention strategy, we highly recommend increasing the plantation of relatively resistant varieties to *X. fastidiosa* and reducing varieties that could be easily damaged, especially that olive plantation in Morocco contributes to a huge part of its agriculture and economy. However, and since it is only a matter of time until *X. fastidiosa* infects olive trees in Morocco, extensive studies on developing an effective cure are needed. A promising target would be the inhibition of the attachment of *X. fastidiosa* to vegetal cells (e.g., the effect of Ca), inhibiting the first stages of its invasion.

Author Contributions: Conceptualization, K.E.H. and E.H.A.; methodology, K.E.H.; software, K.E.H. and M.O.; validation E.H.A. and M.H.; formal analysis, K.E.H.; investigation, E.H.A.; resources, E.H.A.; data curation, K.E.H. and M.S.; writing—original draft preparation, K.E.H.; writing—review and editing, M.E.M.; visualization, K.H. and A.B.; supervision, E.H.A. and M.H.; project administration, Cure-Xf. All authors have read and agreed to the published version of the manuscript.

Funding: This research was funded by CURE-XF, an EU-funded project.

Institutional Review Board Statement: Not applicable.

Informed Consent Statement: Informed consent was obtained from all subjects involved in the study.

Acknowledgments: This research was supported by CURE-XF, an EU-funded project, coordinated by CIHEAM Bari (H2020-Marie Sklodowska-Curie Actions-Research and Innovation Staff Exchange. Reference number: 634353). The authors greatly appreciate the English Language editing and review services supplied by Elvira Lapedota, language consultant (https://www.linkedin.com/in/elvira-lapedota-12a53473/) (accessed on 5 May 2021).

Conflicts of Interest: The authors declare no conflict of interest.

References

1. Bajoub, A.; Medina-Rodríguez, S.; Olmo-García, L.; Ajal, E.A.; Monasterio, R.P.; Hanine, H.; Fernández-Gutiérrez, A.; Carrasco-Pancorbo, A. In-Depth Two-Year Study of Phenolic Profile Variability among Olive Oils from Autochthonous and Mediterranean Varieties in Morocco, as Revealed by a LC-MS Chemometric Profiling Approach. *Int. J. Mol. Sci.* **2017**, *18*, 52. [CrossRef]
2. FAOSTAT. Food and Agriculture Organization Statistical Data-Base. 2019. Available online: http://faostat.fao.org/default.aspx (accessed on 5 May 2021).
3. Bouymajane, A.; El Majdoub, Y.O.; Cacciola, F.; Russo, M.; Salafia, F.; Trozzi, A.; Filali, F.R.; Dugo, P.; Mondello, L. Characterization of Phenolic Compounds, Vitamin E and Fatty Acids from Monovarietal Virgin Olive Oils of "*Picholine marocaine*" Cultivar. *Molecules* **2020**, *25*, 5428. [CrossRef] [PubMed]
4. Barbara, H.; Terral, J.F.; Ater, M. Première Caractérisa-tion Pomologique des Variétés Locales De L'olivier (*Olea Europaea*, L.) Des Oliveraies Tradition-nelles des Agroécosystèmes des Montagnes Du Nord-Ouest Du Maroc. *Eur. Sci. J. ESJ* **2020**, *16*, 556–575.
5. Chliyeh, M.; Touhami, A.O.; Filali-Maltouf, A.; El Modafar, C.; Moukhli, A.; Oukabli, A.; Benkirane, R.; Douira, A. Phytophthora Palmivora: A New Pathogen of Olive Trees in Morocco. *Atlas J. Biol.* **2013**, *2*, 130–135. [CrossRef]
6. Sicard, A.; Zeilinger, A.R.; Vanhove, M.; Schartel, T.E.; Beal, D.J.; Daugherty, M.P.; Almeida, R.P. Xylella fastidiosa: Insights into an Emerging Plant Pathogen. *Annu. Rev. Phytopathol.* **2018**, *56*, 181–202. [CrossRef] [PubMed]
7. White, S.M.; Navas-Cortés, J.A.; Bullock, J.M.; Boscia, D.; Chapman, D.S. Estimating the Epidemiology of Emerging Xylella fastidiosa Outbreaks in Ol-ives. *Plant Pathol.* **2020**, *69*, 1403–1413. [CrossRef]
8. Afechtal, M.; Vicent, A.; Saponari, M.; D'Onghia, A.M. Pest risk analysis on Xylella fastidiosa in Morocco. *J. Plant Prot. Res.* **2018**, *58*.
9. Frem, M.; Chapman, D.; Fucilli, V.; Choueiri, E.; El Moujabber, M.; La Notte, P.; Nigro, F. Xylella fastidiosa invasion of new countries in Europe, the Middle East and North Africa: Ranking the potential exposure scenarios. *NeoBiota* **2020**, *59*, 77–97. [CrossRef]
10. Afechtal, M.; Aitfriha, A.; Bibi, I. A Preliminary Survey on the Presence of Xylella fastidiosa in Olive, Citrus and Grapevine Groves in Morocco. *Rev. Maroc. Sci. Agron. Vétérinaires* **2017**, *6*, 6–9.
11. Martelli, G.P. The current status of the quick decline syndrome of olive in southern Italy. *Phytoparasitica* **2016**, *44*, 1–10. [CrossRef]
12. Oepp, B.; Eppo, B. PM 7/24 (4) Xylella fastidiosa. *EPPO Bull.* **2019**, *49*, 175–227.
13. D'Attoma, G.; Morelli, M.; Saldarelli, P.; Saponari, M.; Giampetruzzi, A.; Boscia, D.; Savino, V.N.; De La Fuente, L.; Cobine, P.A. Ionomic Differences between Susceptible and Resistant Olive Cultivars Infected by Xylella fastidiosa in the Outbreak Area of Salento, Italy. *Pathogens* **2019**, *8*, 272. [CrossRef]
14. Del Coco, L.; Migoni, D.; Girelli, C.R.; Angilè, F.; Scortichini, M.; Fanizzi, F.P. Soil and Leaf Ionome Heterogeneity in Xylella fastidiosa Subsp. Pauca-Infected, Non-Infected and Treated Olive Groves in Apulia, Italy. *Plants* **2020**, *9*, 760. [CrossRef]
15. Cruz, L.F.; Cobine, P.A.; De La Fuente, L. Calcium Increases Xy-lella fastidiosa Surface Attachment, Biofilm Formation, and Twitching Motility. *Appl. Environ. Microbiol.* **2012**, *78*, 1321–1331. [CrossRef] [PubMed]

16. Raffini, F.; Bertorelle, G.; Biello, R.; D'Urso, G.; Russo, D.; Bosso, L. From Nucleotides to Satellite Imagery: Approaches to Identify and Manage the Invasive Pathogen Xylella fastidiosa and Its Insect Vectors in Europe. *Sustainability* **2020**, *12*, 4508. [CrossRef]
17. Hood, M.I.; Skaar, E.P. Nutritional immunity: Transition metals at the pathogen–host interface. *Nat. Rev. Genet.* **2012**, *10*, 525–537. [CrossRef] [PubMed]
18. Baxter, I.; Dilkes, B.P. Elemental Profiles Reflect Plant Adaptations to the Environment. *Science* **2012**, *336*, 1661–1663. [CrossRef]
19. Oliver, J.; Sefick, S.A.; Parker, J.K.; Arnold, T.; Cobine, P.A.; De La Fuente, L.; Oliver, J.; Sefick, S.A.; Parker, J.K.; Arnold, T.; et al. Ionome Changes in Xylella fastidiosa–Infected Nicotiana tabacum Correlate with Virulence and Discriminate Between Subspecies of Bacterial Isolates. *Mol. Plant-Microbe Interact.* **2014**, *27*, 1048–1058. [CrossRef] [PubMed]
20. Navarrete, F.; De La Fuente, L. Zinc Detoxification Is Required for Full Virulence and Modification of the Host Leaf Ionome by Xylella fastidiosa. *Mol. Plant-Microbe Interact.* **2015**, *28*, 497–507. [CrossRef]
21. Cobine, P.A.; Cruz, L.F.; Navarrete, F.; Duncan, D.; Tygart, M.; De La Fuente, L. *Xylella fastidiosa* Differentially Accumulates Mineral Elements in Biofilm and Planktonic Cells. *PLoS ONE* **2013**, *8*, e5493610. [CrossRef] [PubMed]
22. Andersen, P.C.; French, W.J. Biophysical Characteristics of Peach-Trees Infected with Phony Peach Disease. *Physiol. Mol. Plant Pathol.* **1987**, *31*, 25–40. [CrossRef]
23. De La Fuente, L.; Parker, J.K.; Oliver, J.E.; Granger, S.; Brannen, P.M.; van Santen, E.; Cobine, P.A. The bacterial pathogen *Xylella fastidiosa* affects the leaf ionome of plant hosts during infection. *PLoS ONE* **2013**, *8*, e62945. [CrossRef] [PubMed]
24. Rodrigues, C.M.; De Souza, A.A.; Takita, M.A.; Kishi, L.T.; Machado, M.A. RNA-Seq analysis of Citrus reticulata in the early stages of *Xylella fastidiosa* infection reveals auxin-related genes as a defense response. *BMC Genom.* **2013**, *14*, 676. [CrossRef] [PubMed]
25. Vergine, M.; Nicolì, F.; Sabella, E.; Aprile, A.; De Bellis, L.; Luvisi, A. Secondary Metabolites in *Xylella fastidiosa*—Plant Interaction. *Pathogens* **2020**, *9*, 675. [CrossRef] [PubMed]
26. Djelouah, K.; Frasheri, D.; Valentini, F.; D'Onghia, A.M.; Digiaro, M. Direct tissue blot immunoassay for detection of *Xylella fastidiosa* in olive trees. *Phytopathcl. Mediterr.* **2020**, *53*, 559–564.
27. Minsavage, G.V.; Thompson, C.M.; Hopkins, D.L.; Leite, R.M.V.B.C.; Stall, R.E. Development of a polymerase chain reaction protocol for detection of *Xylella fastidiosa* in plant tissue. *Phytopathology* **1994**, *84*, 446–461. [CrossRef]
28. Sanders, T.H.; McMichael, R.W.; Hendrix, K.W. Occurrence of resveratrol in edible peanuts. *J. Agric. Food Chem.* **2020**, *48*, 1243–1246. [CrossRef]
29. Xie, L.; Bolling, B.W. Characterisation of stilbenes in California almonds (Prunus dulcis) by UHPLC–MS. *Food Chem.* **2014**, *148*, 300–306. [CrossRef]
30. Waterhouse, A.L. Determination of total phenolics. *Curr. Protoc. Food Anal. Chem.* **2002**, *6*, 111–118.
31. Barreira, J.C.; Ferreira, I.C.; Oliveira, M.; Pereira, J.A. Antioxidant activities of the extracts from chestnut fower leaf skins and fruit. *Food Chem.* **2018**, *107*, 1106–1113. [CrossRef]
32. European and Mediterranean Plant Protection Organization (EPPO). Diagnostic protocols for regulated pests. In *Xylella fastidiosa*; Bulletin OEPP/EPPO Bulletin: Paris, France, 2004; pp. 187–192. Available online: https://onlinelibrary.wiley.com/doi/epdf/10.1111/j.1365-2338.2004.00718.x (accessed on 5 May 2021).
33. Firrao, G.; Bazzi, C. Specific identification of *Xylella fastidiosa* using the polymerase chain reaction. *Phytopathol. Mediterr.* **1994**, *33*, 90–92.
34. Harper, S.J.; Ward, L.I.; Clover, G.R.G. Development of LAMP and Real-Time PCR Methods for the Rapid Detection of *Xylella fastidiosa* for Quarantine and Field Applications. *Phytopathology* **2010**, *100*, 1282–1288. [CrossRef]
35. Kailis, S.; Harris, D. *Fertiliser Requirements and Monitoring Nutritional Status*; Landlinks Press: Collingwood, Australia, 2007.
36. Pagan, I.; Garcia-Arenal, F. Tolerance to Plant Pathogens: Theory and Experimental Evidence. *Int. J. Mol. Sci.* **2018**, *19*, 810. [CrossRef]
37. El Chami, D.; Galli, F. An assessment of seaweed extracts: Innovation for sustainable agriculture. *Agronomy* **2020**, *10*, 1433. [CrossRef]
38. Saponari, M.; Giampetruzzi, A.; Loconsole, G.; Boscia, D.; Saldarelli, P. *Xylella fastidiosa* in Olive in Apulia:Where We Stand. *Phytopathology* **2019**, *109*, 175–186. [CrossRef] [PubMed]
39. Torres, M.A.; Jones, J.D.G.; Dangl, J.L. Reactive oxygen species signaling in response to pathogens. *Plant Physiol.* **2006**, *141*, 373–378. [CrossRef]
40. Caspi, V.; Droppa, M.; Horvath, G.; Malkin, S.; Marder, J.B.; Raskin, V.I. The effect of copper on chlorophyll organization during greening of barley leaves. *Photosynth. Res.* **1999**, *62*, 165–174. [CrossRef]
41. Eide, D.J. The oxidative stress of zinc deficiency. *Metallomics* **2011**, *3*, 1124–1129. [CrossRef]
42. Cabot, C.; Martos, S.; Llugany, M.; Gallego, B.; Toirà, R.; Poschenrieder, C. A role for zinc in plant defense against pathogens and herbivores. *Front. Plant Sci.* **2019**, *10*, 1171. [CrossRef] [PubMed]
43. Kim, S.H.; Hong, J.K.; Lee, S.C.; Sohn, K.H.; Jung, H.W.; Hwang, B.K. CAZFP1, CYS2/HiS(2)-type zinc-finger transcription factor gene functions as a pathogen-induced early-defense gene in Capsicum annuum. *Plant Mol. Biol.* **2004**, *55*, 883–904. [CrossRef] [PubMed]
44. Gupta, S.K.; Rai, A.K.; Kanwar, S.S.; Sharma, T.R. Comparative analysis of zinc finger proteins involved in plant disease resistance. *PLoS ONE* **2012**, *7*, 8. [CrossRef]

45. Poschenrieder, C.; Tolrà, R.; Barceló, J. Can metals defend plants against biotic stress? *Trends Plant Sci.* **2006**, *11*, 288–295. [CrossRef]
46. Fones, H.N.; Preston, G.M. Reactive oxygen and oxidative stress tolerance in plant pathogenic Pseudomonas. Fems Microbiol. *Letters* **2012**, *327*, 1–8.
47. Scortichini, M. Preliminary results on field trial to control *Xylella fastidiosa* on olive trees in Puglia. *Options Méditerranéennes* **2017**, *121*, 77–78.
48. Sabella, E.; Luvisi, A.; Aprile, A.; Negro, C.; Vergine, M.; Nicolì, F.; Miceli, A.; De Bellis, L. *Xylella fastidiosa* induces differential expression of lignification related-genes and lignin accumulation in tolerant olive trees cv. Leccino. *J. Plant Physiol.* **2018**, *220*, 60–68. [CrossRef] [PubMed]
49. Leite, B.; Ishida, M.; Alves, E.; Carrer, H.; Pascholati, S.; Kitajima, E. Genomics and X-ray microanalysis indicate that Ca2+ and thiols mediate the aggregation and adhesion of *Xylella fastidiosa*. *Braz. J. Med. Biol. Res.* **2002**, *35*, 645–650. [CrossRef] [PubMed]
50. Gutha, L.R.; Casassa, L.F.; Harbertson, J.F.; Naidu, R.A. Modulation of flavonoid biosynthetic pathway genes and anthocyanins due to virus infection in grapevine (*Vitis vinifera* L.) leaves. *BMC Plant Biol.* **2010**, *10*, 187. [CrossRef]
51. El Aabidine, A.Z.; Baissac, Y.; Moukhli, A.; Jay-Allemand, C.; Khadari, B.; El Modafar, C. Resistance of olivetree to Spilocaea oleagina is mediated by the synthesis of phenolic compounds. *Int. J. Agric. Biol.* **2010**. *12*, 61–67.
52. Rusjan, D.; Veberič, R.; Mikulič-Petkovšek, M. The response of phenolic compounds in grapes of the variety 'Chardonnay' (Vitis vinifera L.) to the infection by phytoplasma Bois noir. *Eur. J. Plant Pathol.* **2012**, *133*, 965–974. [CrossRef]
53. Petridis, A.; Therios, I.; Samouris, G.; Tananaki, C. Salinity-induced changes in phenolic compounds in leaves and roots of four olive cultivars (*Olea europaea* L.) and their relationship to antioxidant activity. *Environ. Exp. Bot.* **2012**, *79*, 37–43. [CrossRef]
54. Raman, T.; Muthukathan, G. Field Suppression of Fusarium Wilt Disease in Banana by the Combined Application of Native Endophytic and Rhizospheric Bacterial Isolates Possessing Multiple Functions. *Phytopathol. Mediterr.* **2015**, *54*, 241–252. [CrossRef]
55. Wallis, C.M.; Wallingford, A.K.; Chen, J. Effects of cultivar, phenology, and *Xylella fastidiosa* infection on grapevine Xylem Sap and tissue phenolic content. *Physiol. Mol. Plant Pathol.* **2013**, *84*, 28–35. [CrossRef]

sustainability

MDPI

Article

Mechanical Soil Resistance Influenced by Different Tillage Systems and Tractor Tire Pressures

Robert Benković [1,*], Danijel Jug [2], Luka Šumanovac [2], Irena Jug [2], Krunoslav Mirosavljević [1], Domagoj Zimmer [2] and Teuta Benković-Lačić [1]

1 Biotechnical Department, University of Slavonski Brod, 35000 Slavonski Brod, Croatia; kmirosavljevic@unisb.hr (K.M.); tblacic@unisb.hr (T.B.-L.)

2 Faculty of Agrobiotechnical Sciences Osijek, Josip Juraj Strossmayer University of Osijek, 31000 Osijek, Croatia; djug@fazos.hr (D.J.); lsumanov@fazos.hr (L.Š.); ijug@fazos.hr (I.J.); dzimmer@fazos.hr (D.Z.)

* Correspondence: rbenkovic@unisb.hr

Abstract: Intensive agricultural practices affect soil compaction, and their indirect and direct effects on crop growth and development are an increasingly important focus of scientific research. The objective of this study was to determine the influence of different tillage systems on soil compaction and to observe the influence of tractor tire pressure on penetrometer resistance during sowing. The three-year study was conducted on the heavy pseudogley soil of Brod-Posavina County in the Republic of Croatia. During the research, crops were observed in the following cropping sequence: soybean (*Glycine max* L.) in the first year, maize (*Zea mays* L.) in the second year and winter wheat (*Triticum aestivum* L.) in the third year. The tillage systems as the main study factor were conventional tillage (CT) plowing to a depth of 35 cm, disc tillage (DH) to a depth of 15 cm, loosening (CH) to a depth of 30 cm, and undermining (SS) to a depth of 50 cm. The following pressures were used as a subfactor of this study, namely the pressure of the front and rear tires of the tractor during sowing: p_1 (front 1.0 bar/rear 0.8 bar), p_2 (front 2.0 bar/rear 1.6 bar), and p_3 (front 3.0 bar/rear 2.4 bar). The tillage systems applied resulted in different soil compaction, thus the deepest tillage SS had the lowest resistance and the DH tillage had the highest resistance in all three experimental years. Penetrometer measurements showed the influence of tire pressure p_1 on reducing compaction as early as the first year in 2017, while in the last year of research in 2019, tractor tire pressure p_3 during sowing contributed to a significant increase in soil compaction.

Keywords: tillage systems; soil compaction; tire pressure; penetrometer resistance

Citation: Benković, R.; Jug, D.; Šumanovac, L.; Jug, I.; Mirosavljević, K.; Zimmer, D.; Benković-Lačić, T. Mechanical Soil Resistance Influenced by Different Tillage Systems and Tractor Tire Pressures. *Sustainability* **2023**, *15*, 10236. https://doi.org/10.3390/su151310236

Academic Editors: Daniel El Chami and Maroun El Moujabber

Received: 31 May 2023
Revised: 21 June 2023
Accepted: 26 June 2023
Published: 28 June 2023

1. Introduction

Arable area is a limited natural resource, which, in addition to direct and indirect interaction with climate and sudden climate changes, significantly complicates the sustainability and security of agricultural production. Applied tillage systems can be highlighted as an important way to mitigate the current very unfavorable climatic conditions and ensure regular high yields [1]. Tillage is carried out with the aim of improving the mechanical properties of the soil, increasing the incorporation and mixing of fertilizers and organic residues with the soil, suppressing weeds, plant diseases, and pests, i.e., creating optimal conditions for crop germination, growth, and development [2]. Agricultural machinery is becoming larger, faster, more reliable, more economical, and generally more productive, but this leads to greater compaction and degradation of agricultural soils [3]. A significant increase in soil compaction during soil preparation prior to sowing is influenced by the size, power, and method of locomotion (wheels or tracks) of agricultural tractors [4].

Poor soil structure increases evaporation, reduces infiltration and soil water permeability, has poor aeration, contributes to crust formation, and paves the way for greater erosion [5,6]. Conservation tillage is carried out through a number of different tillage systems and approaches, mainly with the aim of conserving soil moisture and reducing erosion, leaving the surface covered with at least 30% crop residue after tillage and sowing of the next crop [7]. Farmers involved in various aspects of agricultural production are encouraged to use conservation tillage because it can reduce erosion and compaction, nutrient leaching, fuel consumption, and labor hours [8] while increasing microbiological activity and carbon storage [9]. The authors of [10], in their study of different tillage systems and their influence on the physical properties of clayey luvisol soil, found significant differences between tillage systems: density, porosity, water retention capacity, and air capacity of the soil. Conservation tillage systems have the least effect on the tearing off of the surface layer of the soil due to the slip of the tractor wheels, while the reduction in tillage and the reduction in the tillage depth can have significant effects on the fuel consumption of the tractor [8].

The author of [11] investigated the physical changes in clay soils of Slovakia under the influence of two different cultivation systems a "no-till" cultivation system and a cultivation system with conventional tillage. The analysis of the results showed that in the "no-till" system the density increased and the total porosity decreased. Capillarity increased significantly in the system with conventional tillage, while the maximum water capacity of the soil (=field capacity) showed no significant differences regardless of the tillage system. Different tillage systems (conventional plowing, disk tillage, loosening, and no-till) of chernozem in northern Baranja affect soil resistivity, such that the greatest compaction and soil resistance occurred at a depth of 10–15 cm with disk tillage and at a depth of 25–30 cm with plowing [12]. The same authors note that the perennial use of conventional tillage can have negative effects on soil erosion and the formation of "plow bottoms." Proper management of soil as a resource can protect soil from excessive water and erosion, prevent crust formation and compaction of layers, and ensure good and easy germination and root development [13]. The specific location, crop, soil type, and climate are important in selecting a tillage system [2]. According to [14], one of the most important objectives of tillage is to create a favorable soil structure so that seed can be placed at an ideal depth, and therefore tillage is essential for seeding, growth, development, and ultimately yield [15]. In practice, tillage affects the physical, chemical, and biological properties of the soil and ultimately directly influences plant growth and yield [16–18]. Different tillage systems affect soil compaction, soil temperature, and yield of wheat, corn, and soybean, and reduced tillage has a beneficial effect on soil moisture at the time of sowing [19]. Control of agrotechnical traffic on arable land and conservation tillage were found to be acceptable solutions for surface runoff, soil water movement, and yield components on heavy vertisol in Queensland—Australia [20]. The author of [21], in a multi-year study of three tillage systems, found that tillage methods and crop residue cover had a significant effect on yield, compaction, and soil density. The unfavorable condition of soil moisture and tractor tire pressure affect the low wheel resistance factor of the rolling working machine (f = 0.08–0.12), which affects compaction, soil shear, wheel slip, and energy consumption [3,8]. Tillage with a power harrow may be an alternative to good maintenance of favorable soil moisture and wheat yield with conventional systems [22].

The effects of agricultural machinery traffic can be divided into several categories: direct damage to crop yield, effects on compaction due to processing, and long-term damage that occurs after all operations [23]. Proper distribution of ballast on the axles [24] and selection of the correct size and tire pressure can reduce compaction of previously loosened soil [25]. From the research mentioned so far, it is necessary and important to study the influence of tillage before sowing and other utilization parameters during sowing on different soil types. The aim of this research was to determine the influence of tillage system and tractor tire pressure during sowing on soil compaction in the surface layer.

2. Materials and Methods

Three-year research on crop rotation (2017. soybean (*Glycine max* L.); 2018. maize (*Zea mays* L.) and 2018/2019. winter wheat (*Triticum aestivum* L.)) was conducted in Croatia in

Brod-Posavina County with the exact coordinates of the experimental field 45°10′14″ North latitude and 18°6′3″ East longitude. Mechanical soil analysis was performed according to the standard HRN ISO 11464 (2004) [26] and a combination of sieving and sedimentation method according to the standard ISO 11277 (2009) [27]. The agricultural soil in the experimental field has mechanical content as follows: coarse sand 5.36%, fine sand 8.4%, coarse silt 37.77%, silt 44.75%, clay 3.72%. The data on the amount of precipitation and the mean values of air temperature in the month of penetrometry are presented in Figure 1.

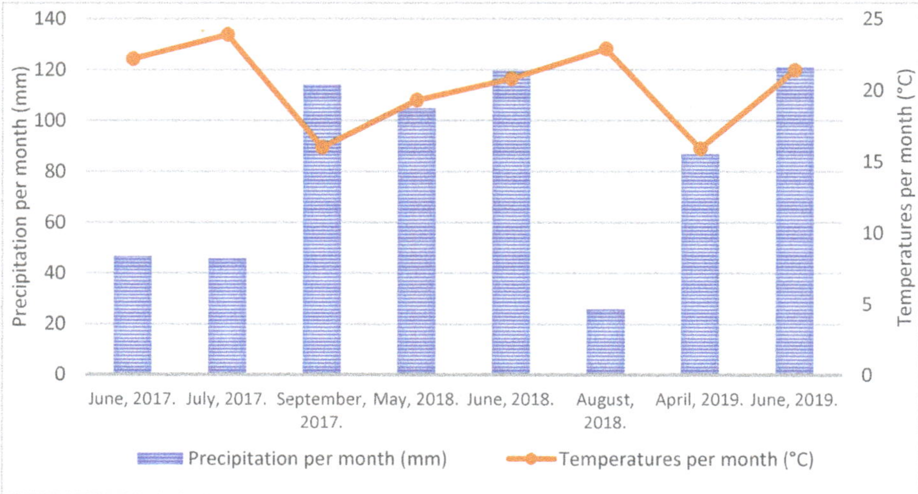

Figure 1. Average precipitation and air temperature by dates of penetrometer measurement period.

2.1. Setting up the Experiment

The experiment was designed as a completely randomized block experiment in four replicates with the main factor "tillage system" and the subfactor "pneumatic pressures at sowing". The size of the experimental tillage area was 10 m × 90 m (900 m^2). The study of the influence of tire pressure at sowing was carried out perpendicular to the tillage direction on areas of 10 m × 30 m. The total size of the area with all replicates was 14,400 m^2.

In 2016, 2017 and 2018, the following tillage systems were used as the main factor: Conventional tillage (CT), plowing to a depth of 35 cm and preparing the soil before sowing; Disc tillage (DH), tilling the soil by discs to a depth of 15 cm and preparing for sowing; Loosening (CH), tilling the soil by loosening to a depth of 30 cm and preparing for sowing; Undermining (SS), tilling the soil by undermining to a depth of 50 cm and preparing for sowing.

The subfactor of the experiment "tire pressure at sowing" was as follows: p_1 for tractor front tire pressure 1.0 bar and rear tire pressure 0.8 bar; p_2 for tractor front tire pressure 2.0 bar and rear tire pressure 1.6 bar; p_3 for tractor front tire pressure 3.0 bar and rear tire pressure 2.4 bar.

A Massey Ferguson 8480 Dyna-VT tractor (Figure A1; Table A1), was used to implement the tillage system and during sowing in all three years of the research, equipped with 600/65 R28 (front) and 710/70 R38 (rear) tires. For the implementation of conventional tillage (CT) it was aggregated with a five-furrow plow Regent Titan 15 (Figure A2; Table A2), for the implementation of reduced tillage by disk tillage (DH) it was aggregated with a disk RAU Rondo XL 44 (Figure A3; Table A3), for the implementation of conservation tillage by loosening (CH) and undermining (SS) it was aggregated with an undermining device Pegoraro MEGA DRAG 7 (Figure A4; Table A4), adjusted to different working depths. The working speeds in the application of the various tillage systems were as follows: 6.5 km h^{-1}

for CT, 9 km h^{-1} for DH, 10 km h^{-1} for CH and 5,5 km h^{-1} for SS tillage. For further tillage, the same tractor was aggregated with a Kongskilde HK 31 (Figure A5; Table A5), power harrow. The other agrotechnical measures such as fertilization, protection and harvesting were the same in all three experimental years.

The experiment begins with the implementation of all the tillage systems studied, which were carried out on 25 November 2016. Penetrometry for each tillage system and tractor tire pressure during sowing, with the aim of determining the influence of the applied agricultural technique on compaction, was performed during the cultivation of soybean (*Glycine max* L.) as follows: on 9 June (1st measurement), 25 July (2nd measurement) and 9 September (3rd measurement) in 2017, during the cultivation of maize (*Zea mays* L.) as follows: on 25 May (1st measurement), 28 June (2nd measurement) and 28 August (3rd measurement) in 2018, and during the cultivation of winter wheat (*Triticum aestivum* L.) as follows: on 9 April (1st Measurement) and 12 June (2nd Measurement) in 2019. On the same day of the start of each penetrometry, the current soil moisture was determined by the gravimetric method. Soil samples were taken every ten centimeters of depth, starting from 10 cm to 40 cm depth. In the laboratory, the samples were weighed on an Ohaus Adventurer pro AV4101 balance and placed in a Memmert Model 100–800 dryer. After drying at a temperature of 105 °C to a constant mass, the samples were weighed again and the moisture content was calculated according to the Equation (A1) Soil in weight %.

Fertilization was carried out according to the recommendations for cultivated plants and was as follows: with 667 kg ha^{-1} 7:20:30 NPK fertilizer for growing soybeans (*Glycine max*, L.), 725 kg ha^{-1} 7:20:30 NPK fertilizer and with 92 kg ha^{-1} UREA N 46 for growing corn (*Zea mays* L.), and with 508 kg ha^{-1} 7:20:30 NPK fertilizer for growing winter wheat (*Triticum aestivum* L.). Soil preparation was carried out before sowing with a Kongskilde HK 31 power harrow to a depth of 7 cm. They were the same for all observed tillage systems and tire pressures in all experimental years. Sowing with changes in tire pressure of the tractor with which it was performed was as follows: soybeans (Sinara variety) on 15 April 2017; maize (Kulak variety) on 18 April 2018; winter wheat (Victoria variety) on 5 November 2018.

2.2. Soil Compaction

The state of soil compaction, possibly caused by the applied treatment systems, was determined with a penetrometer "Eijkelkamp Penetrologger SN" with a conical tip with an area of 2 cm^2 to a depth of 40 cm in the soil profile. Penetrometry on each cultivation system and on each tire pressure of the tractor during sowing (total of 144 measurements per each term) with the aim of determining the influence of the applied agricultural techniques on compaction. The scheme for measuring the penetrometer resistances can be seen in Figure A6.

2.3. Statistical Analysis

Statistical processing of the data was performed using the statistical package for analysis of variance Statistica version 13.5.0.17 (Statistical Package TIBCO Software Inc., 2018, Palo Alto, CA, USA), where the main factor was "tillage system", and the subfactor was "different pressures of the front and rear tractor tires at sowing". In accordance with Fisher's test for significant differences in analysis of variance, least significant differences (LSD) for statistical significance of $p < 0.05$ were calculated by comparing the means. To determine exactly which variants had statistically significant differences, Duncan's test was used. The values of the research results in the tables marked with different capital letters (A, B, C, D ...) in the column have mutually statistically significant differences with 95% probability.

3. Results and Discussion

The condition of soil compaction is shown in Tables 1–9. The results of the influence of the applied soil treatment method can be seen in Tables 1, 4 and 7. Tables 2, 5 and 8 show

how the different tire pressures of the tractors during sowing affected soil compaction. The resistances determined with a conical tip penetrometer every 5 cm from 0 to 40 cm depth of the soil profile are shown in Tables 3, 6 and 9. The average current soil moisture at a depth of 0–40 cm while performing each of the penetrometry in the first year of research was as follows: in the first measurement 17.37%, in the second measurement 17.96%, and in the third measurement 23.26%.

Table 1. The influence of the tillage system on the mechanical resistance of the soil.

2016./2017. Soybean (*Glycine max* L.)			
Tillage System	**1st Measurement (Average Resistance—MPa)**	**2nd Measurement (Average Resistance—MPa)**	**3rd Measurement (Average Resistance—MPa)**
CT	2.01	2.92 $_{BC}$	2.65 $_{B}$
DH	2.04	3.62 $_{A}$	2.97 $_{A}$
CH	1.93	3.06 $_{B}$	2.75 $_{B}$
SS	1.93	2.73 $_{C}$	2.35 $_{C}$
F_O	n.s. ($p < 0.05$, $F = 2.49$)	* ($p < 0.05$, $F = 27.06$)	* ($p < 0.05$, $F = 11.49$)

F_O—F test tillage system; *—statistical significance; n.s.—no statistical significance; values within the same columns, which are marked with different capital letters, differ statistically significantly at the level of $p < 0.05$.

Differently applied tillage systems in the 1st year of observation and at the 1st measurement had no statistically significant effect on soil resistance ($F = 2.49$).

Soil resistance at the 2nd measurement was statistically significantly affected by soil treatment ($F = 27.06$). The average resistance value was 3.08 MPa. The highest resistance was measured in the DH treatment and the lowest in SS. The LSD test revealed statistically significant differences in the value of soil resistance between treatments DH and CT (0.70 MPa), DH and CH (0.56 MPa), DH and SS (0.89 MPa), CH and SS (0.33 MPa).

Soil resistance at the 3rd measurement was statistically significantly affected by soil treatment ($F = 11.49$). The average resistance value was 2.68 MPa. The highest resistance was measured in the DH treatment and the lowest in the SS treatment. Statistically significant differences in soil resistance values were found between all tillage systems except between CT and CH.

The authors of [28,29] found that using a subexcavator with a greater tillage depth reduced soil compaction while improving soil infiltration capacity. Increasing the tillage depth of the same cultivation tool when implementing basic tillage CH and SS resulted in a reduction in soil compaction. Reducing the amount of tillage can affect the reduction in soil degradation and especially the reduction in soil compaction [12].

Table 2. The influence of the tire pressure of a tractor on the mechanical resistance of the soil.

2016./2017. Soybean (*Glycine max* L.)			
Tire Pressure	**1st Measurement (Average Resistance—MPa)**	**2nd Measurement (Average Resistance—MPa)**	**3rd Measurement (Average Resistance—MPa)**
p_1	1.91 $_{B}$	2.78 $_{B}$	2.54 $_{B}$
p_2	2.09 $_{A}$	3.27 $_{A}$	2.40 $_{B}$
p_3	1.93 $_{B}$	3.18 $_{A}$	3.09 $_{A}$
F_P	* ($p < 0.05$, $F = 9.55$)	* ($p < 0.05$, $F = 16.72$)	* ($p < 0.05$, $F = 31.15$)

F_P—F test tire pressure at sowing; *—statistical significance; n.s.—no statistical significance; values within the same columns, which are marked with different capital letters, differ statistically significantly at the level of $p < 0.05$.

Soil resistance in the first year of observation was statistically significantly affected by tire pressure in all three measurements (as follows $F = 9.55$; $F = 16.72$; $F = 31.15$). The average value of soil resistance in the 1st measurement was 1.98 MPa. The highest penetrometer resistance of the 1st measurement was measured at tire pressure p_2, and the lowest at p_1. The differences in resistances between p_2 and p_1 (0.17 MPa) and p_2 and p_3 (0.15 MPa) were statistically justified.

The average resistance value of the 2nd measurement was 3.08 MPa. The highest resistance was measured at p_2 and the lowest at p_1. The LSD test revealed statistically significant differences in soil resistance between variants p_1 and p_3 (0.4 MPa) and between p_1 and p_2 (0.49 MPa). The difference in soil resistance between p_2 and p_3 was not statistically significant.

The average resistance value of the 3rd measurement was 2.68 MPa. The highest resistance was measured at p_3 and at p_2. The LSD test revealed statistically significant differences in soil resistance between p_1 and p_3 (0.55 MPa) and between p_2 and p_3 (0.69 MPa). The difference in soil resistance between p_1 and p_2 was not statistically significant.

In the third measurement of the influence of pressure, an increasingly pronounced compaction can be seen at the highest tire pressure p_3. Similar research results were obtained by [30], and the same authors indicate that by reducing tire pressure from p_{max} = 160 kPa (1.6 bar) to p_{max} = 120 kPa (1.2 bar), especially in agricultural practices with multiple passes, soil compaction can be significantly reduced.

Table 3. Influence of the depth measurement on the mechanical resistance of the soil.

2016./2017. Soybean (*Glycine max* L.)			
Depth of Resistance Measurement	**1st Measurement (Average Resistance—MPa)**	**2nd Measurement (Average Resistance—MPa)**	**3rd Measurement (Average Resistance—MPa)**
0–5 cm	1.62 D	0.72 H	0.79 G
5–10 cm	2.04 B	1.40 G	1.21 F
10–15 cm	2.21 A	2.10 F	2.15 E
15–20 cm	2.17 AB	2.92 E	2.23 ED
20–25 cm	1.90 CB	3.77 D	2.52 D
25–30 cm	1.85 C	4.14 C	3.61 C
30–35 cm	1.99 BC	4.56 B	4.09 B
35–40 cm	2.04 B	5.05 A	4.82 A
F_D	* ($p < 0.05$, F = 14.58)	* ($p < 0.05$, F = 224.30)	* ($p < 0.05$, F = 173.27)

F_D—F test depth of penetrometry; *—statistical significance; n.s.—no statistical significance; values within the same columns, which are marked with different capital letters, differ statistically significantly at the level of $p < 0.05$.

Soil resistances in the 1st year of observation were statistically significantly affected by depth measurement in all three measurements (as follows F = 14.58; F = 224.3; F = 173.27). The highest penetration resistance was found in the 1st measurement at 10–20 cm and 15–20 cm depths without statistical significance, and in the 2nd and 3rd measurements at 35–40 cm depths.

Table 4. The influence of the tillage system on the mechanical resistance of the soil.

2017./2018. Maize (*Zea mays* L.)			
Tillage System	**1st Measurement (Average Resistance—MPa)**	**2nd Measurement (Average Resistance—MPa)**	**3rd Measurement (Average Resistance—MPa)**
CT	1.66 B	2.48 B	2.59 A
DH	1.89 A	2.66 AB	3.30 B
CH	1.82 A	2.72 A	3.19 B
SS	1.47 C	1.78 C	2.83 A
F_O	* ($p < 0.05$, F = 19.96)	* ($p < 0.05$, F = 31.83)	* ($p < 0.05$, F = 12.55)

F_O—F test tillage system; *—statistical significance; n.s.—no statistical significance; values within the same columns, which are marked with different capital letters, differ statistically significantly at the level of $p < 0.05$.

The average current soil moisture at a depth of 0–40 cm during each of the penetrometry measurements in the second year of the research was as follows: in the first measurement 18.53%, in the second measurement 18.27% and in the third measurement 17.29%.

Resistance in the 2nd year of maize cultivation was statistically significantly affected by the tillage system used in all three measurements (as follows $F = 19.96$; $F = 31.83$; $F = 12.55$).

The average resistance value of the 1st measurement was 1.71 MPa. The highest resistances were measured for the DH, and the lowest for the tillage system SS. The LSD test revealed the following statistically significant differences in resistance between: CT and DH (0.23 MPa), CT and CH (0.16 MPa), CT and SS (0.19 MPa), DH and SS (0.41 MPa), and CH and SS (0.35 MPa).

In the 2nd measurement, the average resistance was 2.41 MPa. The highest penetrometer resistance was measured at CH, and the lowest resistance was measured at SS. Statistically significant differences in resistance were found between CT and CH (0.25 MPa), CT and SS (0.70 MPa), DH and SS (0.88 MPa), and CH and SS (0.95 MPa).

The average resistance value of the 3rd measurement was 2.98 MPa. The highest resistances were measured at DH and the lowest at CT processing. The LSD test revealed the following statistically significant differences in resistance between: CT and DH (0.70 MPa), CT and CH (0.60 MPa), DH and SS (0.47 MPa), and CH and SS (0.36 MPa).

In all three study years, the CT processing showed lower compaction compared to the DH and CH processing systems, which is consistent with the results of [31], conducted at similar depths. The deep processing of SS, performed using a submersible with seven working bodies at a working depth of 50 cm, resulted in the lowest penetrometer resistances in all three study years, which is consistent with the results of [28,29].

Table 5. The influence of the tire pressure of a tractor on the mechanical resistance of the soil.

2017./2018. Maize (*Zea mays* L.)			
Tire Pressure	**1st Measurement (Average Resistance—MPa)**	**2nd Measurement (Average Resistance—MPa)**	**3rd Measurement (Average Resistance—MPa)**
p_1	1.62 $_{B}$	2.44 $_{AB}$	2.95
p_2	1.80 $_{A}$	2.53 $_{A}$	3.11
p_3	1.71 $_{AB}$	2.26 $_{B}$	2.87
F_P	* ($p < 0.05$, $F = 6.67$)	* ($p < 0.05$, $F = 4.26$)	n.s. ($p < 0.05$, $F = 2.48$)

F_P—F test tire pressure at sowing; *—statistical significance; n.s.—no statistical significance; values within the same columns, which are marked with different capital letters, differ statistically significantly at the level of $p < 0.05$.

Resistance in the 2nd year of maize planting was statistically significantly affected by tractor tire pressure at sowing in the first two measurements (as follows $F = 6.67$; $F = 4.26$; $F = 2.48$ n.s.).

The average resistance value of the 1st measurement was 1.28 MPa. The highest resistances were measured at p_2, the lowest at p_1, between which the only statistically significant difference was 0.18 MPa.

In the 2nd measurement, the average resistance was 2.41 MPa. The highest resistance of the penetrometer was measured at tire pressure p_2 and the lowest resistance at tire pressure p_3. The difference in resistances was statistically justified only between p_2 and p_3 and was 0.27 MPa.

The resistances at the 3rd measurement were not significantly affected by the tire pressure and averaged 2.98 MPa. The compaction at the third measurement in the second year was the same at all pressures. This uniformity of compaction at all applied tire pressures can be attributed to the optimum condition of soil moisture during tillage, sowing and at the time of penetrometer resistance measurement.

Table 6. Influence of the depth measurement on the mechanical resistance of the soil.

2017./2018. Maize (*Zea mays* L.)			
Depth of Resistance Measurement	**1st Measurement (Average Resistance—MPa)**	**2nd Measurement (Average Resistance—MPa)**	**3rd Measurement (Average Resistance—MPa)**
0–5 cm	1.53 C	1.59 D	0.94 G
5–10 cm	2.13 A	2.13 C	1.66 F
10–15 cm	1.93 B	2.31 BC	2.30 E
15–20 cm	1.79 B	2.42 ABC	2.92 D
20–25 cm	1.63 BC	2.58 AB	3.43 C
25–30 cm	1.54 C	2.66 A	3.95 AB
30–35 cm	1.54 C	2.77 A	4.21 A
35–40 cm	1.60 C	2.80 A	4.40 A
F_D	* ($p < 0.05$, $F = 14.01$)	* ($p < 0.05$, $F = 13.82$)	* ($p < 0.05$, $F = 93.75$)

F_D—F test depth of penetrometry; *—statistical significance; n.s.—no statistical significance; values within the same columns, which are marked with different capital letters, differ statistically significantly at the level of $p < 0.05$.

Soil resistances in the 2nd year of observation was statistically significantly affected by measurement depth for all three measurements (as follows $F = 14.01$; $F = 13.82.3$; $F = 93.75$). The average resistance value was as follows: 1.71 MPa (1st measurement), 2.41 MPa (2nd measurement), 2.98 MPa (3rd measurement). The highest penetrometer resistance was measured at the 1st measurement at a depth of 5–10 cm, and at the 2nd and 3rd measurements at a depth of 35–40 cm.

Various studies have also found soil compaction in shallower layers under the influence of drought [32], which means that conserving moisture and reducing compaction by implementing conservation treatment systems (CH and SS) gains additional importance as a mitigation measure [7].

The average current soil moisture at a depth of 0–40 cm while performing each of the penetrometry in the third year of research was as follows: in the first measurement 15.08%, in the second measurement 18.81%.

Table 7. The influence of the tillage system on the mechanical resistance of the soil.

2018./2019. Winter Wheat (*Triticum aestivum* L.)		
Tillage System	**1st Measurement (Average Resistance—MPa)**	**2nd Measurement (Average Resistance—MPa)**
CT	3.10 C	1.7 C
DH	3.93 A	2.45 A
CH	3.49 B	1.96 B
SS	2.87 C	1.38 D
F_O	* ($p < 0.05$, $F = 29.76$)	* ($p < 0.05$, $F = 110.21$)

F_O—F test tillage system; *—statistical significance; n.s.—no statistical significance; values within the same columns, which are marked with different capital letters, differ statistically significantly at the level of $p < 0.05$.

Resistance in the 3rd year of the experiment was significantly affected by tillage in both measurements (as follows $F = 29.76$; $F = 110.21$).

The average resistance value of the 1st measurement was 3.35 MPa. The highest resistances were measured with the DH and the lowest with the tillage system SS. The LSD test showed that the resistance on CT was 0.83 MPa lower than the resistance measured on DH and 0.40 MPa lower than CH. Soil resistance on the DH variant was 0.44 MPa higher than on CH and 1.06 MPa higher than on the tillage variant SS. The difference in soil resistance for the CH and SS variants was 0.62 MPa and was statistically significant.

The resistances of the 2nd measurement averaged 1.87 MPa. The highest resistance was measured for DH, and the lowest for tillage SS. Resistance differences between DH and

CT (0.76 MPa), CH and CT (0.27 MPa), CT and SS (0.32 MPa), DH and CH (0.49 MPa), DH and SS (1.08 MPa), and CH and SS (0.59 MPa) were statistically significant. The lower soil resistance values in the second measurement were significantly influenced by the current soil moisture, which was 3.73% higher than in the first measurement.

Table 8. The influence of the tire pressure of a tractor on the mechanical resistance of the soil.

2018./2019. Winter Wheat (*Triticum aestivum* L.)		
Tire Pressure	**1st Measurement (Average Resistance—MPa)**	**2nd Measurement (Average Resistance—MPa)**
p_1	3.14 $_B$	1.69 $_C$
p_2	3.35 $_A$	1.89 $_B$
p_3	3.55 $_A$	2.03 $_A$
F_P	* ($p < 0.05$, $F = 7.81$)	* ($p < 0.05$, $F = 20.23$)

F_P—F test tire pressure at sowing; *—statistical significance; n.s.—no statistical significance; values within the same columns, which are marked with different capital letters, differ statistically significantly at the level of $p < 0.05$.

Tire pressure had a significant effect on resistance in both measurements of the 3rd year of research (as follows $F = 7.81$; $F = 20.23$).

The average resistance value of the 1st measurement was 3.34 MPa. The highest resistances were measured at p_3 and the lowest at p_1. The LSD test showed that the resistance differences were statistically significant only between p_3 and p_1 (0.41 MPa) and p_2 and p_1 (0.21 MPa), while the resistance difference between p_2 and p_3 was not statistically significant.

The resistance value of the 2nd measurement of pressure influence averaged 1.87 MPa. The highest resistance was measured at p_3, and the lowest resistance at p_1. The differences in resistance with respect to tire pressure were significant. Soil resistance at p_1 was 0.20 MPa lower than at p_2 and 0.34 MPa lower than at p_3. Soil resistance at variant p_2 was 0.14 MPa lower than resistance at variant p_3.

In both measurements of the third year of research, compaction under the influence of tire pressure was 13.2% higher than at pressure p_1 in the first measurement and 20% higher in the second measurement, which corresponds to the results of the research of Parkhomenko et al. (2019) [29].

Table 9. Influence of the depth measurement on the mechanical resistance of the soil.

2018./2019. Winter Wheat (*Triticum aestivum* L.)		
Depth of Resistance Measurement	**1st Measurement (Average Resistance—MPa)**	**2nd Measurement (Average Resistance—MPa)**
0–5 cm	1.22 $_D$	1.04 $_C$
5–10 cm	2.60 $_C$	1.67 $_B$
10–15 cm	3.48 $_B$	1.94 $_A$
15–20 cm	3.93 $_A$	2.03 $_A$
20–25 cm	4.21 $_A$	2.08 $_A$
25–30 cm	4.16 $_A$	2.06 $_A$
30–35 cm	3.78 $_{AB}$	2.08 $_A$
35–40 cm	3.40 $_B$	2.07 $_A$
F_D	* ($p < 0.05$, $F = 68.37$)	* ($p < 0.05$, $F = 35.51$)

F_D—F test depth of penetrometry; *—statistical significance; n.s.—no statistical significance; values within the same columns, which are marked with different capital letters, differ statistically significantly at the level of $p < 0.05$.

Soil resistances in the 3rd year of observation were statistically significantly affected by depth measurement in both measurements (as follows $F = 68.37$; $F = 35.51$). The average resistance value was as follows: 3.35 MPa (1st measurement) and 1.87 MPa (2nd measurement). The highest penetrometer resistance was measured at a depth of 20–25 cm in the 1st measurement, and at a depth of 20–25 cm and 30–35 cm in the 2nd measurement.

Under the treated layer in the DH treatment system at a depth of more than 15 cm, a sudden increase in resistance was observed in almost all measurements performed. The authors of [33] obtained similar results of penetrometer resistance at a depth of 25–30 cm (3.75 MPa), resulting from several years of tillage at the same depth. The same authors conduct research on the reduction in the depth of the basic tillage and the application of a disc harrow, which leads to an increase in resistance to 4.6 MPa (at a depth of 17.5–27.5 cm) over the multi-year observation period, as a result of the combination of two tillage bases (plowing and disc tillage), which was also the case in this study.

4. Conclusions

The applied tillage systems resulted in different soil compaction. The deepest tillage SS (undermining to a depth of 50 cm) achieved the lowest average resistance or compaction in all three study years. The highest penetrometer resistances were measured for DH tillage in all three years of observation (especially pronounced for the third measurement in the second and third years). The reason for this resistance to DH processing is the formation of a new machining base under the grip of the slab (below 15 cm depth) in combination with the base of conventional processing that was carried out in previous years.

In addition to the tillage system, soil compaction was also affected by differences in tractor tire pressure during sowing. Penetrometer measurements of compaction already in the first year show that the lowest tractor tire pressure p_1 significantly affects the reduction in compaction. In the second year, relatively uniform penetrometer resistances were obtained, indicating that the agrotechnical work and measurements were carried out at optimal soil moisture. In the third year of observation, there is a significant increase in resistance at the highest observed tractor tire pressure p_3, which also indicates that a decrease in tire pressure can affect the reduction in soil compaction.

In arable crops, a positive response of the soil to the tillage systems applied is expected, which was the case here. The use of an appropriate tillage system with regular monitoring of tire pressure during farm operations can be a more effective soil management practice to achieve optimal yields while preserving the soil as a resource.

Author Contributions: Conceptualization, R.B., K.M. and T.B.-L.; methodology, D.J. and L.Š.; investigation, R.B., T.B.-L. and D.Z.; resources, D.J.; data processing I.J.; writing—original draft preparation, R.B.; writing—review and editing, K.M. All authors have read and agreed to the published version of the manuscript.

Funding: University of Slavonski Brod and Brod-Posavina County (project: The influence of different methods of soil tillage and fertilization on the yield of field crops).

Institutional Review Board Statement: Not applicable.

Informed Consent Statement: Not applicable.

Data Availability Statement: Not applicable.

Acknowledgments: Authors express their sincere gratitude to owner of land and agricultural machinery OPG "Luka Kurkutović" for his generous help with the implementation of the research.

Conflicts of Interest: The authors declare no conflict of interest.

Appendix A

Figure A1. Tractor Massey Ferguson 8480 Dyna-VT (source: own photo).

Table A1. Basic technical characteristics of the tractor Massey Ferguson 8480 Dyna-VT.

Dimensions (length × width × height)	523 cm × 307 cm × 309 cm
Mass	9239 kg
Nominal power	216.3 kW

Figure A2. Regent Titan 15 five-furrow plow (source: own photo).

Table A2. Technical characteristics of the plow Regent Titan 150.

Number of plow bodies	5
Operation on the plow body	29–60 cm

Figure A3. Disc harrow RAU Rondo XL 44 (source: own photo).

Table A3. Technical characteristics of the disc harrow RAU Rondo XL 44.

Broj sekcija	2
The diameter of the disc	66 cm
Number of discs	44
Operation procedure	5 m

Figure A4. Underminer Pegoraro MEGA DRAG 7 (source: own photo).

Table A4. Technical characteristics of the underminer Pegoraro MEGA DRAG 7.

Number of working bodies	7
Working width	270–345 cm
Required tractor power	103–190 kW
Depth of loosening/undermining	30–75 cm
Mass	1420 kg

Figure A5. Rotary harrow Kongskilde HK 31 (source: own photo).

Table A5. Technical characteristics of the rotary harrow Kongskilde HK 31.

Operation procedure	300 cm
Mass	900 kg
The number of rotations of the rotor blades—adjustable	$270/360/400\ \mathrm{min}^{-1}$
Number of rotors with blades	12

Equation (A1) Current soil moisture

$$M_v = \frac{T_{\mathrm{mv}} - T_{\mathrm{s}}}{T_{\mathrm{s}}} \cdot 100 \tag{A1}$$

M_v—current soil moisture in weight percent, (%);
T_{mv}—wet soil mass, (kg);
T_{s}—mass of completely dry soil, (kg).

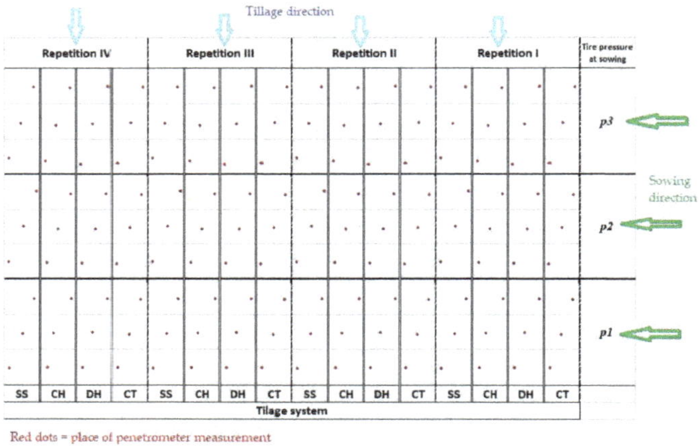

Figure A6. Research scheme.

References

1. Jug, D.; Blažinkov, M.; Redžepović, S.; Jug, I.; Stipešević, B. Effects of different soil tillage systems on nodulation and yield of soybean. *Poljoprivreda* **2005**, *11*, 38–43.
2. Rasmussen, K.J. Impact of ploughless soil tillage on yield and soil quality. *Soil Tillage Res.* **1999**, *53*, 3–14. [CrossRef]
3. Šumanovac, L.; Sebastijanović, S.; Kiš, D. *Transport in Agriculture*, 1st ed.; Faculty of Agriculture: Osijek, Croatia, 2011; pp. 156–178.
4. Normirzayev, A.R.; Nuriddinov, A.D.; Tukhtabayev, M.A. Undercarriages impact on soil of machine-tractor units during tillage and cultivation of agricultural crops. *AIP Conf. Proc.* **2023**, *2612*, 030032. [CrossRef]
5. Belić, M.; Nešić, L.; Ćirić, V. *Practicum in Pedology*; University of Novi Sad, Faculty of Agriculture: Novi Sad, Serbia, 2014.
6. Bluett, C.; Tullberg, J.N.; McPhee, J.E.; Antille, D.L. Soil and Tillage Research: Why still focus on soil compaction? *Soil Tillage Res.* **2019**, *194*, 104282. [CrossRef]
7. Jug, D.; Jug, I.; Đurđević, B.; Vukadinović, V.; Stipesević, B.; Brozović, B. *Conservation Tillage as a Measure to Mitigate Climate Change*, 1st ed.; Croatian Soil and Tillage Research Organization (CROSTRO): Osijek, Croatia, 2017.
8. Benković, R.; Šumanovac, L.; Jug, D.; Jug, I.; Japundžić-Palenkić, B.; Mirosavljević, K.; Popijač, M.; Benković-Lačić, T. Influence of Aggregated Tillage Implements on Fuel Consumption and Wheel Slippage. *Teh. Vjesn.—Tech. Gaz.* **2021**, *28*, 956–962. [CrossRef]
9. Peigné, J.; Ball, B.C.; Roger-Estrade, J.; David, C. Is conservation tillage suitable for organic farming? A review. *Soil Use Manag.* **2007**, *23*, 129–144. [CrossRef]
10. Husnjak, S.; Filipović, D.; Košutić, S. Influence of different tillage systems on soil physical properties and crop yield. *Plant Soil Environ.* **2002**, *48*, 249–254. [CrossRef]
11. Kotorová, D. The changes of clay-loamy soil properties at its different tillage. *Agriculture* **2007**, *53*, 183–190.

12. Jug, D.; Jug, I.; Stipešević, B.; Stošić, M.; Brozović, B.; Đurđević, B. Influence of different soil tillage treatments on soil compaction and nodulation of soybean root. In Proceedings of the 1st International Scientific Conference-CROSTRO, Soil tillage, Osijek, Croatia, 9–11 September 2010; pp. 188–195, ISBN 978-953-6331-83-3.
13. Wright, D.; Small, I.; Mackowiak, C.; Grabau, Z.; Devkota, P.; Paula-Moraes, S. Field Corn Production Guide. UF/IFAS Extension, SS-AGR-85. *EDIS* **2022**, *2022*, 1–13. [CrossRef]
14. Srivastava, A.K.; Goering, C.E.; Rohrbach, R.P.; Buckmaster, D.R. *Engineering Principles of Agricultural Machines*, 2nd ed.; American Society of Agricultural Biological Engineers: St. Joseph, MI, USA, 2006; pp. 169–184.
15. Atkinson, H.S.; Sparkes, D.L.; Mooney, S.J. Using selected soil physical properties of seedbeds to predict crop establishment. *Soil Tillage Res.* **2007**, *97*, 218–228. [CrossRef]
16. Çarman, K. Effect of different tillage systems on soil properties and wheat yield in Middle Anatolia. *Soil Tillage Res.* **1997**, *40*, 201–207. [CrossRef]
17. Ozpinar, S.; Cay, A. Effect of different tillage systems on the quality and crop productivity of a clay-loam soil in semi-arid north-western Turkey. *Soil Tillage Res.* **2006**, *88*, 95–106. [CrossRef]
18. Rashidi, M.; Keshavarzpour, F. Effect of Different Tillage Methods on Grain Yield and Yield Components of Maize (*Zea mays* L.). *Int. J. Agric. Biol.* **2009**, *9*, 274–277.
19. Moraru, P.I.; Rusu, T.; Bogdan, I.; Pop, A.I.; Sopterean, M.L. Effect of diferent tillage systems on soil properties and production on wheat, maize and soybean crop. *Lucr. Stiintifice Ser. Agron.* **2011**, *54*, 258–262.
20. Li, Y.; Tullberg, J.; Freebairn, D. Wheel traffic and tillage effects on runoff and crop yield. *Soil Tillage Res.* **2007**, *97*, 282–292. [CrossRef]
21. Dam, R.; Mehdi, B.; Burgess, M.; Madramootoo, C.; Mehuys, G.; Callum, I. Soil bulk density and crop yield under eleven consecutive years of corn with different tillage and residue practices in a sandy loam soil in central Canada. *Soil Tillage Res.* **2005**, *84*, 41–53. [CrossRef]
22. Zheng, C.-Y.; Yu, Z.-W.; Shi, Y.; Cui, S.-M.; Wang, D.; Zhang, Y.-L.; Zhao, J.-Y. Effects of Tillage Practices on Water Consumption, Water Use Efficiency and Grain Yield in Wheat Field. *J. Integr. Agric.* **2014**, *13*, 2378–2388. [CrossRef]
23. Håkansson, I.; Voorhees, W.B.; Riley, H. Vehicle and wheel factors influencing soil compaction and crop response in different traffic regimes. *Soil Tillage Res.* **1988**, *11*, 239–282. [CrossRef]
24. Keller, T.; Arvidsson, J. Technical solutions to reduce the risk of subsoil compaction: Effects of dual wheels, tandem wheels and tyre inflation pressure on stress propagation in soil. *Soil Tillage Res.* **2004**, *79*, 191–205. [CrossRef]
25. Botta, G.; Jorajuria, D.; Draghi, L. Influence of the axle load, tyre size and configuration on the compaction of a freshly tilled clayey soil. *J. Terramechanics* **2002**, *39*, 47–54. [CrossRef]
26. *HRN ISO 11464*; Soil Qality, Pretreatment of Samples for Physico-Chemical Analyses. Croatian Standards Institute Zagreb: Zagreb, Croatia, 2004.
27. *ISO 11277*; Soil Quality, Determination of Particle Size Distribution in Mineral Soil Material—Method by Sieving and Sedimentation. Polish Committee for Standardization: Warszawa, Poland, 2009.
28. Negev, I.; Shechter, T.; Shtrasler, L.; Rozenbach, H.; Livne, A. The Effect of Soil Tillage Equipment on the Recharge Capacity of Infiltration Ponds. *Water* **2020**, *12*, 541. [CrossRef]
29. Parkhomenko, G.G.; Voinash, S.A.; Sokolova, V.A.; Krivonogova, A.S.; Rzhavtsev, A.A. Reducing the negative impact ofundercarriage systems and agricultural machinery partson soils. *IOP Conf. Ser. Earth Environ. Sci.* **2019**, *316*, 012049. [CrossRef]
30. Jakobsen, B.; Dexter, A. Prediction of soil compaction under pneumatic tyres. *J. Terramechanics* **1989**, *26*, 107–119. [CrossRef]
31. Stošić, M.; Brozović, B.; Vinković, T.; Tkalec, K.M.; Šumanovac, L.; Tadić, V. Long-term tillage and nitrogen fertilization for soybean on gley soil. *Rom. Agric. Res.* **2020**, *37*, 151–160.
32. Birkás, M.; Kisić, I.; Bottlik, L.; Jolánkai, M.; Mesić, M.; Kalmár, T. Subsoil Compaction as a Climate Damage Indicator. *Agric. Conspec. Sci.* **2009**, *74*, 91–97.
33. Birkás, M.; Jolánkai, M.; Gyuricza, C.; Percze, A. Tillage effects on compaction, earthworms and other soil quality indicators in Hungary. *Soil Tillage Res.* **2004**, *78*, 185–196. [CrossRef]

sustainability

MDPI

Article

Assessing Soil Dynamics and Improving Long-Standing Irrigation Management with Treated Wastewater: A Case Study on Citrus Trees in Palestine

Giovanna Dragonetti and Roula Khadra *

Mediterranean Agronomic Institute of Bari (CIHEAM Bari), Via Ceglie 9, 70010 Valenzano, Italy; dragonetti@iamb.it
* Correspondence: khadra@iamb.it; Tel.: +39-080-4606227; Fax: +39-080-4606206

Abstract: Irrigation with Treated Wastewater (TWW) is a well-known agricultural practice in Palestine. The long-term use of irrigation with TWW, a source of water and nutrients, can affect plant development, soil, and groundwater quality. Consequently, the frequency and the intervals of irrigation events should be adequately scheduled, especially when nutrients (TWW-N) cannot be separated from the water. Achieving good water quality implies its immediate reuse in irrigated agriculture. In contrast, long-term soil and groundwater quality conservation is marked by the complex mechanisms that correlate the soil, water, plant, and atmosphere. Therefore, monitoring and modeling (MMA) are combined to retrieve the soil water and nitrate fluxes and identify a proper irrigation management plan in a case study in Beit Dajan-Palestine, where a schedule adapted to conventional water was applied to a 6-year-old citrus orchard continuously irrigated with TWW. Soil nitrogen concentration and water content data were collected from March to August 2021 to calibrate the Hydrus-1D model under the (1) farmer demand (F) scenario, where irrigation volumes are delivered according to the farmer experience, and to define an optimal irrigation management strategy with TWW according to the (2) model demand (M) scenario, based on the irrigation frequency. The latter respects the allowable thresholds of soil solution electrical conductivity, σ_e, assuming an average soil salinity profile and estimated leaf nitrogen concentrations tolerance as reference; 2021 was taken as a calibration year to retrieve water and nitrate fluxes for 2019 and 2020. In 2021, the measured soil electrical conductivity, σ_e, showed no salinity risk with an average value of 1.07 dS m^{-1} (low salinity < 2 dS m^{-1}) but with a leaf nitrogen deficit. Although an acceptable level of available soil nitrogen was observed (ranging between 10 and 35 mg kg^{-1}, whereas the standard value is 10–40 mg kg^{-1}), critical concentrations were observed in the leaves (below 1%) in scenario (F) compared to scenario (M) (ranging between 1.7 and 1.9%). The latter also showed a decrease in nitrate leaching by 33% compared to the former. Overall, the comparison between the simulated and measured soil variables shows that the 1D-Hydrus model could follow the temporal variation in the monitored data, with some overestimation of the measured data during the simulation period. The simulations demonstrate that by modulating the salt tolerance threshold, the M scenario achieved better results in terms of root water and N uptake despite the stress inevitably experienced by citrus with long-term TWW irrigation. Moreover, the optimum threshold values used to assess the soil quality and citrus response under conventional water irrigation were inadequate for TWW practices. Therefore, MMA could be an alternative strategy to schedule proper TWW irrigation.

Keywords: Treated Wastewater (TWW); irrigation scheduling; soil water dynamics; monitoring; modeling; 1D-hydrus model; citrus

check for updates

Citation: Dragonetti, G.; Khadra, R. Assessing Soil Dynamics and Improving Long-Standing Irrigation Management with Treated Wastewater: A Case Study on Citrus Trees in Palestine. *Sustainability* **2023**, *15*, 13518. https://doi.org/10.3390/su151813518

Academic Editors: Daniel El Chami and Maroun El Moujabber

Received: 27 July 2023
Revised: 6 September 2023
Accepted: 7 September 2023
Published: 9 September 2023

1. Introduction

Many geographic areas are already suffering from severe droughts and water resource shortages due to climate change [1], particularly affecting irrigation water requirements [2].

Improving water use efficiency is more than ever an urgent need and entails an effective and sustainable exploitation and management of the irrigation infrastructure [3]. Furthermore, adopting irrigation practices that do not deplete freshwater or damage aquatic ecosystems is vital for adapting irrigated agriculture to climate change conditions [4].

Domestic effluents may be one of the adaptation practices because they contain essential nutrients for agricultural crop development, and Treated Wastewater (TWW) reuse may reduce the freshwater demand and groundwater depletion while decreasing effluent discharges into water bodies [5]. However, the high concentration of salts that characterizes TWW may entail changes in the soil quality ascribable to the soil–water interactions that may lead to long-term soil salinization and yield reduction [6,7]. In other words, sequential TWW irrigation seasons may compromise the soil quality, such as by modifying soil infiltration and aeration and reducing root uptake of nutrients [8]. Moreover, regular TWW irrigation events induce cytotoxicity in plants caused by an excessive uptake of ions and a nutritional imbalance [9–11].

Overall, these effects stem from (1) the applied irrigation schedule and (2) the soil hydrological behavior. So far, most of the adopted TWW irrigation management plans are commonly defined according to the irrigation water quality and the conventional crop water requirement estimates, regardless of the soil–water interactions. However, the latter is significantly relevant and necessitates the consideration of soil water content and soil physical properties when selecting the irrigation schedule to be adopted. In addition, the availability of nutrients and the organic matter content in the soil aggregates are affected, which in turn affects the capacity of the soil to retain water, soil water infiltration, transport of nutrients to the roots and drainage processes, and soil fertility.

All the above confers the inevitable necessity to account for the soil hydraulic characteristics and the water–soil dynamics, among other aspects, in TWW irrigation management plans, since they are basically involved in the replenishment of groundwater through deep percolation fluxes and recharge [12,13]. For this purpose, accounting for only water quality is no longer enough as the soil, along with water and salts, triggers complex transport processes, giving place to fluxes within and below the root zone that may compromise the plant growth, soil, and groundwater quality [14].

Research supports that once the irrigation volumes are supplied to the soil, the infiltration and redistribution processes occur and contribute to regulating the water exchange within the soils, plants, and atmospheric system [15]. Thus, the adopted irrigation management strategy, which defines the frequency and interval of the interventions, does nothing but steadily interfere with the soil–water interactions, modifying the available soil water and the water and salt fate [16,17].

Since the soil's hydraulic properties largely influence the water and salt movement, their assessment is necessary. One example is observed in clay soils, where preferential flows often occur after tillage practices, induce salt transfer below the root zone, and reduce soil volume, contributing to water and nutrient uptake [18–20]. In addition, water moving through the soil brings along salts and may generate fluxes beyond the root zone, thus promoting groundwater pollution [21]. Among the different pollutants, nitrates are a cause of concern, given their capability to move with water [22,23].

In detail, water fluxes defined as an oversupply abandoning the root zone may become a potential source of leaching. Therefore, investigating the fate of water and nitrogen is of paramount importance for designing a proper TWW irrigation management plan.

Nonetheless, the task is not simple, as the mechanisms that govern the movement of water and transport of nutrients in the soils are rather complex, depending on physical laws, and are further convoluted by the soil spatial variability. Moreover, the temporal variability of the water supply and the uniformity of distribution by an irrigation system exacerbate the complexity and imply a high performance of the distribution system to enhance water infiltration [24,25].

To this end, modeling unsaturated soils may be a worthwhile strategy to adopt [26,27], and flow and transport models are the most appropriate to support a thorough plan of irrigation practices with TWW as they consider both water and nitrate fluxes [28]. Moreover, if the physical-based models are also calibrated with in situ measures, the outputs will be pretty accurate. In this regard, Refs. [10,29] observed more robust model results in a citrus orchard irrigated with TWW, where the soil water and salt distribution predictions were compared with detailed numerical analyses. This means that combining modeling and monitoring allows for determining robust outcomes [30], where the model provides estimates of the water and nitrate fluxes reasonably close to the real conditions' aftermath of long-term TWW irrigation management, and the field data allow the model to be addressed for reliable predictions. Likewise, considering that the accuracy of the models depends on the outputs' goodness implies collecting data measurements as well [31]. For this reason, it is pivotal to collect robust data from large-scale experiments rather than small-scale ones, which are not adequate to detect real TWW risks, as they fail to represent soil variability, which is fundamental for the proper characterization of TWW dynamics.

Based on these considerations, the present research developed and adopted an approach combining monitoring and modeling (MMA) to address irrigation management with TWW, merely based so far on the commonly applied strategies for conventional water. MMA allowed the assessment of model accuracy based on data measurements and enabled the provision of outputs, such as water and nitrogen–nitrate fluxes, nitrogen uptake, and deep percolation. These outputs are fundamental to defining an irrigation schedule with TWW and are difficult and complex to ascertain only from measurements. MMA is permitted to retrieve the evolution of water and nutrient dynamics and detect eventual deep percolation fluxes. These were then used to remodulate irrigated agricultural management [32,33].

MMA implies considering (1) soil hydraulic characterization, (2) water infiltration and redistribution process estimation, and (3) soil quality in terms of salinity. Consequently, field techniques to characterize the soil hydraulic properties and measure the soil water content and salt distribution, combined with using a model that provides appropriate estimates of the soil dynamics induced by one irrigation practice rather than another, are integral to the approach.

As previously stated, to account for the long-term effects of TWW use in irrigation, it is essential to consider the soil status since these waters are also rich in nutrients, and their continued use in agriculture may contribute to soil salinization, reduction in nutrient uptake, and the groundwater pollution [7,10,34]. After several irrigation seasons using TWW, mismanagement exposes the crops to stress conditions, such as nutrient deficit, which may become irreversible. Monitoring the soil salinity is also relevant. This supports the definition of TWW irrigation intervals during the growing season, which consequently curbs water oversupply and salt accumulation. In other words, monitoring soil solution electrical conductivity is relevant to the TWW strategy.

Since the salinity impact cannot be represented only by soil solution electrical conductivity measurements, σ_e, due to the soil water and salt complex processes, the soil nitrogen concentration might be a potential parameter to consider and make more robust irrigation management strategies. However, since the latter is a component rarely predictable with modeling, it can be approximated as the nitrogen content uptake by roots.

Finally, for selecting the best TWW irrigation management strategy, the outputs of the models are often compared with the optimum threshold values, which refer to irrigation with conventional water. However, threshold values must be tailored to the long-term reuse of TWW, resizing them, for example, to include plant nutrient deficiency since it is an unrecoverable condition.

Three components should be considered to identify a robust TWW irrigation schedule that accounts for the effects of a simultaneous source of water and nitrogen on soil and plant quality: soil water content, soil solution electrical conductivity, and leaf nitrogen concentration.

Based on all the above, this research proposes to apply MMA on the citrus plantations in Beit Dajan, Palestine, to estimate the soil water content and nitrogen concentration and enable the improvement of irrigation management strategies with TWW based on adequate threshold values.

More specifically, the 1D vadose zone flow model is used to predict and retrieve the long-term effects of TWW on soil quality from 2019 to 2021 based on the collected spot soil water content and soil quality data measured in 2021. A demand scenario is also built to identify the best TWW irrigation management strategy and define the leaf nitrogen threshold levels.

The deep percolation fluxes, instead, are estimated to predict the endangering of the groundwater quality and thus curb the TWW oversupplies.

The selection of a case study in Palestine stems from multiple factors related to water scarcity due to climate change in a vulnerable area with an evident gap in adaptation [35] and a complex geopolitical context that limits the use of available groundwater resources and exerts extreme pressure on its use in agriculture [36,37]. The study allowed the combination of water and soil quality parameters with crop growth to improve citrus's long-standing TWW irrigation management. However, the model can be calibrated for other crops and in different areas where water shortages due to climate change pose an existential threat to agriculture.

2. Materials and Methods

The study area was in Beit Dajan, located 10.74 km east of Nablus Governorate, approximately 80 km away from the Mediterranean Sea (600 m a.s.l). The mean annual rainfall was 343.3 mm, the average annual temperature was 20 °C, and the average annual humidity was approximately 56% [38].

The selected experimental field (Figure 1) had a surface area of 0.35 Ha and was cropped with citrus (6 years old), with an estimated water requirement of 750 mm yr^{-1}. The citrus orchards in the area were irrigated following a 3-day interval with drip systems that consisted of one drip line per tree row with emitters discharging 2 L h^{-1} each and spaced 0.25 m apart. The WW treated at the tertiary level was an exclusive resource for irrigation.

Figure 1. Beit Dajan WW treatment plant and citrus orchard site.

The common soil type is Terra Rossa (Xerochrepts, Rhodoxeralfs) according to the USDA classification [39]. The soil is red-brown and finely textured as clay loam with a percentage distribution of 35, 36.5, and 28.5% among clay, silt, and sand, respectively.

2.1. Monitoring Campaign: Collection of Soil Data

At the beginning of the irrigation season (3 March 2021), a survey allowed us to assess soil quality and hydraulic properties. Samples were collected below and between the two trees at three different depths per spot: 0–25; 25–50, and 50–75 cm. Overall, 6 undistributed soil samples were collected to determine the soil hydraulic properties as well as bulk density (ρ_b), whereas 6 further campaigns were carried out on 29 April, 25 May, 7 and 28 June, 28 July, and 12 August of the same year to collect disturbed samples.

The data acquired during these campaigns were used to determine the soil water content, soil electrical conductivity, and nitrogen concentration at three depths to calibrate the model and then retrieve the scenarios of two previous years: 2019 and 2020, where soil and crop measurements were not available, and assuming the same soil hydraulic properties and crop characteristics as for 2021.

2.1.1. Soil Hydraulic Properties Estimation

An expeditious field method was used to determine both the water retention curve, h(θ), and the hydraulic conductivity function, K(θ) [40]. The method consisted of estimating soil hydraulic properties based on the soil water infiltration process (the curves are not shown in this paper).

2.1.2. Soil Water Content, Soil Salinity, and Nitrogen Measurements

A series of measurement campaigns were performed to monitor the soil quality below and between two trees at three depths. To estimate the gravimetric water content, soil solution electrical conductivity, and available nitrogen 6 disturbed soil samples were collected. Moreover, many undisturbed samples were collected during the first campaign to determine the bulk density, ρ_b (M L^{-3}), and the corresponding volumetric water content of all samples. Chemical analyses were also performed to determine the concentrations of nitrogen using absorption spectrophotometry. In detail, available nitrogen was determined with an automated segmented flow analyzer using the cadmium reduction method to quantify nitrate-nitrogen [41] and ammonium-nitrogen. In addition, leaf nitrogen measurements were carried out by collecting 5 leaves from 10 representative trees during 6 field campaigns. The leaf samples were analyzed according to the Kjeldahl method [42].

2.2. Modeling Soil Water Movement and Solute Transport

HYDRUS-1D software package 4.17 [27] was used to simulate one-dimensional water and multiple solute movements in variably saturated porous media.

Using the finite element method, HYDRUS-1D numerically solved the Richards equation for unsaturated soils and described the water flow and the advection–dispersion equations. For simplification, it neglects water, solute fluxes, pressure head, and concentration gradients in the horizontal direction.

2.2.1. Water Movement

Richards' equation describes the soil water flow:

$$\frac{\partial \theta}{\partial t} = \frac{\partial}{\partial z}\left[K(h)\frac{\partial h}{\partial z} - K(h)\right] - S(z,t) \tag{1}$$

where θ is the volumetric soil water content (L^3 L^{-3}), t is time (T), z is the vertical space coordinate (L), h is the pressure head (L) expressing the potential in terms of the water column's height, K is the hydraulic conductivity (LT^{-1}), and S is the sink term accounting for water uptake by plant roots (T^{-1}).

van Genuchten–Mualem [43] empirical model was used to determine the unsaturated hydraulic properties of the soil.

$$\theta(h) = \begin{cases} \theta_r + (\theta_0 - \theta_r)\left[1 + |\alpha_{VG}\, h|^n\right]^{-m} & h \le 0 \\ \theta_0 & h > 0 \end{cases}$$
$$K(S_e) = \begin{cases} K_0 S_e^{\tau}\left[1 - \left(1 - S_e^{\frac{1}{m}}\right)^m\right]^2 & h \le 0 \\ K_0 & h > 0 \end{cases} \tag{2}$$

In Equation (2), θ_0 and θr are the saturated and residual water content ($cm^3\ cm^{-3}$), respectively; θ is the soil water content at a given pressure head (h); n and m = 1 − 1/n are shape parameters; S_e is the effective saturation (cm^{-1}) obtained as $(\theta - \theta_r)/(\theta_0 - \theta_r)$; τ is a parameter that considers the tortuosity; and K_0 is the saturated hydraulic conductivity.

2.2 2. Solute Transport

The equations governing the one-dimensional advective-dispersive chemical transport under transient flow in the soil are defined as follows:

$$\frac{\partial \theta c_k}{\partial t} + \rho \frac{\partial \overline{c_k}}{\partial t} = \frac{\partial}{\partial z}\left(\theta D \frac{\partial c_k}{\partial z}\right) - \frac{\partial q c_k}{\partial z} + \varnothing_k - Sc_{r,k} \tag{3}$$

where θ is the volumetric water content ($L^3\ L^{-3}$); c, $\overline{c}$ and c_r are solute concentrations in the liquid phase ($M\ L^{-3}$), solid phase ($M\ M^{-1}$), and sink term ($M\ L^{-3}$), respectively; ρ is the soil bulk density ($M\ L^{-3}$); q is the volumetric flux density ($L\ T^{-1}$); D is the hydrodynamic dispersion coefficient ($L^2\ T^{-1}$); $\varnothing$ represents chemical reactions of solutes involved in a sequential first-order decay chain, such as the nitrification of nitrogen forms ($M\ L^{-3}\ T^{-1}$); and subscript k represents the chemical forms as σ_e, $N - NO_3^-$, and $N - NH_4^+$.

The last term in Equation (3) represents a passive root nutrient uptake [44].

The adsorption and cation exchange, which are relatively complex processes, are considered in the solute transport module of the standard HYDRUS using empirical linear or nonlinear adsorption isotherms. The following linear equation describes the adsorption isotherm that relates c and $\overline{c}$ in Equation (3):

$$\overline{c}_k = K_{d,\ k} \cdot c_k \tag{4}$$

where $K_{d,k}$ ($L^3\ M^{-1}$) is the distribution coefficient of the chemical form k.

2.2.3. Root Water Uptake

A macroscopic approach [45] was used to determine the sink term, S. It assumes that the rate of the potential transpiration, T_p ($L\ T^{-1}$), distributed according to the normalized root density function, $\beta(z,\ t)$ (L^{-1}), over the root zone, and multiplied by, $\alpha(h, h_\varnothing, z, t)$, the dimensionless stress response function, accounts for osmotic and water stresses [44–46]:

$$S(h, h_\varnothing, z, t) = S_p(z,t) \times \alpha(h, h_\varnothing, z, t) = T_p(t) \times \beta(z,t) \times \alpha(h, h_\varnothing, z, t) \tag{5}$$

where $S(h, h_\varnothing, z, t)$ and $S_p(z,t)$ are, respectively, the actual and potential volumes of water taken from a unit volume of soil per unit of time ($L^3 L^{-3} T^{-1}$), and $\alpha(h, h_\varnothing, z, t)$ is a dimensionless function of the pressure (h) and the osmotic (h_ϕ) head. Then, by integrating Equation (5) over the roots' domain L_R, the actual transpiration rate, T_a ($L\ T^{-1}$), is derived as

$$T_a = \int_{L_R}^{0} S(h, h_\varnothing, z, t)dz = T_p(t)\int_{L_R}^{0} \alpha(h, h_\varnothing, z, t) \times \beta(z,t)dz \tag{6}$$

The reduction in the water uptake by the roots induced by water stress, $\alpha_1(\mathrm{h})$, was described [45] in the following model:

$$\alpha_1(h) = \begin{cases} 0, \; h > h_1 \; or \; h \leq h_4 \\ \dfrac{h - h_1}{h_2 - h_1}, \; h_2 < h \leq h_1 \\ 1, \; h_3 < h \leq h_2 \\ \dfrac{h - h_4}{h_3 - h_4}, \; h_4 < h \leq h_3 \end{cases} \quad (7)$$

where $\mathrm{h_1}, \mathrm{h_2}, \mathrm{h_3}$, and $\mathrm{h_4}$ are the threshold parameters, and the water uptake is at the potential rate when $\mathrm{h_3} < \mathrm{h} \leq \mathrm{h_2}$. It linearly drops off for $\mathrm{h} > \mathrm{h_2}$ or $\mathrm{h} < \mathrm{h_3}$, and is null for $\mathrm{h} < \mathrm{h_4}$ or $\mathrm{h} > \mathrm{h_1}$. The internal database of HYDRUS-1D provides the soil water pressure head parameters for citrus [47].

The reduction in the water uptake by the roots induced by salinity stress, $\alpha_2(\mathrm{h})$, was described using the threshold and slope functions [48]. The model was implemented in HYDRUS as follows:

$$\alpha_2(h) = \begin{cases} 1, & \sigma_e \leq \sigma_{threshold} \; or \; h_\varnothing \geq h_{\varnothing threshold} \\ 1 - (\sigma_e - \sigma_{threshold})0.01\; slope, & \sigma_e > \sigma_{threshold} \\ or & or \\ 1 + (h_\varnothing - h_{\varnothing threshold})slope^*, & h_\varnothing < h_{\varnothing threshold} \end{cases} \quad (8)$$

where $\sigma_{\mathrm{threshold}}$ is the salinity threshold (dS m^{-1}) corresponding to the electrical conductivity (σ_e) value below which no reduction in the root water uptake occurs. It is expressed in terms of osmotic head (L), *slope*, and *slope**. The slopes determine the decline in root water uptake per unit increase in salinity or osmotic head, above or below the threshold, respectively.

2.3. Data Input

A geometric soil domain depth of 100 cm was set in Hydrus 1D, subdivided into two layers: 0–70 and 70–100 cm to include the variability of the soil hydraulic properties.

Moreover, the following data were inputted:

- Daily meteorological data were recorded at the meteorological station in Nablus and averaged for 2010–2021;
- Citrus crop coefficients along the developmental stages were collected from the literature;
- TWW irrigation volumes applied by the farmers during the irrigation seasons of 2019, 2020, and 2021;
- Soil and water quality: initial electrical conductivity (σ_e), nitrate, and ammonium, measured in both TWW and soil solution, in addition to soil water content;
- Soil hydraulic and hydro-dispersive properties: The former was obtained according to the method described above, whereas the latter was collected from citrus experiments proposed in the literature [49,50].

2.4. The Setup

The geometry domain was drawn in terms of the initial, top, and boundary conditions (Figure 2). The model was run twice to simulate water content, soil solution electrical conductivity, σ_e, nitrogen concentrations, and root water uptake (below and between the two trees), where the average of both outputs was considered to represent the study area. The water fluxes outputs were used to estimate the deep percolation below 75 cm, whereas the root water uptake and soil nitrogen concentrations were used to estimate the leaf nitrogen content. To select the best model demand (M), the acceptable soil solution electrical conductivity values, σ_e, were set in the range of 2.0–2.2 dS m^{-1}. At the same time, the estimated leaf nitrogen concentrations (LNC) were considered to be the acceptable values that accounted for the citrus nutrient deficit since the threshold values observed under conventional water irrigation were not representative of long-term TWW reuse.

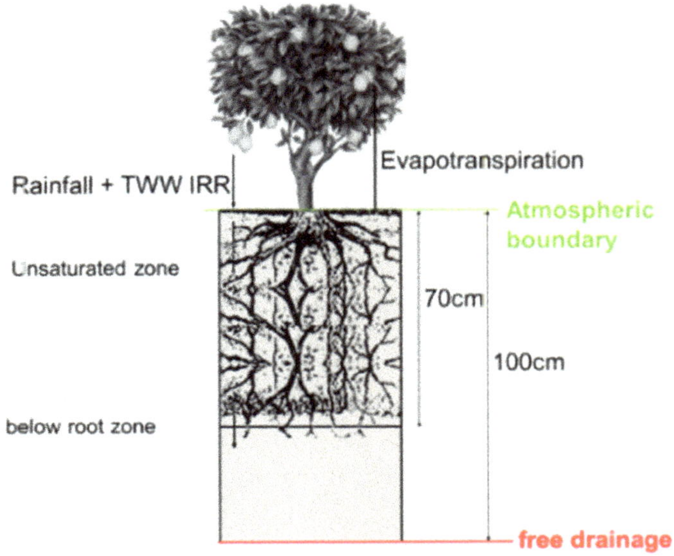

Figure 2. Geometry Domain setup in Hydrus 1D.

The input modules were prepared to report the following initial conditions parameters: soil solution electrical conductivity (σ_e), nitrate ($N-NO_3^-$), ammonium ($N-NH_4^+$), and initial soil water content (θ_{init}) at three depths: 25, 50, and 75 cm as listed in Table 1, while the initial pressure head was assumed uniform along the two soil profiles and set equal to −100 cm.

Table 1. Initial soil parameters.

Soil Layer (cm)	θ_{init} ($cm^3\ cm^{-3}$)	σ_e ($dS\ m^{-1}$)	$N-NH_4^+$ ($mg\ cm^{-3}$)	$N-NO_3^-$
0–25	0.105	1.267	0.0019	0.0174
25–50	0.105	1.127	0.0034	0.00956
50–75	0.105	1.333	0.0030	0.0239

The boundary conditions were defined at the soil's top and bottom (Figure 2). Meteorological data and daily crop evapotranspiration (ET_c) (where daily reference evapotranspiration (ET_0) collected by the Nablus meteorological station and calculated using the Penman-Monteith method [51] and K_c values of citrus as per [51] were considered) were used to define the former, and the free drainage (unitary hydraulic gradient) to set the latter.

The root water uptake was computed using the HYDRUS-1D model and assumed as the fraction, T_r, of water transpired by the crop and obtained by splitting ET_c daily values into two components (i) crop transpiration (T_r) and (ii) soil evaporation (E) rates, according to the authors of [52].

The hydraulic parameters, such as the residual water content θ_r, the saturated water content θ_s, van Genuchten shape parameters (α, n, and λ), and the saturated hydraulic conductivity (K_s), were obtained as described in Section 2.1. The estimated soil hydraulic parameters used are listed in Table 2.

Table 2. Soil hydraulic parameters at three depths.

Soil Layer (cm)	θ_s (cm^3 cm^{-3})	θ_r (cm^3 cm^{-3})	α (cm^{-1})	n	K_s (cm d^{-1})	λ
0–25	0.540	0	0.025	1.35	7.7	0.5
25–50	0.530	0	0.033	1.23	6.7	0.5
50–75	0.480	0	0.046	2.33	5.9	0.5

To describe the transport and transformations of *N* forms, the convection-dispersion equation and the first-order decay chain were used, respectively [27]. Since ammonium and nitrate were considered as available nitrogen provided by TWW, the adsorption of ammonium to the soil particles and nitrification ($N-NH_4^+$ transformation into NO_2^- and then further into NO_3^-) were considered as the main reaction processes.

The parameters of solute transport, i.e., dispersity, λ, and molecular diffusion coefficient, D_w, σ_e, $N-NH_4^+$, and $N-NO_3^-$, were initially set based on [50,53,54], and the data are listed in Table 3. The HYDRUS-1D model, nitrogen-nitrate $N-NO_3^-$ and soil solution electrical conductivity σ_e were assumed only in the dissolved phase (adsorption coefficient, $K_d = 0$ cm^3 d^{-1}), while nitrogen-ammonium $N-NH_4^+$ was assumed to adsorb to the solid phase using a distribution coefficient K_d of 3.5 cm^3 d^{-1}.

Table 3. Soil hydro-dispersive parameters.

Soil Layer (cm)	λ (cm)	D_w (cm^2 d^{-1})		
		σ_e	$N-NH_4^+$	$N-NO_3^-$
0–25	8	0.864	2.32	2.32
25–50	8	0.864	2.32	2.32
50–75	10	0.864	2.32	2.32

The initial ammonium and nitrate concentrations were set to a uniform zero concentration according to the authors of [55]. Mineralization and immobilization processes were neglected [56]. In this case, the denitrification process can be neglected due to the unsaturated and aerobic conditions in the soil under drip irrigation [57]. Ref. [55] states that denitrification losses may be ignored in silty clay, loam, and sandy loam soils where a micro-irrigation system was adopted, which was the case in the present study. Unlimited passive nutrient uptakes were considered for N forms according to the authors of [27,50]. The hydro-dispersive parameters are listed in Table 3.

As for the water supply, irrigation water volumes stemming from the treatment plant and rainfall were accounted for. The TWW irrigation volumes were collected from farmers and then resized following the model demand (M), whereas the rainfall data were collected from the meteorological station, as explained above.

To monitor the TWW quality, three effluent samples were collected monthly from the outlet between 3 March and 30 August 2021, and their physical and chemical properties were analyzed following the Standard Methods of Water and Wastewater Analysis [58]. Table 4 shows the analysis results as an average of the three collected samples.

Table 4. Chemical parameters of TWW.

Parameter	Value
σ_w (dS m^{-1})	1.916
$N-NH_4^+$(mg L^{-1})	32
$N-NO_3^-$ (mg L^{-1})	0.768
P (mg PO_4 L^{-1})	22.9
K (mg L^{-1})	270

Table 4 shows the high values of σ_w (electrical conductivity of water) and ammonium $N-NH_4^+$. In particular, the high concentration of $N-NH_4^+$ can be attributed to the high organic matter percentage. Having assumed that the available nitrogen (N − AVA) is the sum of $N-NO_3^-$ and $N-NH_4^+$, it is evident that $N-NH_4^+$ plays a significant role in the soil nitrogen balance. Being positively charged, it is attracted and retained by the soil particles, which may favor either root uptake or fast nitrification. The latter scenario causes acidification of the soil and increases $N-NO_3^-$ availability, inducing losses through leaching and denitrification [59].

2.5. Modeling Optimal Irrigation Management Strategy with TWW

The demand (M) set in the Hydrus 1D model considered the simultaneous supply of salt and water and adjusted the irrigation frequency accordingly during the growing season. The threshold values of σ_e and LNC (measured monthly from March to August 2021) were considered.

In practice, the irrigation frequency was defined based on σ_e, while the soil water storage was calculated along the soil profile (0–75 cm) by summing the water content of the 3 compartments to define the irrigation volumes.

A frequency of 4–5 days was scheduled by the demand (M) to reach both an optimum LNC (between 1.7 and 1.9%) and a σ_e, ranging between 2.0 and 2.2 dS m^{-1} and assumed as the average soil salt tolerance of the root zone 0–75 cm. A diagram of the applied approach is illustrated in Figure 3.

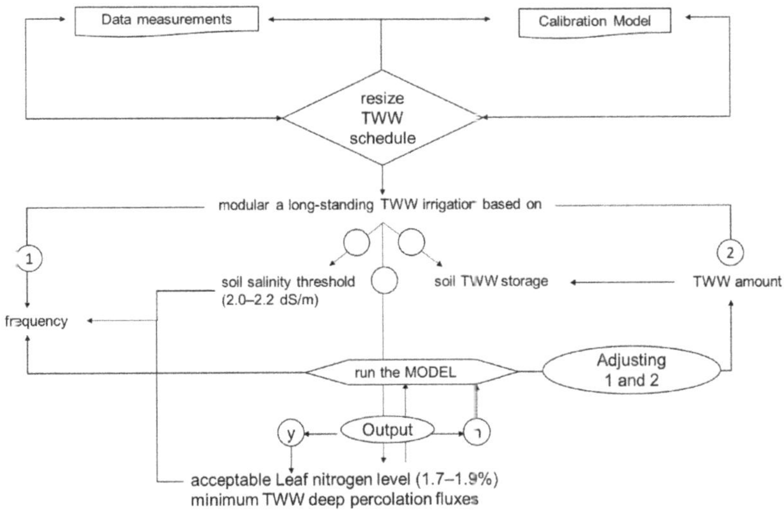

Figure 3. Flow chart showing the approach used to resize a long-standing TWW irrigation management (y = yes; n = no).

3. Results and Discussion

Six field campaigns (29 April, 25 May, 7 and 28 June, 28 July, and 12 August) were carried out during the citrus irrigation season 2021 to predict the evolution of water and salt behavior using the Hydrus 1D model. The collected data in 2021 allowed the realistic simulation of the field conditions and the reproduction of the soil water and salt movement observed in the field. Once the model was calibrated, the same geometry domain was used to retrieve the soil water and salt fluxes related to the two previous irrigation seasons: 2019 and 2020. The results are shown as a comparison between the farmer (F) and model (M) scenarios.

3.1. Monitoring Phase

The results depicted in the subsequent figures show the trend of water content, θ, soil solution electrical conductivity, σ_e, and soil nitrogen concentrations at three different depths (25, 50, and 75 cm) were obtained as an average of the data collected below and between the two trees.

Available soil nitrogen (N − AVA) was obtained as a sum of the nitrate and ammonium concentrations.

3.1.1. Soil Water Content

Figure 4 shows the trend of soil water content as per the collected data during the six field campaigns. A fluctuation likely due to a steady sequential TWW irrigation frequency period (every 2–3 days and for 9 months) influencing the soil quality is clearly shown, underlining the relevant impact of the soil hydraulic properties on soil–water dynamics. The case study area, characterized by a fair percentage of clay and loam particles, had a low capacity of the soil to release water. Moreover, the low hydraulic conductivity slows the water movement among and inside the pores, thus affecting the water distribution and availability to the plant. In this regard, the researchers observed that the suspended materials of the TWW, such as salts, can occupy even the spaces surrounding the smaller pores [60]. This may lead the mesopores to disconnect from the soil matrix, inhibiting any of their contributions to the pore distribution. This relevant soil structure property expresses the amount of water and where the water is distributed. If water is not well allocated among different pore classes, it reduces available water and compromises the roots' uptake activity. The authors of [61] showed that the percentage of the soil clay particles and salt content of TWW used for irrigation reduced soil porosity and available water, inducing root uptake stress. In Figure 4a, the measured water content at the three soil depths shows a slight variability, with values varying between 0.150 and 0.240 and a significant difference observed on 12 August 2021, likely due to the high temperatures recorded in the summer and provoking the quick evaporation of water from the shallow soil layers, without influencing the overall soil water storage (W) 0–75 cm, as shown in Figure 4b. Moreover, two peaks can be observed in Figure 4b, with a down-peak recorded on 7 June, showing an overestimation of water demand and greater storage compared to 28 July, where an up-peak likely induced the movement of part of the water below the root zone. This would also explain why a citrus nitrogen deficit was observed; the water storage at 0–50 cm, considered the volume widely explored by active roots, was smaller than that at 0–75 cm. This is due to two main reasons: (1) the roots, while up-taking water, bring salts toward the aerial part of the plant, instigating stress and nutrient deficit over time; (2) the overestimated TWW volumes allow the redistribution of water along the soil profile, not only in the root zone. The low peak recorded on 28 June may be explained by the sequence of TWW irrigation events adopted by the farmer, overestimating the citrus requirements and the fraction of water withheld by the small pores of the clay soil, which is not easily accessible to the plant. It is likely that the water oversupply was redistributed in the deep layers, contributing to the leaching of the salts introduced with TWW through the preferential paths produced by agricultural practices and often manifested in clay soils. This behavior may also be explained by the soil hydraulic conductivity, whose values were collected along the soil profile and are shown in Table 2. These low values, as discussed above, demonstrate that the soil holds more water in the small pores that are poorly accessible to the plant, causing stress to the roots.

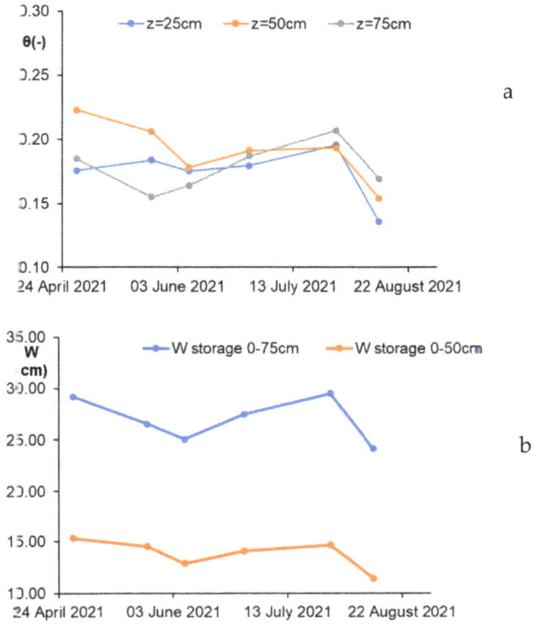

Figure 4. (**a**) The trend of water content measured at three depths: 0–25, 25–50, and 50–75 cm; (**b**) the estimated soil water storage (W) at two soil depths, 0–50 and 0–75 cm, during 2021 citrus TWW irrigation season.

3.1.2. Soil Solution Electrical Conductivity (σ_e)

Soil solution σ_e showed a similar trend to the water content (Figure 5), considering that the salinity measured at three different depths did not show any critical values that may suggest that TWW would have contributed to soil salinity. Soil nitrogen concentrations were within the optimum range. Nevertheless, the low frequency of irrigation events and the high contribution of TWW in terms of ammonium (32 mg L^{-1}) induced a leaf nitrogen imbalance. The clay particles absorbed part of the ammonium [62], and another part was subjected to nitrogen transformation processes. This aspect is better discussed later.

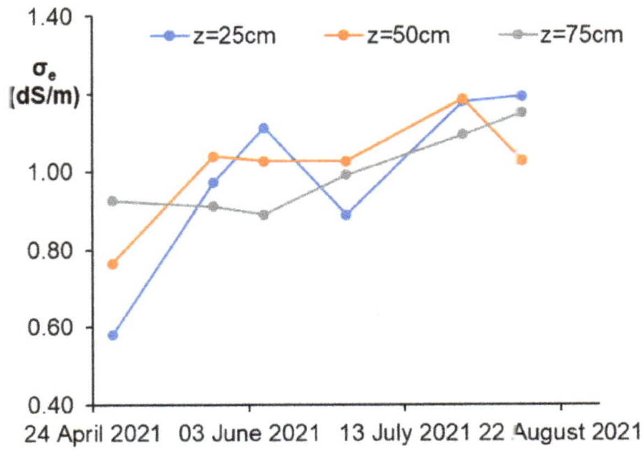

Figure 5. The trend of soil solution σ_e measured at three depths during the 2021 citrus irrigation season.

Despite the high electrical conductivity of the TWW averaging about 1.90 dS m^{-1} (Table 4), no correspondence was found in soil salinity, suggesting that the water and salt distribution along the soil profile is governed by the soil hydraulic properties. Nitrate, being the available part of nitrogen and moving with the water, may be conveyed below the root zone and leached, partly explaining the nitrogen deficit observed in the leaves (Table 5).

Table 5. Comparison between measured (F scenario) and simulated (predicted M scenario) in 2021: leaf nitrogen concentration measurements, LNC.

Date	LNC (%)	
	F	M
	Measured	**Predicted**
29 April	0.95	1.87
25 May	1.06	1.75
7 June	0.77	1.65
28 June	0.89	1.94
28 July	0.81	1.90
12 August	0.88	1.97

According to [63,64], the salinity risk for citrus is considered high for σ_e values greater than 1.9 dS m^{-1}, low for values lower than 1.7 dS m^{-1}, and medium for values ranging between 1.7 and 1.9 dS m^{-1}. Overall, the nutrient deficit is explained by the simultaneous uptake of water and salt citrus plants, which inevitably induce stress. Therefore, irrigation management with TWW should account for the salt supply and find alternative strategies that do not harm the soil quality or compromise the yield. To this end, the irrigation events' frequency should be adjusted considering leaf nitrogen and soil salinity; consequently, assessing the soil salinity becomes fundamental.

Refs. [10,29], Assouline et al. (2015), and Russo et al. (2015) observed a 20–30% yield drop in citrus cultivated on clay soil, after ten years of TWW irrigation supply. Therefore, resizing the acceptable thresholds of salinity, including the nutrient deficit induced by stress and imposed by TWW is primordial, while improving irrigation management with TWW remains a valid option to control leaf salinity levels.

3.1.3. Available Nitrogen ($N - AVA$)

Figure 6 shows that $N - AVA$ was around 10 mg kg^{-1} at the beginning of the monitoring phase, which corresponds to the optimum minimum [65,66], and that an increase in N concentration occurred after the sequential TWW irrigation events. Nevertheless, the contribution of TWW was around 1 mg L^{-1}, while most of the increase was due to the nitrogen–ammonium distributed along the soil profile during the 2021 irrigation season. The observed ammonium levels in the soil were between 4 and 5 mg kg^{-1} over time, which is an order of magnitude lower than the nitrate concentrations. This also means that the high ammonium level of TWW (32 mg kg^{-1}) contributed to the nitrogen transformation processes in the soil [67]. The salinity stress can be explained by the fact that the salts supplied with TWW induced an osmotic effect that deteriorated the soil's physical and hydraulic properties [68] and, in turn, reduced the water uptake and the roots' activity [29,69,70]. Furthermore, osmotic stress induced the loss of elasticity in roots and thus reduced the uptake of water and nutrients.

Moreover, soil nitrate–nitrogen concentration reached 10 mg kg^{-1}, a value ranging around the acceptable minimum threshold, in the first half of the irrigation period and for two depths (25 and 50 cm), and then slowly increased and ranged between 15 and 40 mg kg^{-1} after 23 June, considering that the optimum threshold range is 10–40 mg kg^{-1} under conventional water [66]. Further experiments on citrus registered levels of total N equal to 1.4% in TWW irrigated soils [71] that may not be translated into availability to the plant. In fact, another study reported, under the same conditions, higher accumulation of nitrogen (N) only in the soils but not in the leaves [72].

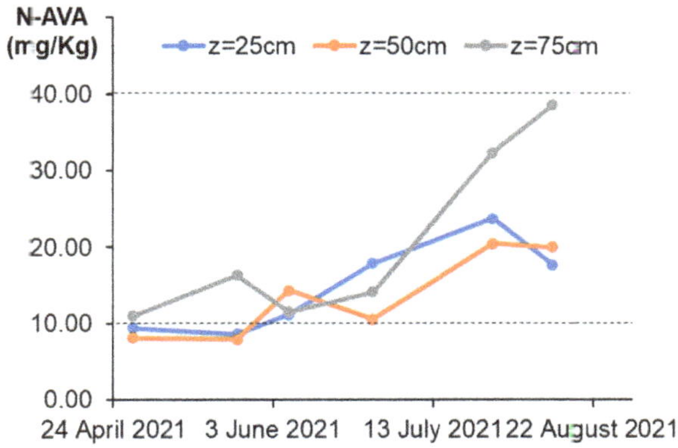

Figure 6. The trend of nitrogen–nitrate measured at three depths during the irrigation season of citrus.

3.2. Simulation Results

To improve the TWW irrigation strategy, simulations were generated based on the soil quality measurements carried out in 2021. Two runs were performed to compare the farmer's irrigation schedule (three times per week) (F) and the model demand (M). By weighing the water and salt supply and adjusting the irrigation frequency, (M) could also improve the leaf nitrogen content that reached acceptable values.

The two graphs in Figure 7a,b show the trend of upward (ET_c) and downward (TWW supply and rainfall) fluxes during three irrigation seasons. ET_c, and rainfall, were input data to the model, while the cumulative TWW irrigation volumes were calculated for both (F) and (M). It is evident how the irrigation frequency set by the model slightly reduced the amount of TWW supplied. The cumulative volume of supplied TWW was reduced by 15% compared to (F) during the three irrigation seasons (2019–2021).

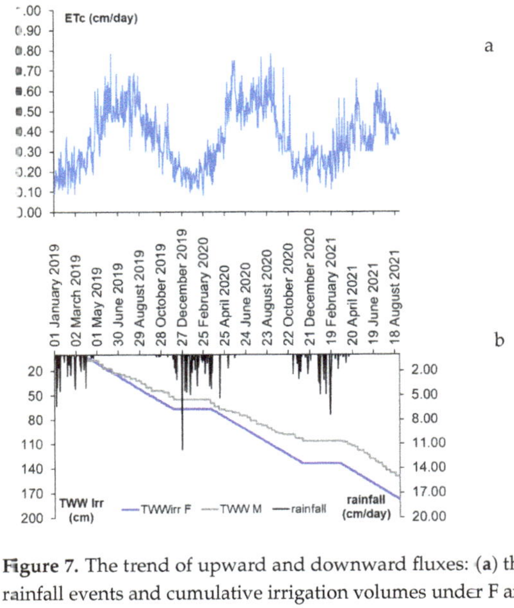

Figure 7. The trend of upward and downward fluxes: (**a**) the crop evapotranspiration ET_c; (**b**) the rainfall events and cumulative irrigation volumes under F and M scenarios over three years.

3.3. Comparison between Measured and Simulated Outputs

The simulated water content, the concentration of available nitrogen N − AVA, and the soil solution, σ_e, at 25, 50, and 75 cm depths were compared with the measured data and are illustrated in Figures 8–11. The field campaigns from 29 April to 12 August 2021 were used to calibrate the model and retrieve the evolution of water and nitrate fluxes where no measured data were available.

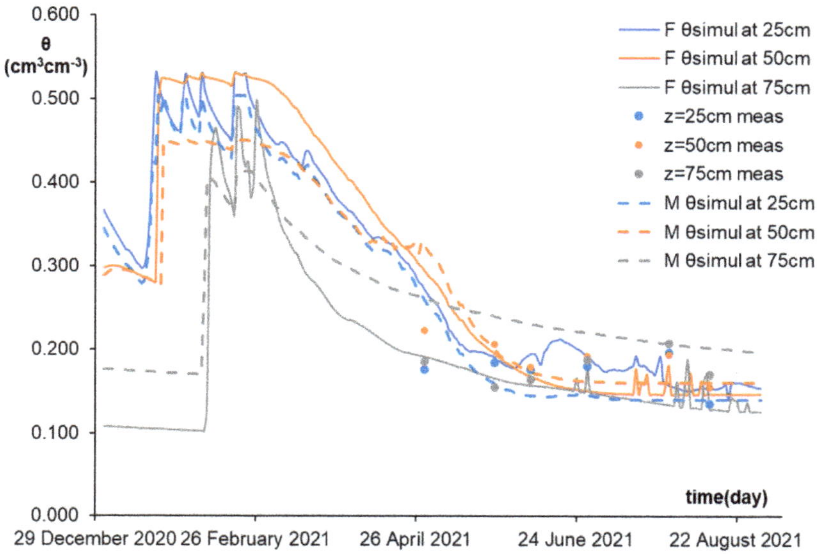

Figure 8. The trend of the soil water content simulated (1 January to 30 August 2021) for two scenarios: farmer (F) and the model (M).

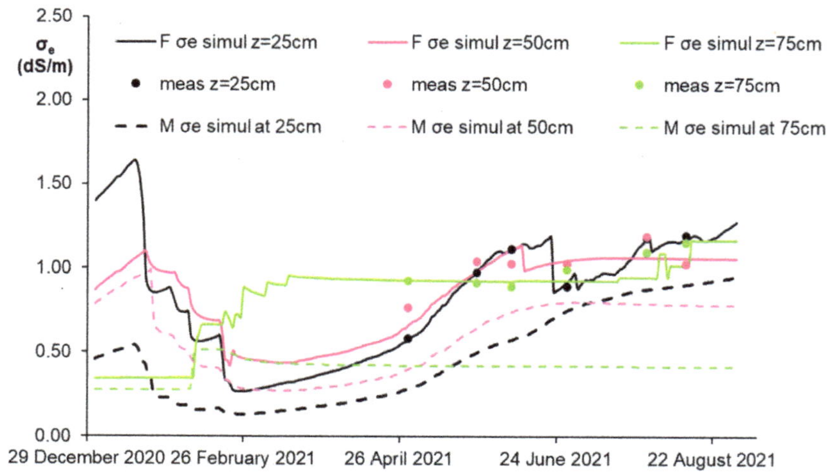

Figure 9. The trend of soil solution electrical conductivity σ_e at three depths z = 25, 50, and 75 cm during the 3rd irrigation season (1 January to 30 August 2021) for two simulation scenarios: farmer (F) and the model (M).

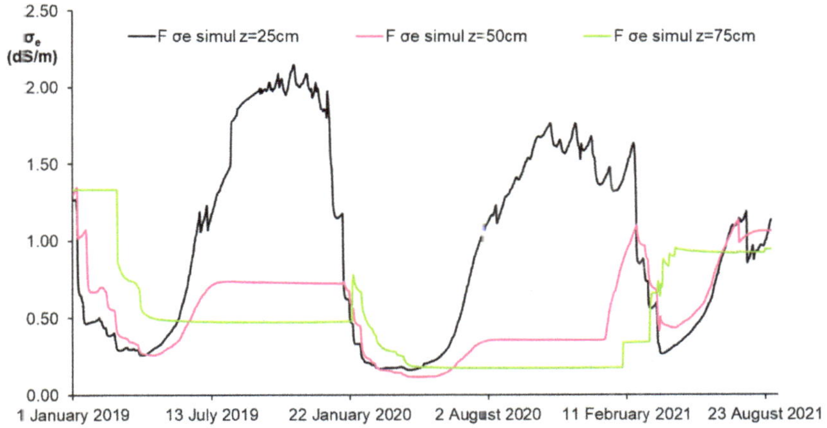

Figure 10. The trend of soil solution electrical conductivity σ_e at three depths z = 25, 50, and 75 cm from 2019 to 2021.

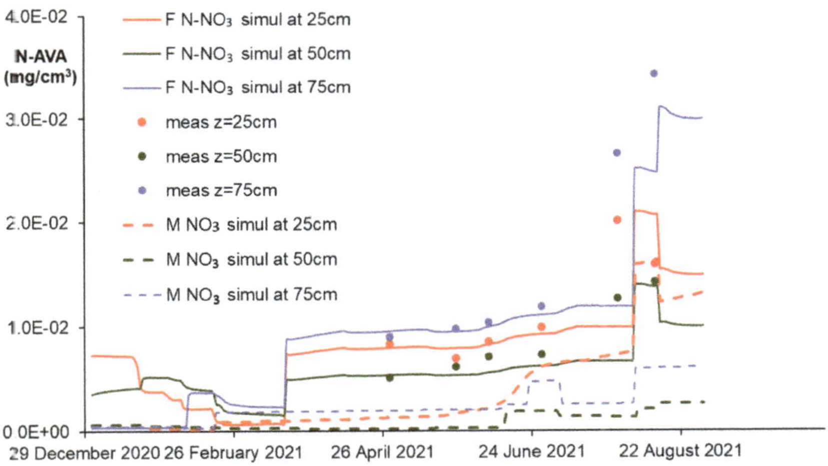

Figure 11. The trend of nitrogen nitrate $N-NO_3^-$ at three depths z = 25, 50, and 75 cm during the 3rd irrigation season (1 January to 30 August 2021) for two simulation scenarios: farmer (F) and the model (M).

Soil water content and soil solution σ_e: The simulation closely matched the measured local values, but at the deep layer (50 and 75 cm), the values were underestimated by the model that smoothed the local soil variabilities at each site.

The nitrogen root uptake calculated by the model was assumed to be equal to that of LNC. Accordingly, (M) shows better LNC (Table 5) and σ_e values compared to (F) (Figure 9), which ranges around 2 dS m^{-1}. Even though this value is acceptable, it is not enough to assess the adequacy of the farmer's management (Figure 10). It is a point measurement that cannot provide an overall indication of the combined processes and their impact on the soil and crop quality. The measured nitrogen concentrations showed unbalanced plant nitrogen, although the nitrogen available in the soil was within an optimum threshold range.

The leaf nitrogen analyses showed a nutrient deficiency since the values were below 1.0%. This demonstrates that the plant nutrient deficit is not recoverable and cannot be compared with the optimum values proposed for citrus under conventional water conditions, ranging between 2.1 and 2.4% [73]. This range was not verified under TWW irrigation management, which induces, in the long run, stress hardly improvable even with subsequent proper management due to the hysteresis effect that characterizes the roots [74]. Other studies reported higher leaf N concentrations in citrus plants (2.8–3.0%) due to an imbalanced supply with sequential TWW irrigation events [75].

Figure 11 compares the trend of available nitrogen (N − AVA) in the two scenarios F and M, and clearly shows that an acceptable nitrogen concentration in the soil does not imply its immediate availability for root uptake. The soil–water interactions, the soil hydraulic characteristics, and the overestimated TWW supply reduced the roots' elasticity and, thus, their uptake capacity. These parameters and processes were mainly caused by the salts introduced with TWW continuous irrigation events, which may be partly conveyed below the root zone, generating deep percolation fluxes.

Under scenario (F), the σ_e trend was acceptable, but since citrus experienced a long stand and improper TWW irrigation management, the capability of roots to uptake nitrogen was significantly reduced, rendering it often unavailable to the crop. Moreover, most of the water is held in the small pores, requiring excessive energy from the roots to be taken up. Consequently, part of the nitrogen available in the nitrate form may be leached, and only a small portion is kept in the shallow soil layers (0–35 cm), even though less available to the plant, with water being mostly distributed in the small pores. The authors of [61] state that these mechanisms reduce soil porosity and hydraulic conductivity, decreasing soil infiltration. In [76], it was also reported that using TWW in agriculture decreased soil infiltration after 4 years, accompanied by macropore reduction. Moreover, the suspended solids in the TWW reduce the soil infiltration capacity as they replace water and occupy the soil pores [70].

(M) suggests an adequate TWW irrigation management, orientating the simulations to real field conditions, where citrus has experienced a nutrient deficit under stress conditions, as shown by the observed low leaf nitrogen values in Table 5. Since stress is almost irreversible, modulation of the threshold values for irrigation with conventional water is necessary. In fact, (M), which accounts for stress conditions, shows leaf nitrogen values lower than the ones observed with conventional methods.

3.4. Water Fluxes below the Root Zone

The water fluxes below the root zone (at 75 cm), simulated according to the (M) scenario (Figure 12), show how the management imposed by the model reduced the potential downward fluxes and nitrate leaching compared to (F) management that overestimated TWW requirements and contributed to reducing nitrogen, potentially promoting leaching.

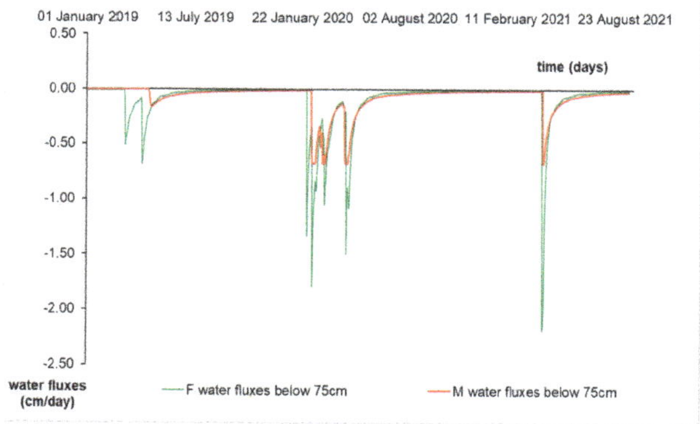

Figure 12. Water fluxes simulated at 75 cm under M and F scenario.

Assuming that these fluxes represent the deep percolation (Figure 13) and that, at the considered depth, root activity is low and, therefore, nitrate abandons the root zone, the nitrate leaching, NL, can be computed at 75 cm following (Equation (9)).

$$NL = J_w \times NC_{75} \tag{9}$$

where J_w is the water flux and NC_{75} the nitrate concentration at 75 cm depth.

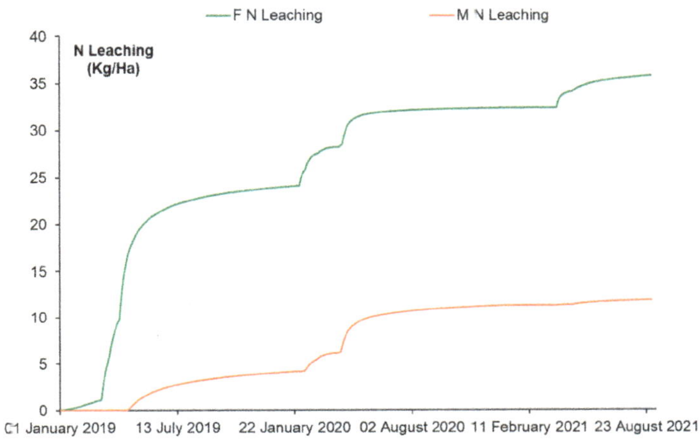

Figure 13. Simulated cumulative nitrogen leaching for both F and M scenarios.

The overestimation of TWW supply volumes observed under the farmer management induced an NL equal to 33% (35 kg ha^{-1}) compared to the scenario (M) (12 kg ha^{-1}). This result demonstrates how the frequency of TWW irrigation events and the correlated soil hydraulic properties influence the movement of water and salt.

In summary, to curb the environmental impacts, the irrigation management with TWW should simultaneously consider soil and water quality and the citrus nitrogen deficit induced by the long-term use of TWW.

3.5. Model Testing

The model was evaluated by comparing the simulated and measured values over time and at three depths. Statistical analysis was performed, and the average error (AE), the root mean square error (RMSE), and the modeling efficiency expressed by the Nash–Sutcliffe criterion (EF) in terms of water content and available nitrogen were calculated.

AE is the average difference between the simulated and the measured values, which may be affected by a positive or negative sign, indicating whether the model overestimates or underestimates a certain output. The RMSE is a statistical index that calculates the mean difference between the simulated and observed data. In an ideal situation, the RMSE should be nil, indicating that the model estimates have the highest possible accuracy. As for the Nash-Sutcliffe criterion (EF), a value of one indicates the complete conformity of the simulated and observed data [77].

The average error and root mean square error were calculated as follows:

$$AE = \frac{\sum_{i=1}^{n} S_i - O_i}{n} \tag{10}$$

$$RMSE = \sqrt{\frac{\left(\sum_{i=1}^{n} S_i - O_i\right)^2}{n}} \tag{11}$$

The Nash–Sutcliffe efficiency is defined as follows:

$$EF = 1 - \frac{\sum_{i=1}^{n} (S_i - O_i)^2}{\sum_{i=1}^{n} (O_i - \overline{O_i})^2} \quad (12)$$

where n is the number of observations, $\overline{O}_i$ is the average of the observed values, and S_i and O_i are the simulated and measured values, respectively [78].

Table 6 summarizes the statistical analysis for the model evaluation, which demonstrates a good correlation between the model outputs and the field data. The EF values are very close to 1, whereas the AE and RMSE values are close to zero [79].

Table 6. Statistical criteria for the simulated and measured data of soil water content, nitrate, and ammonium in 2021.

Depth	θ			σ_e			N−AVA		
	AE	RMSE	EF	AE	RMSE	EF	AE	RMSE	EF
0–25	0.00013	0.005	0.921	0.027	0.059	0.924	0.00007	0.0007	0.979
25–50	−0.0021	0.005	0.936	−0.018	0.057	0.790	−0.00094	0.0019	0.692
50–75	−0.0019	0.040	0.918	−0.0035	0.033	0.890	−0.00092	0.0018	0.967

The performance of the Hydrus model is satisfactory, especially at depths of 0–25 cm and 50–75 cm, and the efficiency indexes demonstrate the goodness of statistical fit and the accuracy of the model for the soil water and salt processes.

4. Conclusions

The reuse of TWW in agriculture is an appealing and practical solution that significantly relieves pressure on water resources and, as such, is a consolidated adaptation strategy for the current and future sustainability of agriculture in water-scarce countries; nonetheless, it necessitates a sizable re-calibration for local conditions and specific crops.

In this regard, a Monitoring and Modeling Approach (MMA) was developed to improve the current irrigation management of a citrus crop, regularly and chronically irrigated with TWW, in Beit Dajan, Palestine. MMA considered the soil solution's electrical conductivity and the leaf nitrogen concentration to characterize the water–soil–plant interactions and compared the outcomes of conventional and TWW-tailored irrigation practices.

MMA enabled the conservation of the soil salinity level between 2.0 and 2.2 dS m^{-1} and estimated an acceptable leaf nitrogen range between 1.7 and 1.9%, considering that a nitrogen deficit is usually induced by the stress conditions established under TWW supply. These results directly correlate with the re-scaling of the frequency of irrigation events based on comparing the farmer (F) and the model (M) routines. They demonstrate that a robust TWW irrigation management cannot but consider the quality of water and soil and the crop nitrogen deficit.

The HYDRUS-1D model was used to estimate the water flow and nitrogen transport in a citrus orchard under TWW irrigation using data collected in 2021. Based on the 2021 calibration, the same geometry domain was set up to retrieve the water and salt behavior during the irrigation seasons of 2019–2020. The simulations also allowed for the prediction of the deep percolation and leaf nitrogen optimum values. The results show that at 75 cm depth, the (M) strategy generated 33% less nitrate leaching than (F), which induced an increase in the available soil nitrogen N − AVA in the long run. This is due to the high frequency of the irrigation events (2–3 times per week) scheduled by the farmer, coupled with a reduction in the root uptake.

Despite the TWW quality being characterized by high salinity (1.9 dS m^{-1}), soil salinity was acceptable, but leaves suffered from nitrogen deficit experienced under inadequate long-standing practices. The high concentration of soil N − AVA did not translate into

Citrus N − AVA since citrus root uptake was reduced under water and salt stresses, in addition to the part of nitrogen that was pushed beyond the root zone.

Irrigation scheduling with TWW based on the combination of water and soil parameters showed an improvement in nitrogen root uptake and a reduction in deep percolation fluxes compared to the F practice.

All the above results show that irrigation management with TWW should be re-scaled to consider both supplies of water and salts and deficits witnessed by the citrus crop.

At the time being, the irrigation management follows a business-as-usual plan that is most adapted for and applies optimum thresholds that fit practices with conventional water. This behavior does not account for the different mechanisms induced when TWW is used and their impact on the soil and plant.

The main contribution of this study is, therefore, a new perspective in managing irrigation with TWW, where sustainability stands and no harm irrigation plans apply, especially in those areas where TWW is the only source, and its reuse is consolidated to ensure irrigation. These results demonstrate the potential of MMA in this regard, although further studies are still required to consolidate the methodology.

Author Contributions: Conceptualization, R.K.; Methodology, G.D.; Software, G.D.; Validation, R.K.; Formal analysis, R.K.; Resources, R.K.; Data curation, G.D.; Writing—original draft, G.D.; Writing—review & editing, R.K.; Supervision, R.K. All authors have read and agreed to the published version of the manuscript.

Funding: This research received no external funding.

Institutional Review Board Statement: Not applicable.

Data Availability Statement: The data presented in this study are available on request from the corresponding author. The data are generated in collaboration with stakeholders and are not publicly available.

Acknowledgments: This research was carried out due to the partnership established in the framework of the MENAWARA project. The authors express their deep gratitude toward the colleagues of We World, Annajah University, and the farmers in Beit Dajan.

Conflicts of Interest: The authors declare no conflict of interest.

References

1. Caretta, M.A.; Mukherji, A.; Arfanuzzaman, M.; Betts, R.A.; Gelfan, A.; Hirabayashi, Y.; Lissner, T.K.; Liu, J.; Lopez, E.; Gunn, E.; et al. Water. In *Climate Change 2022: Impacts, Adaptation, and Vulnerability. Contribution of Working Group II to the Sixth Assessment Report of the Intergovernmental Panel on Climate Change*; Pörtner, H.-O., Roberts, D.C., Tignor, M., Poloczanska, E.S., Mintenbeck, K., Alegría, A., Craig, M., Langsdorf, S., Löschke, S., Möller, V., et al., Eds.; Cambridge University Press: Cambridge, UK; New York, NY, USA, 2022; pp. 551–712. [CrossRef]
2. Soares, D.; Paço, T.A.; Rolim, J. Assessing climate change impacts on irrigation water requirements under Mediterranean conditions—A review of the methodological approaches focusing on maize crop. *Agronomy* **2023**, *13*, 117. [CrossRef]
3. Khadra, R.; Sagardoy, J.A.; Taha, S.; Lamaddalena, N. Participatory irrigation management and transfer: Setting the guiding principles for a sustaining monitoring & evaluation system–a focus on the mediterranean. *Water Resour. Manag.* **2017**, *31*, 4227–4238.
4. Rosa, L. Adapting agriculture to climate change via sustainable irrigation: Biophysical potentials and feedbacks. *Environ. Res. Lett.* **2022**, *17*, 063008. [CrossRef]
5. Nikolaou, G.; Neocleous, D.; Christou, A.; Kitta, E.; Katsoulas, N. Implementing sustainable irrigation in water-scarce regions under the impact of climate change. *Agronomy* **2020**, *10*, 1120. [CrossRef]
6. Gao, Y.; Shao, G.; Wu, S.; Xiaojun, W.; Lu, J.; Cui, J. Changes in soil salinity under treated wastewater irrigation: A meta-analysis. *Agric. Water Manag.* **2021**, *255*, 106986. [CrossRef]
7. Hashem, M.S.; Qi, X. Treated wastewater irrigation—A review. *Water* **2021**, *13*, 1527. [CrossRef]
8. Rusan, M.J.M.; Hinnawi, S.; Rousan, L. Long term effect of wastewater irrigation of forage crops on soil and plant quality parameters. *Desalination* **2007**, *215*, 143–152. [CrossRef]
9. Isayenkov, S.V. Genetic sources for the development of salt tolerance in crops. *Plant Growth Regul.* **2019**, *89*, 1–17. [CrossRef]
10. Russo, D.; Laufer, A.; Bardhan, G.; Levy, G.J. Salinity control in a clay soil beneath an orchard irrigated with treated waste water in the presence of a high water table: A numerical study. *J. Hydrol.* **2015**, *531*, 198–213. [CrossRef]

11. Zavadil, J. The effect of municipal wastewater irrigation on the yield and quality of vegetables and crops. *Soil Water Res.* **2009**, *4*, 91–103. [CrossRef]
12. Ochoa-Herrera, V.; Banihani, Q.; León, G.; Khatri, C.; Field, J.A.R. Sierra-Alvarez. Toxicity of fluoride to microorganisms in biological wastewater treatment systems. *Water Res.* **2009**, *43*, 3177–3186. [CrossRef] [PubMed]
13. Seyfried, M.S.; Schwinning, S.; Walvoord, M.A.; Pockman, W.T.; Newman, B.D.; Jackson, R.B.; Phillips, F.M. Ecohydrological control of deep drainage in arid and semiarid regions. *Ecology* **2005**, *86*, 277–287. [CrossRef]
14. Russo, D.; Laufer, A.; Shapira, R.H.; Kurtzman, D. Assessment of solute fluxes beneath an orchard irrigated with treated sewage water: A numerical study. *Water Resour. Res.* **2013**, *49*, 657–674. [CrossRef]
15. Lal, R.; Shukla, M.K. *Principles of Soil Physics*; CRC Press: Boca Raton, FL, USA, 2004. [CrossRef]
16. Strudley, M.W.; Green, T.R.; Ascough, J.C., II. Tillage effects on soil hydraulic properties in space and time: State of the science. *Soil Tillage Res.* **2008**, *99*, 4–48. [CrossRef]
17. Green, T.R.; Ahuja, L.R.; Benjamin, J.G. Advances and challenges in predicting agricultural management effects on soil hydraulic properties. *Geoderma* **2003**, *116*, 3–27. [CrossRef]
18. Mossadeghi-Björklund, M.; Arvidsson, J.; Keller, T.; Koestel, J.; Lamandé, M.; Larsbo, M.; Jarvis, N. Effects of subsoil compaction on hydraulic properties and preferential flow in a Swedish clay soil. *Soil Tillage Res.* **2016**, *156*, 91–98. [CrossRef]
19. Zhang, Z.B.; Zhou, H.; Zhao, Q.G.; Lin, H.; Peng, X. 2014. Characteristics of cracks in two paddy soils and their impacts on preferential flow. *Geoderma* **2014**, *228*, 114–121. [CrossRef]
20. Bouma, J. Field measurement of soil hydraulic properties characterizing water movement through swelling clay soils. *J. Hydrol.* **1980**, *45*, 149–158. [CrossRef]
21. Gavrilescu, M. Water, Soil, and Plants Interactions in a Threatened Environment. *Water* **2021**, *13*, 2746. [CrossRef]
22. Shukla, S.; Saxena, A. Sources and leaching of nitrate contamination in groundwater. *Curr. Sci.* **2020**, *118*, 883–891. [CrossRef]
23. Huan, H.; Hu, L.; Yang, Y.; Jia, Y.; Lian, X.; Ma, X.; Jiang, Y.; Xi, B. Groundwater nitrate pollution risk assessment of the groundwater source field based on the integrated numerical simulations in the unsaturated zone and saturated aquifer. *Environ. Int.* **2020**, *137*, 105532. [CrossRef] [PubMed]
24. Fouial, A.; Khadra, R.; Daccache, A.; Lamaddalena, N. Modelling the impact of climate change on pressurised irrigation distribution systems: Use of a new tool for adaptation strategy implementation. *Biosyst. Eng.* **2016**, *150*, 182–190. [CrossRef]
25. Lamaddalena, N.; Khadra, R.; Fouial, A. Use of localized loops for the rehabilitation of on-demand pressurized irrigation distribution systems. *Irrig. Sci.* **2015**, *33*, 453–468. [CrossRef]
26. Coppola, A.; Dragonetti, G.; Sengouga, A.; Lamaddalena, N.; Comegna, A.; Basile, A.; Noviello, N.; Nardella, L. Identifying optimal irrigation water needs at district scale by using a physically based agro-hydrological model. *Water* **2019**, *11*, 841. [CrossRef]
27. Šimůnek, J.; van Genuchten, M.T.; Šejna, M. Development and Applications of the HYDRUS and STANMOD Software Packages and Related Codes. *Vadose Zone J.* **2008**, *7*, 587–600. [CrossRef]
28. Letey, J.; Hoffman, G.J.; Hopmans, J.W.; Grattan, S.R.; Suarez, D.; Corwin, D.L.; Oster, J.D.; Wu, L.; Amrhein, C. Evaluation of soil salinity leaching requirement guidelines. *Agric. Water Manag.* **2011**, *98*, 502–506. [CrossRef]
29. Assouline, S.; Russo, D.; Silber, A.; Or, D. Balancing water scarcity and quality for sustainable irrigated agriculture. *Water Resour. Res.* **2015**, *51*, 3419–3436. [CrossRef]
30. Fernández-Cirelli, A.; Arumí, J.L.; Rivera, D.; Boochs, P.W. Environmental effects of irrigation in arid and semi-arid regions. *Chil. J. Agric. Res.* **2009**, *69* (Suppl. S1), 27–40. [CrossRef]
31. Russo, D. Alternating irrigation water quality as a method to control solute concentrations and mass fluxes below irrigated fields: A numerical study. *Water Resour. Res.* **2016**, *52*, 3440–3456. [CrossRef]
32. Doltra, J.; Muñoz, P. Simulation of nitrogen leaching from a fertigated crop rotation in a Mediterranean climate using the EU-Rotate_N and Hydrus-2D models. *Agric. Water Manag.* **2010**, *97*, 277–285. [CrossRef]
33. Hanson, B.R.; Šimůnek, J.; Hopmans, J.W. Evaluation of urea–ammonium–nitrate fertigation with drip irrigation using numerical modeling. *Agric. Water Manag.* **2006**, *86*, 102–113. [CrossRef]
34. Dragonetti, G.; Khadra, R.; Daccache, A.; Oubelkacem, A.; Choukr-Allah, R.; Lamaddalena, N. Development and application of a predictive model for treated wastewater irrigation management in a semiarid area. *Integr. Environ. Assess. Manag.* **2020**, *16*, 910–919. [CrossRef]
35. El Chami, D.; Trabucco, A.; Wong, T.; Abdel Monem, M.; Mereu, V. Costs and effectiveness of climate change adaptation in agriculture: A systematic review from the NENA region. *Clim. Policy* **2022**, *22*, 445–463. [CrossRef]
36. Hamdan, M.; Abu-Awwad, A.; Abu-Madi, M. Willingness of farmers to use treated wastewater for irrigation in the West Bank, Palestine. *Int. J. Water Resour. Dev.* **2021**, *38*, 497–517. [CrossRef]
37. Palestinian Water Authority (PWA). Water Supply Report 2010. 2012. Available online: http://www.pwa.ps/page.aspx?id=0PHprpa2520241944a0PHprp (accessed on 18 January 2021).
38. ARIJ Applied Research Institute—Jerusalem. *Village Profile and Needs Assessment in the West Bank Governorates—Database*; ARIJ: Jerusalem, Palestine, 2014. Available online: https://www.arij.org/publications/books-atlases/books-atlases-2014/locality-profiles-and-needs-assessment-in-jerusalem-governorate/ (accessed on 14 October 2021).

39. NRCS-USDA. *Soil Taxonomy a Basic System of Soil Classification for Making and Interpreting Soil Surveys*, 2nd ed.; Agriculture Handbook, Number 436; NRCS-USDA: Washington, DC, USA, 1999. Available online: https://www.nrcs.usda.gov/sites/default/files/2022-06/Soil%20Taxonomy.pdf (accessed on 14 October 2021).
40. Lassabatère, L.; Angulo-Jaramillo, R.; Soria Ugalde, J.M.; Cuenca, R.; Braud, I.; Haverkamp, R. Beerkan estimation of soil transfer parameters through infiltration experiments–BEST. *Soil Sci. Soc. Am. J.* **2006**, *70*, 521. [CrossRef]
41. Henriksen, A.; Selmer-Olsen, A.R. Automatic methods for determining nitrate and nitrite in water and soil extracts. *Analyst* **1970**, *95*, 514–518. [CrossRef]
42. Kalra, Y. (Ed.) *Handbook of Reference Methods for Plant Analysis*; CRC Press: Boca Raton, FL, USA, 1997. [CrossRef]
43. Van Genuchten, M.T. A closed-form equation for predicting the hydraulic conductivity of unsaturated soils. *Soil Sci. Soc. Am. J.* **1980**, *44*, 892–898. [CrossRef]
44. Šimůnek, J.; Hopmans, J.W. Modeling compensated root water and nutrient uptake. *Ecol. Model.* **2009**, *220*, 505–521. [CrossRef]
45. Feddes, R.A.; Kowalik, P.; Zaradny, H. Simulation of field water use and crop yield. In *Simulation Monographs*; PUDOC: Wageningen, The Netherlands, 1978; 189p, ISSN 0924-8439.
46. Van Genuchten, M.T. *A Numerical Model for Water and Solute Movement in and below the Root Zone*; United States Department of Agriculture Agricultural Research Service US Salinity Laboratory: Riverside, CA, USA, 1987.
47. Wesseling, J.G.; Kabat, P.; van den Broek, B.; Feddes, R.A. Simulation function of water balance of a cropped soil with different types of boundary conditions including the possibility of drainage and irrigation and the calculation of crop yield (SWACROP). Instruction for input. WSC for Integrated Land. *Soil Water Res.* **1989**.
48. Maas, E.V.; Nieman, R.H. Physiology of plant tolerance to salinity. In *Crop Tolerance to Suboptimal Land Conditions*; John Wiley & Sons: Hoboken, NJ, USA, 1978; Volume 32, pp. 277–299. [CrossRef]
49. Lyu, S.; Chen, W.; Wen, X.; Chang, A.C. Integration of HYDRUS-1D and MODFLOW for evaluating the dynamics of salts and nitrogen in groundwater under long-term reclaimed water irrigation. *Irrig. Sci.* **2019**, *37*, 35–47. [CrossRef]
50. Ramos, T.B.; Šimůnek, J.; Gonçalves, M.C.; Martins, J.C.; Prazeres, A.; Castanheira, N.L.; Pereira, L.S. Field evaluation of a multicomponent solute transport model in soils irrigated with saline waters. *J. Hydrol.* **2011**, *407*, 129–144. [CrossRef]
51. Allen, R.G.; Pereira, L.S.; Raes, D.; Smith, M. *Crop Evapotranspiration-Guidelines for Computing Crop Water Requirements-Fao Irrigation and Drainage Paper 56*; FAO: Rome, Italy, 1998; Volume 300, p. D05109. Available online: http://www.climasouth.eu/sites/default/files/FAO%2056.pdf (accessed on 14 October 2021).
52. Ritchie, J.T. Model for predicting evaporation from a row crop with incomplete cover. *Water Resour. Res.* **1972**, *8*, 1204–1213. [CrossRef]
53. Ye, W.; Wang, H.X.; Gao, J.; Liu, H.J.; Yan, L. Simulation of salt ion migration in soil under reclaimed water irrigation. *J. Agro-Environ. Sci.* **2014**, *33*, 1007–1015.
54. Chen, W.; Lu, S.; Pan, N.; Jiao, W. Impacts of long-term reclaimed water irrigation on soil salinity accumulation in urban green land in Beijing. *Water Resour Res.* **2013**, *49*, 7401–7410. [CrossRef]
55. Gärdenäs, A.I.; Hopmans, J.W; Hanson, B.R.; Šimůnek, J. Two-dimensional modeling of nitrate leaching for various fertigation scenarios under micro-irrigation. *Agric. Water Manag.* **2005**, *74*, 219–242. [CrossRef]
56. Azad, N.; Behmanesh, J.; Rezaverdinejad, V.; Abbasi, F.; Navabian, M. An analysis of optimal fertigation implications in different soils on reducing environmental impacts of agricultural nitrate leaching. *Sci. Rep.* **2020**, *10*, 7797. [CrossRef] [PubMed]
57. Ravikumar, V.; Vijayakumar, G.; Šimůnek, J.; Chellamuthu, S.; Santhi, R.; Appavu, K. Evaluation of fertigation scheduling for sugarcane using a vadose zone flow and transport model. *Agric. Water Manag.* **2011**, *98*, 1431–1440. [CrossRef]
58. Greenberg, A.E.; Clesceri, L.S.; Eaton, A.D. *Standard Methods for the Examination of Water and Wastewater*, 18th ed.; American Public Health Association: Washington, DC, USA; American Water Works Association: Denver, CO, USA; Water Environment Federation: Alexandria, VA, USA, 1992.
59. Wrage, N.; Velthof, G.L.; van Beusichem, M.L.; Oenema, O. Role of nitrifier denitrification in the production of nitrous oxide. *Soil Biol. Biochem.* **2001**, *33*, 1723–1732. [CrossRef]
60. Albalasmeh, A.A.; Gharaibeh, M.A.; Alghzawi, M.I.Z.; Morbidelli, R.; Saltalippi, C.; Ghezzehei, T.A.; Flammini, A. Using wastewater in irrigation: The effects on infiltration process in a clayey soil. *Water* **2020**, *12*, 968. [CrossRef]
61. Viviani, G.; Iovino, M. Wastewater reuse effects on soil hydraulic conductivity. *J. Irrig. Drain. Eng.* **2004**, *130*, 476–484. [CrossRef]
62. Abdulgawad, F.; Bockelmann, E.B.; Sapsford, D.; Williams, K.P.; Falconer, R. Ammonium ion adsorption on clays and sand under freshwater and seawater conditions. In *Advances in Water Resources and Hydraulic Engineering*; Springer: Berlin/Heidelberg, Germany, 2009; pp. 656–661.
63. Swietlik, D. USA–Texas. Responses of citrus trees in Texas to foliar and soil Zn applications. In Proceedings of the International Society of Citriculture, VIII International Citrus Congress, Sun City, South Africa, 12–17 May 1996; Volume 2, pp. 772–776 and 1265–1268.
64. Maas, E.V.; Hoffman, J.G. Crop salt tolerance-current assessment. *J. Irrig. Drain. Div. Amer. Soc. Civ. Eng.* **1977**, *103*, 115–134. [CrossRef]
65. Apal Agricultural Laboratory–Soil Test Interpretation Guide. 2014. Available online: https://www.apal.com.au/SoilTesting.aspx (accessed on 21 May 2021).
66. Malik, D.M.; Khan, M.A.; Chaudhry, T.A. *Analysis Manual for Soils, Plants and Waters*; Soil Fertility Survey and Soil Testing Institute, Department of Agriculture: Lahore, Pakistan, 1984.

67. Chen, J.F. Adsorption and diffusion of ammonium in soils. In *Nitrogen in Soils of China*; Springer: Dordrecht, The Netherlands, 1997; Volume 74, pp. 87–111.
68. Carden, D.E.; Walker, D.J.; Flowers, T.J.; Miller, A.J. Single-cell measurements of the contributions of cytosolic Na+ and K+ to salt tolerance. *Plant Physiol.* **2003**, *131*, 676–683. [CrossRef] [PubMed]
69. Paudel, I.; Bar-Tal, A.; Levy, G.J.; Rotbart, N.; Ephrath, J.E.; Cohen, S. Treated wastewater irrigation: Soil variables and grapefruit tree performance. *Agric. Water Manag.* **2018**, *204*, 126–137. [CrossRef]
70. Bardhan, G.; Russo, D.; Goldstein, D.; Levy, G. Changes in the hydraulic properties of a clay soil under long-term irrigation with treated wastewater. *Geoderma* **2016**, *264*, 1–9. [CrossRef]
71. Albdaiwi, R.N.; Al-Hawadi, J.S.; Al-Rawashdeh, Z.B.; Al-Habahbeh, K.A.; Ayad, J.Y.; Al-Sayaydeh, R.S. Effect of treated wastewater irrigation on the accumulation and transfer of heavy metals in lemon trees cultivated in arid environment. *Horticulturae* **2022**, *8*, 514. [CrossRef]
72. ZekrI, M.; Parsons, L.R. Salinity tolerance of citrus rootstocks: Effects of salt on root and leaf mineral concentrations. *Plant Soil* **1992**, *147*, 171–181. [CrossRef]
73. Plessis, S.F.; Koen, T.J. Relationship between mineral nutrition and postharvest fruit disorders of 'Fuerte' avocados. In Proceedings of the World Avocado Congress II, Orange, CA, USA, 21–26 April 1991; Lovatt, C.J., Ed.; pp. 395–402.
74. Coppola, A.; Chaali, N.; Dragonetti, G.; Lamaddalena, N.; Comegna, A. Root uptake under non-uniform root-zone salinity. *Ecohydrology* **2015**, *8*, 1363–1379. [CrossRef]
75. Pereira, B.F.F.; He, Z.L.; Stoffella, P.J.; Melfi, A. Reclaimed wastewater: Effects on citrus nutrition. *Agric. Water Manag.* **2011**, *98*, 1828–1833. [CrossRef]
76. Bedbabis, S.; Rouina, B.; Boukhris, B.; Ferrara, M. Effect of irrigation with treated wastewater on soil chemical properties and infiltration rate. *J. Environ. Manag.* **2014**, *133*, 45–50. [CrossRef]
77. Olinski, A.J. Verification of Drainmod ver. 5.1 for Estimating Water Balance and Nitrogen Transport through Soils in Southern Ontario. Ph.D. Thesis, University of Guelph, Guelph, ON, Canada, 2005.
78. Moriasi, D.N.; Arnold, J.G.; Van Liew, M.W.; Bingner, R.L.; Harmel, R.D.; Veith, T.L. Model evaluation guidelines for systematic quantification of accuracy in watershed simulations. *Trans. ASABE* **2007**, *50*, 885–900. [CrossRef]
79. Agah, A.E.; Meire, P.; de Deckere, E. Simulation of Phosphorus Transport in Soil Under Municipal Wastewater Application Using Hydrus-1D. In *Soil Contamination-Current Consequences and Further Solutions*; IntechOpen: London, UK, 2016.

sustainability

MDPI

Article

Pepper Growing Modified by Plasma Activated Water and Growth Conditions

Božica Japundžić-Palenkić [1], Robert Benković [1], Teuta Benković-Lačić [1], Slavica Antunović [1], Matija Japundžić [2], Nataša Romanjek Fajdetić [1] and Krunoslav Mirosavljević [1,*]

[1] Biotechnical Department, University of Slavonski Brod, 35000 Slavonski Brod, Croatia
[2] Independent Researcher, 35252 Sibinj, Croatia
* Correspondence: kmirosavljevic@unisb.hr

Abstract: Plasma-activated water (PAW) is a novel and promising technique in the agricultural field that has the potential to improve vegetable growth and yield. The objective of this study was to determine the effect of plasma-activated water seeds treatment and growth conditions on pepper plant growth parameters and fruit quality. A factorial design of three factors (C = cultivar, GC = growth condition, and PAW = plasma activated water treatment seeds) was established, with two variants for each one: Cultivar 1 (C1) and Cultivar 2 (C2); greenhouse (G) and open field (F); PAW seeds treatment (PAW) and seeds without treatment with PAW (C). Four replicates with fifty seeds were taken for each variety. Growth and fruit quality parameters were measured in the three month period during 2021 and 2022, respectively. The significant influence of cultivar, growing condition, and PAW on fruit quality and pepper plant growth parameters were determined. The lowest values of measured parameters were obtained in the open field without PAW treatment. Pepper growth in a greenhouse from PAW-treated seeds had a higher canopy height (17.85%), weight (10.57%), number of leaves (10.5%), nodes (18.94%), and buds (37.83%). Moreover, dry matter content was higher (33.73%) as well as fruit quality: fruit weight (50.19%), diameter (24.3%), length (20.88%), and pericarp weight (49.49%). Results indicate that PAW treatment of peppers seeds can lead to production and yield improvement under different climates and growing conditions.

Keywords: *Capsicum annuum* L.; greenhouse; field; PAW; seed

Citation: Japundžić-Palenkić, B.; Benković, R.; Benković-Lačić, T.; Antunović, S.; Japundžić, M.; Romanjek Fajdetić, N.; Mirosavljević, K. Pepper Growing Modified by Plasma Activated Water and Growth Conditions. *Sustainability* **2022**, *14*, 15967. https://doi.org/10.3390/su142315967

Academic Editors: Daniel El Chami and Maroun El Moujabber

Received: 20 October 2022
Accepted: 24 November 2022
Published: 30 November 2022

1. Introduction

Pepper (*Capsicum annuum* L.) has economic importance and is a widely produced agricultural crop with nutritionally valuable fruits. The impact of the quality of fruits is not only a matter of local but also the global economy, so the response to environmental changes is crucial for that production [1]. Production of pepper is increasing annually due to the possibility of production in most areas worldwide. Spice and sweet peppers are economically relevant and worldwide cultivated species, and the area used for production is over 1.5 million hectares [2]. Sweet pepper is a very important part of cultivation in protected areas [3]. Favorable microclimate for growth and continuous all-year production increase the overall yield in the greenhouse [4]. Ventilated greenhouse as a place for sweet pepper production can be profitable due to the longer crop season achieved with controlled growing conditions in comparison with open field production [5]. Influence of the greenhouse microenvironment results in increased fruit quality, higher crop growth, and yield [6]. Evaluation of all environmental conditions, such as temperature [7], humidity, and plant response on it [8,9], is very important in sweet pepper cultivation. Low temperature is one of the most important abiotic factors that restrict the optimal production of warm-season vegetables [10]. Additionally, climate changes in recent years have increased the number and duration of extremely high or low temperatures. Challenges for pepper producers in Croatia during the spring period are low temperatures and high humidity.

Therefore, there is a need for an alternative model of safe and efficient cultivation [11]. Although greenhouse cultivation makes plants stronger to stress [12,13], the growth of sweet pepper in the greenhouse and open field is under the influence of biotic and abiotic factors [14] because global warming increases abiotic stress and impair plant growth, yield, and product quality [15]. Developing sustainable methods such as non-thermal plasma (NTP) to alleviate plant stress since plasma-generated reactive species trigger the activity of stress-responsive genes in plants [16] is necessary. Besides the fact that non-thermal or cold plasma improves resistance to abiotic stress, NTP technology is becoming increasingly popular in agriculture, in particular for seed treatment, since it improves seeds germination, growth and development of plants, and yield [17–21]. Plasma-activated water (PAW) or plasma-treated water (PTW) influences plant growth and development. The pathway for that is the transformation of reactive nitrogen and oxygen into ROS and NS (reactive oxygen and nitrogen species), including H_2O_2 and NO biomolecules [22–26]. Signaling molecules such as liquid ROS and NS (H_2O_2, NO_2^-, NO_3^-) could be nutrients for a plant. In contact with plasma and water, these species dissolve and form a plasma-activated water which is a mixture that can be, among others, a clean and sustainable alternative to nowadays widely used chemical fertilizers. Reactive species increase plant growth, but their ratio has to be optimized. Several properties were improved by using PAW, but the results are limited. Therefore, the use of PAW in agriculture demands optimization of PAW content [27].

Several authors reported the positive influence of nonthermal plasma [28–33] and plasma-activated water (PAW) [34] on the growth of pepper. However, despite studies carried out in the field of pre-sawing plasma treated seeds, reports about the effects of plasma-activated water on the morphology and phenology of resulting plants, yield, and quality of fruits are absent or very limited [35].

In this study, we set four treatments combined with two different cultivars of pepper (Bibic and Bernita) and two different growth conditions (open field and greenhouse). The combined treatments of cultivar and growing conditions were carried out during twol vegetation seasons. Additionally, we investigated the effect of PAW seed treatments on the growth and fruit quality of pepper. The objective of this study was to determine the effect of plasma-activated water seeds treatment and growth conditions on pepper plant growth parameters and fruit quality.

2. Material and Methods

2.1. Preparing of Plasma Activated Water (PAW)

Purified water (commercially purified water of Pharmaceutical degree (Pharmacopoeia Europea, Ph. Eur. 9)) was used for making active species with a conductivity of 0.98 $\mu S\ cm^{-1}$ and pH value of 6.5. The PAW was created by establishing contact of the liquid surface with the plasma plume of a kHz plasma jet. The outer and inner diameters of the quartz tube of the atmospheric pressure plasma jet were 1.5 and 1 mm, and a copper wire of 100 µm diameter served as an electrode inserted in the capillary. The sinusoidal voltage waveform of 28 kHz with 12 kV maximum voltage powered the electrode. This was performed on the grounds of preliminary optimization of the used appliance, which delivers about 15 W of power. The generated discharge was conducted from N_2 (99.996% purity) with a flow rate in a capillary of 500 sccm. At the distance of 5 mm from the liquid surface, the Berzelius with 215 mL sample was placed in front of the plasma jet, which resulted in H_2O_2 and NO_2^- in the PAL. Overall treatment time was 40. Analyses of samples were conducted after treatment and a few times while storage. The QUANTOFIX test strips (nitrate/nitrite 500, 10–500 mg l^{-1} NO_3^-, 1–80 mg l^{-1} NO_2^-, nitrate/nitrite 100, 5–100 mg l^{-1} NO_3^-, 0.5–50 mg l^{-1} NO_2^-, Peroxide H_2O_2 25, 0.5–25 mg l^{-1}, Peroxide H_2O_2 100, 0.5–25 mg l^{-1}) were used for those analyses. The results on strips had the accurate and quantitative evaluation by QUANTOFIX Relax unit optical reader (by Macherey-Nagel, GmbHha). The calibration of nitrate/nitrite strips was conducted by $NaNO_3$ and NH_4NO_3 solutions with determined concentrations. UV-VIS absorption spectroscopy was used for the verification of calibration. Monitoring the aging of samples by strips in 1 min time resolution allows determination without significant consumption of samples, and the measuring error has been less than 10%. The pH value is crucial for the aging dynamics of PAW, and it is usually achieved by using metal ion concentration, as demonstrated by Kutasi et al. [36]. Therefore, the Mg ions were used in the form of a 5 g piece of solid Mg during the plasma treatment and kept 1 h inserted in liquid. The PAW conductivity was also measured after the process and while storage with the use of a conductivity probe (Metrohm 914pH/DO/Conductometer) Pt 1000/B/O. Optical emission spectroscopy (OES) was used for monitoring each treatment. The presence of NO, and N_2 was determined during the treatment and monitored in 10 min intervals, as well as peroxide, nitrate/nitrite concentrations, and pH to keep consistency. The volume of 215 mL was used during three PAW treatments, first mixed in 2000 mL bottles, and separated in 200 mL containers for the purpose of the experiment. The 15 mL was kept for aging tests within 2 weeks (Table 1). Plasma-activated water was prepared at the Institute of Physics Zagreb according to Kutasi et al. (2019) [36], Gierczik et al. (2020) [37], and Kutasi et al. (2021) [38].

Table 1. Plasma-activated water parameters.

	H_2O_2	NO_2^-	NO_3^-	pH
40 min treatment (t1, t2, t3)	4.1 mg L^{-1}	3.3 mg L^{-1}	11.2 mg L^{-1}	6.1
For seeds treatment (experiment)	3.1 mg L^{-1}	2.8 mg L^{-1}	5.5 mg L^{-1}	6.1
Two weeks later	0 mg L^{-1}	1.4 mg L^{-1}	5.4 mg L^{-1}	5.8

2.2. Experimental Site

Experiments were conducted in the climate chamber, greenhouse, and open field of the Biotechnical Department, University of Slavonski Brod, between March and July of 2021 and 2022. The greenhouse and open field are geographically located at 45°09′57″ N and 17°57′08″ E and have an altitude of 87 m. The climate of Brod-Posavina County is moderately warm and humid, with warm summers (Cfb according to Köppen) and moderately cold winters. The location is characterized by winter temperatures that can fall below 0 °C and summer temperatures that can rise above 30 °C. Greenhouse indoor temperatures and relative humidity values in 2021 and 2022 were measured daily with H560 DewPoint Pro placed 1 m above the ground in the middle of the greenhouse. Outdoor temperature and relative humidity during the experimental period were recorded by a meteorological station situated nearby the experimental field. In this experiment, a naturally ventilated passive solar greenhouse was used, and to allow air circulation on the hottest days, the greenhouse was ventilated passively through flap roof windows and lateral side panels. The greenhouse dimensions were 30 m x 8 m with a height of 8 m, and the structure consisted of polycarbonate walls and a trilaminate low-density polyethylene (LDPE) film roof (200 µm thickness) with approximately 60% photosynthetically active radiation (PAR) transmittance. It had no heating or artificial light. Average monthly (April–July 2021 and 2022, respectively) indoor greenhouse temperature (°C) and relative humidity (%) values and open field temperature (°C) and relative humidity (%) are shown in Figure 1.

Four replicates (50 seeds each) for a single treatment were used. Samples of 50 seeds were placed in Petri dishes. Seed samples were divided into two groups: seeds poured with PAW and control (seeds poured with water without plasma treatment). The samples were immersed (soaked) with 20 mL of PAW and controlled water in 24 h intervals.

After imbibition, seeds were sowed in PVC containers (Pöpellmann TEKU (BP 3153/60) 53 cm × 31 cm × 5.6 cm) with 60 sowing places (volume 76 mL), filled with Potgrond P Klasmann Deilman substrate (Rp No 002). Trays with sowed seeds have been placed in controlled conditions within the growth chamber (Memmert ICH260L) with a 16/8 h photoperiod, a temperature 22/18 °C (day/night), and relative humidity of 60%.

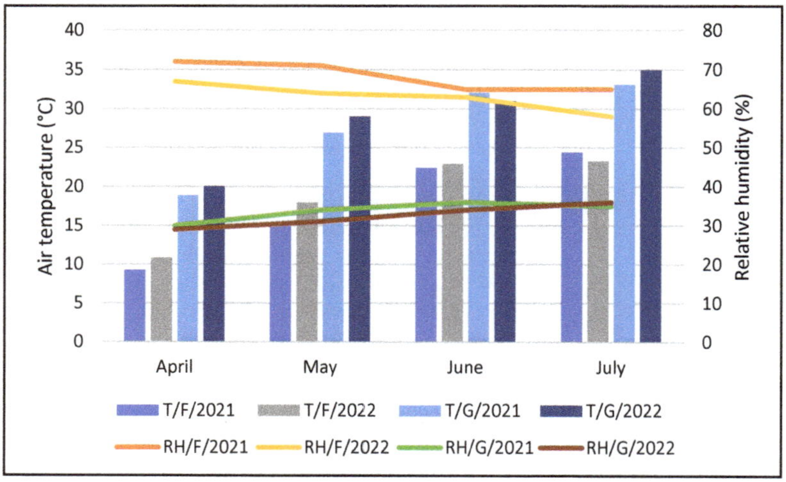

Figure 1. Air temperature (*T*) and relative humidity (*RH*) measured in field (F) and greenhouse (G) conditions in the vegetative period (April-July) 2021 and 2022.

After four weeks (19 April 2021 and 19 April 2022), seedlings were transplanted in pots (Pöpellmann TEKU VTG 9) 9 × 6.8 cm (0.27 L) in a greenhouse using ProLine Herb substrate (Rp No 693) suited for ecological production of pepper in small pots with added organic fertilizer Stallatico Pellettato (1800–2300 kg/ha). Six-weeks-old pepper transplants were planted on 10 May 2021 and 10 May 2022 in pots (Pöpellmann TEKU VCG 19) 19 × 14.9 cm (3 L) fulfilled with ProLine Herb Substrate 70% TerrAktiv®/coir + 30% GreenFibre® with added fertilizer. After the transplantation in pots, the plants were divided into two groups, one for a greenhouse and the other for open field cultivation (same area as the greenhouse). Irrigation was performed after sowing, when necessary, in accordance with precipitation in an open field experiment.

2.3. Experimental Design

Sweet pepper seeds: F1 seeds (Bibic (Cultivar 1) and Bernita (Cultivar 2)), produced in 2020 and 2021, respectively, were used to investigate the effect of PAW seed treatment on plant growth in the greenhouse and open fields. A factorial design of three factors (C = cultivar, GC = growth condition, and PAW = plasma-activated water treatment seeds) was established, with two variants for each one: Cultivar Bibic (C1) and Cultivar Bernita (C2); greenhouse (G) and open field (F); PAW seeds treatment (PAW) and seeds without treatment with PAW (C). Four replicates with fifty seeds were taken for each variety.

2.4. Measurements

In both years, harvests were practiced in five terms: 10 May 2021, 17 May 2021, 24 May 2021, 7 June 2021, 21 June 2021, and 10 May 2022, 17 May 2022, 24 May 2022, 7 June 2022, 21 June 2022. Plants in each replication were used to assess canopy height (cm), canopy weight (g), number of leaves, number of nodes, and number of buds. Four plants were randomly selected for each replication. Samples were transferred to the Agroecological laboratory, Biotechnical department, University of Slavonski Brod, Croatia, and stored in a cooler until used. All analyses were performed within 24 h. Mature fruits were harvested on 7 July 2021 and 7 July 2022, and then fruit weight (g), fruit diameter (cm), fruit length (cm), pericarp weight (g), and residual weight (g) (seeds, placenta, and calix) of fruits were determined. Ten mature fruits were randomly selected for each replication, and samples were transferred to the Agroecological laboratory, Biotechnical department, University of Slavonski Brod, Croatia, and stored in a cooler until used within 24 h.

Plant samples were oven dried at 70 °C for 48 h to the constant dry weight for the plant, which was weighed on an electronic scale with a precision of 0.01 g (PL3002, Mettler–Toledo International Inc., Greifensee, Switzerland) and sliding scale DIGI-MET 1226932-D (Helios Preisser, Gammertingen, Germany).

2.5. *Statistical Analysis*

According to the previous statements, the experiment was conducted as a three-factorial trial: two varieties of pepper (C1 and C2), treated seeds (PAW and control), and growth conditions (greenhouse and open field) in randomized blocks design with four replications. Data were analyzed using RStudio, and statistically significant differences were determined by the program MS Excell, 2019.

According to the Fisher test on the significance of variance analysis, the least significant differences (LSD) for $p < 0.05$ were calculated by comparing the mean values. Duncan's test was used to determine which values have statistically significant differences precisely.

3. Results

The evaluation of measured parameters is shown in Tables 2–13. It is obvious that differences between the cultivars, growing conditions, and PAW treatment were confirmed. However, even though differences in the number of measurement parameters were evident, there was no statistical difference between parameters in each sampling date.

Although canopy length was higher in cultivar C1 in both years of research (Tables 2 and 3), a statistically significant influence of cultivar was established in 2022 (Table 3). Determining the difference between cultivars regarding length of the canopy was 2.25% and 4.05% in 2021 and 2022, respectively (Tables 2 and 3).

Also, in both years included in this research, higher canopy length in plants was determined among seeds treated with PAW. The significance of the impact of the PAW seed treatment was determined in 2021 (Table 2). Canopy length was higher than in control at 11.81% and 6.35% in 2021 and 2022, respectively (Tables 2 and 3). In this study, the impact of growing conditions on canopy length was obvious. The influence of growing conditions was statistically significant in both research years (Tables 2 and 3). Pepper grown in the greenhouse had a higher canopy length, 13.04%, and 6.7%, respectively. Likewise, in both years of research, interactions between cultivars and growing conditions were determined ($p < 0.05$) (Tables 2 and 3).

Table 2. Canopy length of pepper cultivars developed under different growth conditions and seeds treatment in vegetative season 2021.

Pepper Component		Canopy Length (cm)									
Sampling Date		10 May 2021		17 May 2021		24 May 2021		7 June 2021		21 June 2021	
		Seeds Treatment									
Cultivar	Growing Conditions	C	PAW	C	PAW	C	PAW	C	PAW	C	PAW
C1	G	16.00	18.13	22.33 A	24.00 A	27.00 AC	29.38 A	27.13 bB	37.75 aB	35.75 aB	47.00 bA
C1	F			14.63 bD	19.63 aB	19.25 B	19.63 B	24.50 B	27.00 C	34.25 bBC	46.13 aA
C2	G	15.50	15.70	19.88 bB	23.50 aA	26.25 A	26.13 A	30.00 bA	42.00 aA	44.50 A	44.00 B
C2	F			17.13 C	17.00 C	20.25 B	18.75 B	26.50 B	27.00 C	32.75 C	33.50 C
F_C		n.s. (F = 3.61)		n.s. (F = 0.64)		n.s. (F = 1.12)		n.s. (F = 1.83)		n.s. (F = 4.14)	
F_{GC}				* (F = 30.05)		* (F = 71.0)		* (F = 22.3)		* (F = 35.8)	
F_{PAW}		n.s. (F = 2.28)		n.s. (F = 0.54)		n.s. (F = 0.09)		* (F = 14.4)		* (F = 32.2)	
ANOVA—F test											
GC × C				n.s. (F = 0.54)		n.s. (F = 1.27)		n.s. (F = 0.58)		* (F = 23.3)	
GC × PAW				n.s. (F = 0.01)		n.s. (F = 0.85)		* (F = 8.45)		n.s. (F = 0.21)	
C × PAW		n.s. (F = 1.56)		n.s. (F = 0.64)		n.s. (F = 1.43)		n.s. (F = 0.01)		* (F = 30.87)	
GC × C × PAW				n.s. (F = 3.34)		n.s. (F = 0.03)		n.s. (F = 0.25)		n.s. (F = 0.02)	

F_C—F test cultivar; F_{GC}—F test growing conditions; F_{PAW}—F test PAW seeds treatment; *—statistical significance; n.s.—no statistical significance; Values inside same columns with different upper case differ with statistical significance (LSD) $p < 0.05$; values inside the same row with different lower case differ with statistical significance (LSD) $p < 0.05$.

Table 3. Canopy length of pepper cultivars development in 2022 under different growth conditions and seeds treatment in vegetative season 2022.

Pepper Component		Canopy Length (cm)									
Sampling Date		10 May 2022		17 May 2022		24 May 2022		7 June 2022		21 June 2022	
Cultivar	Growing Conditions	Seeds Treatment									
		C	PAW	C	PAW	C	PAW	C	PAW	C	PAW
C1	G	15.63	16.00	26.50 bA	28.13 aA	31.88 A	33.13 A	32.75 B	33.63 B	43.00 A	43.75 A
C1	F			21.75 bB	25.13 aB	31.63 A	32.25 A	30.88 B	33.38 B	34.00 B	36.00 B
C2	G	15.88	19.38	20.88 B	21.38 C	31.38 A	35.00 A	38.00 A	40.00 A	46.25 A	46.50 A
C2	F			13.13 bC	15.38 aD	24.88 B	25.50 B	31.75 B	33.88 B	33.50 B	39.00 B
F_C		n.s. (F = 2.23)		* (F = 106.48)		* (F = 5.65)		* (F = 7.48)		n.s. (F = 1.52)	
F_{GC}				* (F = 31.99)		* (F = 11.27)		* (F = 9.3)		* (F = 28.85)	
F_{PAW}		n.s. (F = 2.55)		* (F = 5.26)		n.s. (F = 0.51)		n.s. (F = 0.54)		n.s. (F = 1.35)	
ANOVA—F test											
GC × C				* (F = 40.5)		* (F = 8.51)		* (F = 4.65)		n.s. (F = 0.26)	
GC × PAW				* (F = 47.53)		n.s. (F = 1.44)		n.s. (F = 1.46)		n.s. (F = 1.03)	
C × PAW		n.s. (F = 1.66)		n.s. (F = 0.37)		n.s. (F = 0.22)		n.s. (F = 0.47)		n.s. (F = 1.31)	
GC × C × PAW				n.s. (F = 3.65)		n.s. (F = 0.22)		n.s. (F = 0.28)		n.s. (F = 0.43)	

F_C—F test cultivar; F_{GC}—F test growing conditions; F_{PAW}—F test PAW seeds treatment; *—statistical significance; n.s.—no statistical significance; Values inside same columns with different upper case differ with statistical significance (LSD) $p < 0.05$; values inside the same row with different lower case differ with statistical significance (LSD) $p < 0.05$.

The influence of the cultivar on the canopy weight was significant in the initial periods of measurement in both years of the study (Tables 4 and 5). However, the impact of cultivar is not consistent. Higher canopy weight was determined in the C1 pepper cultivar in 2021 (6.61%), while in 2022, cultivar C2 had a higher canopy weight (5.61%) (Tables 4 and 5). Plants of pepper developed from seeds treated with PAW had a higher canopy weight in both years (Tables 4 and 5). Determined differences were 9.06% and 9.01% in 2021 and 2022, respectively. Although statistically significant influence was not determined in all sampling dates, in both years of research, a significant influence of growing conditions (greenhouse) was established (Tables 4 and 5). Pepper plant growth in the greenhouse developed a 2.49% higher canopy weight than in the field in 2021 (Table 4). In 2022, a statistically significant interaction between cultivar, growing conditions, and PAW treatment on canopy weight was determined ($p < 0.05$) (Table 5).

Table 4. The canopy weight of pepper cultivars developed in 2021 under different growing conditions and seeds treatment in vegetative season 2021.

Pepper Component		Canopy Weight (g)									
Sampling Date		10 May 2021		17 May 2021		24 May 2021		7 June 2021		21 June 2021	
Cultivar	Growing Conditions	Seeds Treatment									
		C	PAW	C	PAW	C	PAW	C	PAW	C	PAW
C1	G	9.93	10.82 A	20.27 A	20.90 A	35.83 A	39.63 A	62.77 B	73.59 A	96.22	106.81
C1	F			14.60 B	17.00 C	26.11 B	27.76 B	43.03 C	58.52 B	89.90	105.16
C2	G	9.58	8.65 B	15.69 B	18.98 B	36.54 A	36.14 A	73.28 A	87.81 A	83.87	87.16
C2	F			15.71 B	15.24 C	26.7 B	25.44 B	51.99 C	52.38 B	79.30	80.14
F_C		* (F = 4.81)		* (F = 4.04)		n.s. (F = 0.56)		n.s. (F = 1.74)		n.s. (F = 4.15)	
F_{GC}				* (F = 15.25)		* (F = 48.2)		* (F = 19.2)		n.s. (F = 0.35)	
F_{PAW}		n.s. (F = 1.88)		n.s. (F = 2.96)		n.s. (F = 0.39)		n.s. (F = 3.9)		n.s. (F = 0.82)	

Table 4. *Cont.*

Pepper Component		**Canopy Weight (g)**									
Sampling Date		**10 May 2021**		**17 May 2021**		**24 May 2021**		**7 June 2021**		**21 June 2021**	
		Seeds Treatment									
Cultivar	**Growing Conditions**	C	PAW	C	PAW	C	PAW	C	PAW	C	PAW
		ANOVA—F test									
GC × C				n.s. (F = 2.97)		n.s. (F = 0.03)		n.s. (F = 1.1)		n.s. (F = 0.01)	
GC × PAW				n.s. (F = 0.34)		n.s. (F = 0.25)		n.s. (F = 0.21)		n.s. (F = 0.0)	
C × PAW		n.s. (F = 1.42)		n.s. (F = 0.0)		n.s. (F = 1.37)		n.s. (F = 0.3)		n.s. (F = 0.43)	
GC × C × PAW				n.s. (F = 2.66)		n.s. (F = 0.04)		n.s. (F = 0.81)		n.s. (F = 0.05)	

F_C—F test cultivar; F_{GC}—F test growing conditions; F_{PAW}—F test PAW seeds treatment; *—statistical significance; n.s.—no statistical significance; Values inside same columns with different upper case differ with statistical significance (LSD) $p < 0.05$; values inside the same row with different lower case differ with statistical significance (LSD) $p < 0.05$.

Table 5. The canopy weight of pepper cultivars developed under different growing conditions and seeds treatment in vegetative season 2022.

Pepper Component		**Canopy Weight (g)**									
Sampling Date		**10 May 2022**		**17 May 2022**		**24 May 2022**		**7 June 2022**		**21 June 2022**	
		Seeds Treatment									
Cultivar	**Growing Conditions**	C	PAW	C	PAW	C	PAW	C	PAW	C	PAW
C1	G	9.06	9.09 B	23.34 bA	27.78 aB	36.31 B	37.55 B	67.08	72.61	91.14 A	96.21 A
C1	F			16.04 bB	23.46 aC	34.28 B	36.82 B	65.08	67.66	66.33 B	69.59 C
C2	G	10.36	12.43 A	22.85 bA	30.74 aA	40.94 bA	57.62 aA	73.17	75.19	90.92 A	91.48 AB
C2	F			15.25 bB	19.57 aD	30.37 B	33.47 B	68.12	68.82	73.57 B	86.55 B
F_C		* (F = 5.88)		n.s. (F = 0.39)		n.s. (F = 3.25)		n.s. (F = 1.43)		n.s. (F = 2.43)	
F_{GC}				* (F = 75.98)		* (F = 15.03)		n.s. (F = 2.91)		* (F = 35.65)	
F_{PAW}		n.s. (F = 1.21)		* (F = 18.99)		n.s. (F = 2.74)		n.s. (F = 1.01)		n.s. (F = 3.14)	
		ANOVA—F test									
GC × C				n.s. (F = 4.2)		* (F = 10.93)		n.s. (F = 0.17)		* (F = 5.57)	
GC × PAW				* (F = 5.66)		n.s. (F = 2.36)		n.s. (F = 0.16)		n.s. (F = 0.74)	
C × PAW		n.s. (F = 1.13)		* (F = 7.0)		* (F = 5.94)		n.s. (F = 0.25)		n.s. (F = 0.18)	
GC × C × PAW				* (F = 19.59)		n.s. (F = 1.62)		n.s. (F = 0.02)		n.s. (F = 1.33)	

F_C—F test cultivar; F_{GC}—F test growing conditions; F_{PAW}—F test PAW seeds treatment; *—statistical significance; n.s.—no statistical significance; Values inside same columns with different upper case differ with statistical significance (LSD) $p < 0.05$; values inside the same row with different lower case differ with statistical significance (LSD) $p < 0.05$.

Cultivar C1 in 2021 and 2022 developed a greater number of leaves, 3.7%, and 2.01%, respectively (Tables 6 and 7). Moreover, seed treatment with PAW resulted in a higher number of leaves in 2022 (3.13%) (Table 7). The significant influence of PAW seed treatment on the number of leaves was determined in 2021 (Table 6). In that year, the number of leaves developed from seeds treated with PAW were 16.85% higher than the control (Table 6). In both years of the research, a significant influence of growing conditions on the number of leaves was determined (Tables 6 and 7). In 2021 the higher number of leaves (4.24%) in the greenhouse than in the field was determined (Table 6). The interaction between the cultivar and growing conditions was determined in both years of the study (Tables 6 and 7), and cultivar and PAW treatment, as well as growth condition and PAW treatment in 2021 ($p < 0.05$) (Table 6).

Table 6. The number of leaves pepper cultivars developed under different growing conditions and seed treatment in vegetative season 2021.

Pepper Component		Number of Leaves									
Sampling Date		10 May 2021		17 May 2021		24 May 2021		7 June 2021		21 June 2021	
Cultivar	Growing Conditions	Seeds Treatment									
		C	PAW	C	PAW	C	PAW	C	PAW	C	PAW
C1	G	11.50	12.00	26.50 A	27.50 A	29.25 bB	51.00 aA	72.00 A	82.50 A	107.25 bB	156.50 aA
C1	F			17.25 B	18.75 B	25.00 B	28.50 B	52.50 C	55.75 C	10.005 bB	149.75 aA
C2	G	11.25	10.75	20.75 B	26.50 A	35.00 bA	46.75 aA	65.25 B	74.25 B	134.75 A	143.50 A
C2	F			17.50 B	20.50 B	29.00 B	31.75 B	58.00 BC	63.00 C	98.00 B	106.75 B
F_C		n.s. (F = 2.57)		n.s. (F = 0.26)		n.s. (F = 0.63)		n.s. (F = 0.02)		n.s. (F = 1.97)	
F_{GC}				* (F = 8.56)		* (F = 18.80)		* (F = 16.7)		* (F = 10.6)	
F_{PAW}		n.s. (F = 0)		n.s. (F = 1.46)		* (F = 13.0)		* (F = 3.06)		* (F = 19.4)	
ANOVA—F test											
GC × C				n.s. (F = 0.88)		n.s. (F = 0.27)		n.s. (F = 3.06)		* (F = 6.5)	
GC × PAW				n.s. (F = 0.06)		* (F = 6.12)		n.s. (F = 0.5)		n.s. (F = 0.03)	
C × PAW		n.s. (F = 0.31)		n.s. (F = 0.45)		n.s. (F = 0.95)		n.s. (F = 0.0)		* (F = 9.15)	
GC × C × PAW				n.s. (F = 0.12)		n.s. (F = 0.71)		n.s. (F = 0.04)		n.s. (F = 0.03)	

F_C—F test cultivar; F_{GC}—F test growing conditions; F_{PAW}—F test PAW seeds treatment; *—statistical significance; n.s.—no statistical significance; Values inside same columns with different upper case differ with statistical significance (LSD) $p < 0.05$; values inside the same row with different lower case differ with statistical significance (LSD) $p < 0.05$.

Table 7. The number of leaves pepper cultivars developed under different growing conditions and seed treatment in vegetative season 2022.

Pepper Component		Number of Leaves									
Sampling Date		10 May 2022		17 May 2022		24 May 2022		7 June 2022		21 June 2022	
Cultivar	Growing Conditions	Seeds Treatment									
		C	PAW	C	PAW	C	PAW	C	PAW	C	PAW
C1	G	28.00	30.00	38.75 A	40.00 A	63.75	68.50	87.50	93.75	129.00 B	122.50 A
C1	F			30.25 B	39.00 A	55.75	57.25	85.50	87.50	121.25 B	118.00 A
C2	G	30.25	33.25	32.75 B	41.00 A	58.25	63.00	93.50	98.25	143.50 A	127.00 A
C2	F			26.75 B	27.75 B	50.25	52.25	79.00	94.50	105.00 C	103.00 B
F_C		n.s. (F = 1.10)		n.s. (F = 2.76)		n.s. (F = 1.40)		n.s. (F = 0.24)		n.s. (F = 0.32)	
F_{GC}				* (F = 5.84)		n.s. (F = 0)		n.s. (F = 0.99)		* (F = 11,82)	
F_{PAW}		n.s. (F = 0.91)		n.s. (F = 1.98)		n.s. (F = 0.3)		n.s. (F = 0.01)		n.s. (F = 1.69)	
ANOVA—F test											
GC × C				n.s. (F = 0.67)		* (F = 4.36)		n.s. (F = 0.38)		* (F = 5.34)	
GC × PAW				n.s. (F = 0.05)		n.s. (F = 0.04)		n.s. (F = 1.59)		n.s. (F = 0.67)	
C × PAW		n.s. (F = 0.04)		n.s. (F = 0.02)		n.s. (F = 0.04)		n.s. (F = 0.7)		n.s. (F = 0.16)	
GC × C × PAW				n.s. (F = 2.1)		n.s. (F = 0.24)		n.s. (F = 0.28)		n.s. (F = 0.27)	

F_C—F test cultivar; F_{GC}—F test growing conditions; F_{PAW}—F test PAW seeds treatment; *—statistical significance; n.s.—no statistical significance; Values inside same columns with different upper case differ with statistical significance (LSD) $p < 0.05$; values inside the same row with different lower case differ with statistical significance (LSD) $p < 0.05$.

Cultivar C1 developed a 14.04% higher number of nodes in 2021, and C2 cultivar developed an 8.65% higher number of nodes in 2022 (Tables 8 and 9). Pepper plants with preliminary PAW seed treatments developed a higher number of nodes, 64.29%, and 10.92%, in 2021 and 2022, respectively (Tables 8 and 9). In the present study, the significant influence of growing conditions on the number of nodes was proven. The plants grown in the greenhouse developed a higher number of nodes, 10.94%, and 6.6%, respectively (Tables 8 and 9). Regarding the number of nodes, the interaction between growing conditions and cultivar has been proven in 2022 ($p < 0.05$) (Table 9).

Table 8. The number of nodes pepper cultivars developed under different growing conditions and seed treatment in vegetative season 2021.

Pepper Component		**Number of Nodes**									
Sampling Date		**10 May 2021**		**17 May 2021**		**24 May 2021**		**7 June 2021**		**21 June 2021**	
		Seeds Treatment									
Cultivar	**Growing Conditions**	**C**	**PAW**	**C**	**PAW**	**C**	**PAW**	**C**	**PAW**	**C**	**PAW**
C1	G	5.00	5.50 A	5.75 A	6.25 B	7.50	8.00	13.50 A	14.00 A	13.75 A	14.25 A
C1	F			4.75 B	5.00 C	7.25	7.25	9.25 B	12.25 B	10.00 B	13.00 A
C2	G	4.75	4.75 B	5.00 bB	7.00 aA	7.75	7.50	9.50 B	10.50 BC	12.25 A	14.00 A
C2	F			4.00 bC	4.75 aC	6.50	7.50	8.00 B	8.75 C	7.75 C	9.75 B
F_C		* (F = 4.8)		n.s. (F = 0.8)		n.s. (F = 0.18)		* (F = 5.31)		n.s. (F = 3.13)	
F_{GC}				* (F = 24.2)		n.s. (F = 1.59)		n.s. (F = 3.03)		* (F = 11.2)	
F_{PAW}		n.s. (F = 1.2)		* (F = 9.8)		n.s. (F = 0.49)		n.s. (F = 0.97)		n.s. (F = 3.13)	
		ANOVA—F test									
GC × C				n.s. (F = 0.8)		n.s. (F = 0.02)		n.s. (F = 0.27)		n.s. (F = 0.84)	
GC × PAW				n.s. (F = 1.8)		n.s. (F = 0.17)		n.s. (F = 0.18)		n.s. (F = 0.45)	
C × PAW		n.s. (F = 1.2)		n.s. (F = 3.2)		n.s. (F = 0.02)		n.s. (F = 0.11)		n.s. (F = 0.0)	
GC × C × PAW				n.s. (F = 0.8)		n.s. (F = 0.96)		n.s. (F = 0.27)		n.s. (F = 0.3)	

F_C—F test cultivar; F_{GC}—F test growing conditions; F_{PAW}—F test PAW seeds treatment; *—statistical significance; n.s.—no statistical significance; Values inside same columns with different upper case differ with statistical significance (LSD) $p < 0.05$; values inside the same row with different lower case differ with statistical significance (LSD) $p < 0.05$.

Table 9. The number of nodes pepper cultivars developed under different growing conditions and seed treatment in vegetative season 2022.

Pepper Component		**Number of Nodes**									
Sampling Date		**10 May 2022**		**17 May 2022**		**24 May 2022**		**7 June 2022**		**21 June 2022**	
		Seeds Treatment									
Cultivar	**Growing Conditions**	**C**	**PAW**	**C**	**PAW**	**C**	**PAW**	**C**	**PAW**	**C**	**PAW**
C1	G	6.25	8.00	11.00 B	12.25 B	14.25 A	15.25 B	15.25	17.00	19.25	20.25
C1	F			7.75 C	12.00 B	10.75 B	14.00 B	14.00	15.75	17.00	17.50
C2	G	8.25	10.75	13.00 A	14.25 A	16.25 A	17.00 A	18.00	20.75	20.00	20.50
C2	F			11.75 AB	11.25 B	11.00 B	13.25 B	16.25	17.50	16.05	16.50
F_C		n.s. (F = 4.21)		* (F = 4.65)		n.s. (F = 0.18)		n.s. (F = 3.23)		n.s. (F = 2.41)	
F_{GC}				n.s. (F = 0.14)		* (F = 7.05)		n.s. (F = 1.65)		n.s. (F = 2.09)	
F_{PAW}		n.s. (F = 3.37)		n.s. (F = 3.45)		n.s. (F = 0.63)		n.s. (F = 0.01)		n.s. (F = 0.65)	
		ANOVA—F test									
GC × C				* (F = 4.65)		n.s. (F = 0.49)		n.s. (F = 0.18)		n.s. (F = 0.49)	
GC × PAW				n.s. (F = 2.44)		n.s. (F = 1.82)		n.s. (F = 0.07)		n.s. (F = 0.65)	
C × PAW		n.s. (F = 0.75)		n.s. (F = 1.99)		n.s. (F = 0.11)		n.s. (F = 1.65)		n.s. (F = 0.35)	
GC × C × PAW				n.s. (F = 0.27)		n.s. (F = 0.63)		n.s. (F = 0.07)		n.s. (F = 0.03)	

F_C—F test cultivar; F_{GC}—F test growing conditions; F_{PAW}—F test PAW seeds treatment; *—statistical significance; n.s.—no statistical significance; Values inside same columns with different upper case differ with statistical significance (LSD) $p < 0.05$; values inside the same row with different lower case differ with statistical significance (LSD) $p < 0.05$.

Cultivar C1 developed a higher number of buds (1.54%) in 2021 (Table 10), and cultivar C2 had a higher number of buds (2.92%) in 2022 (Table 11). Statistically significant impact of cultivar and preliminary PAW seed treatment was proven in both years (Tables 10 and 11). A higher number of buds (35.54%) were developed in pepper plants from PAW seeds in 2021, and the number of buds was 10.75% higher in plants developed in 2022 (Tables 10 and 11). The variability in the number of buds among growth conditions with different temperatures and relative humidity was statistically significant (Tables 10 and 11). The plants grown in the greenhouse had a higher number of buds (32.13%) in 2021 and 8.44% in 2022 (Tables 10 and 11). There was an interaction between growing conditions and cultivar in 2022 ($p < 0.05$) (Table 11) and growing conditions and seed treated with PAW (Table 10).

Table 10. The number of buds of pepper cultivars developed under different growing conditions and seed treatment in vegetative season 2021.

Pepper Component		**Number of Buds**							
Sampling Date		**17 May 2021**		**24 May 2021**		**7 June 2021**		**21 June 2021**	
		Seeds Treatment							
Cultivar	**Growing Conditions**	**C**	**PAW**	**C**	**PAW**	**C**	**PAW**	**C**	**PAW**
C1	G	2.75 bA	5.25 aA	5.00 b	12.25 aA	20.50 B	26.50 A	25.25 bB	59.25 aA
C1	F	1.75 B	2.25 C	4.75	5.25 C	12.50 C	17.25 C	18.25 bB	57.00 aA
C2	G	3.00 A	3.25 B	6.25	9.00 B	29.00 A	28.25 A	38.75 bA	51.75 aA
C2	F	1.25 B	1.75 C	6.00	4.25 C	21.00 B	22.25 B	18.50 B	27.25 B
F_C		n.s. (F = 2.2)		n.s. (F = 0.16)		* (F = 9.03)		n.s. (F = 1.42)	
F_{GC}		* (F = 15.29)		* (F = 7.77)		* (F = 15.6)		* (F = 7.51)	
F_{PAW}		n.s. (F = 4.09)		* (F = 3.96)		n.s. (F = 2.03)		* (F = 23)	
		ANOVA—F test							
GC × C		n.s. (F = 0.16)		n.s. (F = 0.26)		n.s. (F = 0.17)		n.s. (F = 3.25)	
GC × PAW		n.s. (F = 0.89)		* (F = 6.55)		n.s. (F = 0.01)		n.s. (F = 0.0)	
C × PAW		n.s. (F = 1.47)		n.s. (F = 2.36)		n.s. (F = 1.68)		n.s. (F = 6.7)	
GC × C × PAW		n.s. (F = 1.47)		n.s. (F = 0.26)		n.s. (F = 0.17)		n.s. (F = 0.21)	

F_C—F test cultivar; F_{GC}—F test growing conditions; F_{PAW}—F test PAW seeds treatment; *—statistical significance; n.s.—no statistical significance; Values inside same columns with different upper case differ with statistical significance (LSD) $p < 0.05$; values inside the same row with different lower case differ with statistical significance (LSD) $p < 0.05$.

Table 11. The number of buds of pepper cultivars developed under different growing conditions and seed treatment in vegetative season 2022.

Pepper Component		**Number of Buds**									
Sampling Date		**10 May 2022**		**17 May 2022**		**24 May 2022**		**7 June 2022**		**21 June 2022**	
		Seeds Treatment									
Cultivar	**Growing Conditions**	**C**	**PAW**	**C**	**PAW**	**C**	**PAW**	**C**	**PAW**	**C**	**PAW**
C1	G	4.50	4.75	14.25 A	14.25 A	20.25	23.00	16.50	19.50	16.75	18.00
C1	F			6.75 bC	11.50 aB	15.25	15.50	16.50	15.75	14.05	14.25
C2	G	5.75	6.50	9.50 bB	12.50 aA	15.75	21.50	21.25	18.50	16.50	17.50
C2	F			5.00 bC	8.75 aC	13.25	15.75	15.00	18.50	15.75	16.00
F_C		n.s. (F = 1.82)		* (F = 5.01)		n.s. (F = 0.73)		n.s. (F = 0.26)		n.s. (F = 0.4)	
F_{GC}				* (F = 14.16)		n.s. (F = 0.22)		n.s. (F = 0.06)		n.s. (F = 1.1)	
F_{PAW}		n.s. (F = 0.2)		* (F = 5.42)		n.s. (F = 0.4)		n.s. (F = 0.09)		n.s. (F = 0.4)	
		ANOVA—F test									
GC × C				n.s. (F = 0.69)		* (F = 5.24)		n.s. (F = 1.03)		* (F = 5.33)	
GC × PAW				n.s. (F = 1.25)		n.s. (F = 0.48)		n.s. (F = 1.03)		n.s. (F = 0.04)	
C × PAW		n.s. (F = 0.83)		n.s. (F = 0.17)		n.s. (F = 1.41)		n.s. (F = 0.02)		n.s. (F = 0.0)	
GC × C × PAW				n.s. (F = 0.66)		n.s. (F = 0.0)		n.s. (F = 0.06)		n.s. (F = 0.4)	

F_C—F test cultivar; F_{GC}—F test growing conditions; F_{PAW}—F test PAW seeds treatment; *—statistical significance; n.s.—no statistical significance; Values inside same columns with different upper case differ with statistical significance (LSD) $p < 0.05$; values inside the same row with different lower case differ with statistical significance (LSD) $p < 0.05$.

Fruit quality components were in the present study under the significant influence of cultivars in 2021 (Table 12). In the first year of research, in cultivar C1, the following higher values have been proven: fruit weight (8.84%), pericarp weight (3.27%), and residual weight (20.28%) (Table 12). However, fruit diameter and fruit length were higher in cultivar C2 (2.86% and 1.55%, respectively) (Table 12). In the second year of research, all measured parameters were higher in cultivar C2: fruit weight (13.62%), fruit diameter (6%), fruit length (9.43%), pericarp weight (12.72%), and residuals weight (17%) (Table 13). The cause for this could be stress caused by low temperatures that occasionally occurred during the cultivation period.

Table 12. Fruit components of pepper cultivars developed under different growing conditions and seed treatment in the vegetative season 2021.

Fruit Component, 2021		Fruit Weight (g)		Fruit Diameter (cm)		Fruit Length (cm)		Pericarp Weight (g)		Residuals Weight (g)	
		Seeds Treatment									
Cultivar	Growing Conditions	C	PAW	C	PAW	C	PAW	C	PAW	C	PAW
C1	G	31.27 bA	39.48 aA	4.93 A	4.94 B	5.01 bA	5.14 aB	23.35 bA	24.35 aB	7.92 bA	15.12 aA
C1	F	22.04 bC	24.55 aC	3.69 bB	4.10 aC	4.02 bC	4.85 aC	15.70 C	15.70 C	6.33 bC	8.85 aD
C2	G	29.52 bB	34.93 aB	4.96 bA	5.28 aA	5.06 bA	5.35 aA	22.85 bB	25.57 aA	6.67 bB	9.36 aB
C2	F	18.88 bD	23.64 aD	3.90 bB	4.04 aC	4.44 B	4.47 D	13.63 bD	14.46 aD	5.26 bD	9.18 aC
F_C		* (F = 303.51)		* (F = 6.65)		n.s. (F = 1.37)		* (F = 43.25)		* (F = 1269.04)	
F_{GC}		* (F = 5999.94)		* (F = 471.94)		* (F = 114.19)		* (F = 8667.08)		* (F = 1884.84)	
F_{PAW}		* (F = 1233.42)		* (F = 19.05)		* (F = 24.4)		* (F = 133.89)		* (F = 5634.35)	
ANOVA—F test											
GC × C		* (F = 14.11)		n.s. (F = 1.19)		n.s. (F = 0.72)		* (F = 104.79)		* (F = 831.95)	
GC × PAW		* (F = 114.09)		n.s. (F = 1.19)		n.s. (F = 2.86)		* (F = 53.86)		* (F = 250.76)	
C × PAW		n.s. (F = 0.84)		n.s. (F = 0.04)		* (F = 5.96)		* (F = 42.09)		* (F = 204.36)	
GC × C × PAW		* (F = 71.9)		* (F = 8.28)		* (F = 13.62)		* (F = 5.09)		* (F = 739.23)	

F_C—F test cultivar; F_{GC}—F test growing conditions; F_{PAW}—F test PAW seeds treatment; *—statistical significance; n.s.—no statistical significance; Values inside same columns with different upper case differ with statistical significance (LSD) $p < 0.05$; values inside the same row with different lower case differ with statistical significance (LSD) $p < 0.05$.

Table 13. Fruit components of pepper cultivars developed under different growing conditions and seed treatment in the vegetative season of 2022.

Fruit Component, 2022		Fruit Weight (g)		Fruit Diameter (cm)		Fruit Length (cm)		Pericarp Weight (g)		Residuals Weight (g)	
		Seeds treatment									
Cultivar	Growing conditions	C	PAW	C	PAW	C	PAW	C	PAW	C	PAW
C1	G	57.13 B	79.23 A	5.50 A	5.73 A	6.53 A	7.16 A	45.09 AB	64.87 A	12.03 A	14.37 A
C1	F	28.90 B	43.72 B	4.31 B	4.81 B	4.77 B	5.36 B	22.77 B	33.59 B	6.14 B	10.13 B
C2	G	70.49 A	72.95 A	5.63 A	5.99 A	7.18 A	7.21 A	55.65 A	58.98 A	13.97 A	14.85 A
C2	F	39.02 B	59.48 B	4.73 B	5.30 B	5.32 B	6.59 A	29.58 B	46.35 AB	9.45 B	13.14 A
F_C		n.s. (F = 0.94)		n.s. (F = 1.41)		n.s. (F = 1.85)		n.s. (F = 0.73)		n.s. (F = 2.21)	
F_{GC}		* (F = 10.24)		* (F = 11.22)		* (F = 10.96)		* (F = 10.66)		* (F = 7.72)	
F_{PAW}		n.s. (F = 0.21)		n.s. (F = 1.17)		n.s. (F = 0.44)		n.s. (F = 0.16)		n.s. (F = 0.58)	
ANOVA—F test											
GC × C		n.s. (F = 0.31)		n.s. (F = 0.22)		n.s. (F = 0.35)		n.s. (F = 0.28)		n.s. (F = 0.44)	
GC × PAW		n.s. (F = 2.62)		n.s. (F = 0.71)		n.s. (F = 1.91)		n.s. (F = 2.43)		n.s. (F = 3.41)	
C × PAW		n.s. (F = 0.79)		n.s. (F = 0.36)		n.s. (F = 0.49)		n.s. (F = 1.06)		n.s. (F = 0.04)	
GC × C × PAW		n.s. (F = 0.31)		n.s. (F = 0.22)		n.s. (F = 0.0)		n.s. (F = 0.37)		n.s. (F = 0.1)	

F_C—F test cultivar; F_{GC}—F test growing conditions; F_{PAW}—F test PAW seeds treatment; *—statistical significance; n.s.—no statistical significance; Values inside same columns with different upper case differ with statistical significance (LSD) $p < 0.05$; values inside the same row with different lower case differ with statistical significance (LSD) $p < 0.05$.

In both years of the research, a significant difference has been found in the fruit quality between the growing conditions (Tables 12 and 13). Better fruit quality was found for plants grown in the greenhouse in 2021 and 2022: fruit weight (34.09% and 38.84%), fruit diameter (21.78% and 16.19%), fruit length (13.52% and 21.51%), pericarp weight (38.11% and 41.1%) and residuals weight (24.19% and 29.63%). The influence of PAW seed treatment on fruit quality was statistically significant in 2021 (Table 12). Preliminary seeds PAW treatment resulted in higher: fruit weight (17% and 23.43%), fruit diameter (4.79% and 7.6%), fruit length (6.46% and 9.57%), pericarp weight (5.68% and 24.88%) and residuals weight (38.41% and 20.77%) (Tables 12 and 13). Moreover, in 2021 an interaction of all treatments on pericarp and residuals weight has been found (Table 12). Likewise, the interaction between growing conditions, cultivar and PAW seed treatment were found regarding fruit weight, diameter, and fruit length (Table 12).

Dry matter content in pepper plant organs was different between cultivars, growing conditions, and seed treatment with PAW, as well as between years (Figure 2). The content of dry matter C1 was 7.82 g, and C2 had a dry matter content was 6.90 g. Comparing the dry matter of plants grown in different growing conditions, it was found that the dry matter of plants grown in the greenhouse was higher (8.10 g) than for those grown in the open field (6.62 g). On average, plants grown from the seeds treated with PAW have a dry matter of 8.93 g, and control has 5.78 g. Moreover, the difference between years has been established: in 2022, dry matter content was 8.63 g, and in 2021 it was 7.36 g (Figure 2).

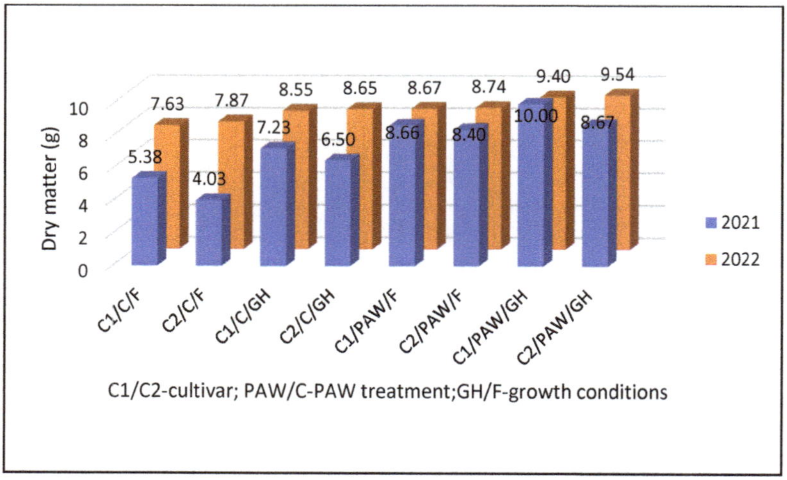

Figure 2. Dry matter content in pepper plant organs developed under different growing conditions and seed treatment in vegetative seasons 2021 and 2022.

4. Discussion

The product quality of greenhouse vegetables [39], as well as yield and fruit weight [40], depend on plants' genetic composition of species and cultivars (Tables 2–13).

Temperature [7] and relative humidity [8] are the basic environmental factors that affect sweet pepper growth. The minimum temperature for pepper growth is about 8–12 °C [41], and the optimum temperature for pepper growth is 20–25 °C [10]. Although higher air temperature (32 °C) has improved plant height, fresh weight, dry weight, and enhanced development of leaves [7], pepper are sensitive to temperature extremes [10]. In the present study, growth conditions varied in temperature (Figure 1). A low temperature in the first hours of the day was measured in an open field (Figure 1), which could influence the low biomass and fruit quality of pepper (Tables 2–13). Moreover, the spring period of 2021 had a long period of low temperatures below 0 °C, which could have an influence on low biomass of pepper plants and fruit quality. Low temperature is one of the most important abiotic factors that restrict the optimal production of warm-season vegetables [10], which can be a reason for lower values determined in the open field in comparison with production in the greenhouse in the present study (Tables 2–13). The reduction in dry mass accumulation could be a result of the restriction of sucrose synthesis by low temperatures [7]. Low temperature coupled with high humidity (Figure 1) reduces the yield under open field conditions [5]. However, high night-time humidity could result in higher production of pepper biomass (Tables 2–13) and fruit weight in the greenhouse [42]. Previous research [5,8,10,43] determined an increase in plant height, the number of leaves, dry weight of shoot and root, as well as fruit number and fruit size of pepper and flowering in the greenhouse in order to field conditions which are in accordance with results determined in the present study (Tables 2–13) (Figure 2).

Several authors stated that PAW changes the metabolism of seeds which improves plant development [18,19,31,34]. However, the results of research on the influence of PAW on the development of pepper plants and fruits are limited. Most of the previous research investigated germination and seedlings characteristics of pepper [32]. According to provided research, nonthermal plasma improved stem length [33], shoot length, total fresh weight and leaf area, and biomass [28,30]. In the present study, higher canopy height, the number of buds, nodes, leaves, and canopy weight in plants with seeds treated with PAW were determined (Tables 2–13). In the research of Nalwa et al. [29], pepper seeds exposed to plasma achieved maximum ripe fruit yield/plant. This is in accordance with the achievement in the first vegetative season in this research (Table 12). Additionally, in the present research, pepper was cultivated during two vegetative seasons. However, in the second vegetative season significant impact of PAW on fruit quality was not determined (Table 13). Moreover, in the research of Shapira et al. [35], the yield parameters of pepper (fruit yield and average mass of fruit) were not significantly different, comparing plasma treated to non-treated pepper.

The highest average of dry pepper weight under greenhouse conditions was proven [43]. Since low temperature could be the cause of the restriction of sucrose synthesis, dry weight rapidly increases under higher air temperatures [7,39]. In this research, the dry matter was increased in the greenhouse compared to the field, probably as a result of higher temperature [7] and lower humidity (Figure 2). As the authors reported, the plant's dry weight was increased with the application of PAW [44]. In this study, the PAW treatment also increased dry matter weight in both growth conditions and vegetative seasons (Figure 1).

PAW-treated seeds developed pepper with higher growth parameters in the open field and greenhouse in both vegetative seasons. Pepper development in the field was exposed to low temperatures and high relative humidity, which caused stress to them. However, pepper produced by plasma treatments probably had a better adaptive response against stress which can be explained by enhanced total soluble sugar concentration [45]. The action of various stressors on plants usually causes the formation and accumulation of ROS at some point, which is vital for stress signaling [46]. Plasma-activated water contains ROS that acts as a mild stressor that prepares the plant for the occurrence of stronger stress [24]. This is probably the reason for the higher parameters of pepper developed from treated seeds in this research. Higher growth parameters of pepper developed in the greenhouse could be the consequence of nitrogen content in PAW. The NO_3^- is the main source of nitrogen for plant production of proteins and nucleic acids; therefore, it can be considered the main component of the PAW responsible for biomass increase. Another species (H_2O_2) can also take part in weight increase through the process of plant tissue lignification. Moreover, PAW had a positive effect on both the chlorophyll content and the photosynthetic rate. High levels of chlorophyll can be attributed to increased physiological activity and photosynthesis in plants and plant growth [47].

The response to PAW treatment depended on the plant cultivar, growing conditions as well as their interaction [17,20,21]. Higher growth parameters were determined in both vegetable seasons in PAW-developed plants. However, results concerning fruit quality were different in the vegetative season 2021 and 2022, which demand further investigation in specific conditions.

5. Conclusions

Contemporary agriculture requires new approaches in production related to the use of fertilizers and the application of substances that will increase the resistance of plants to the stress caused by climate change. From this point of view, the present study evaluated the performance of PAW, a new promising eco-friendly stimulant in terms of plant productivity. PAW seeds were compared to water-soaked seeds for four treatments, comprising two different cultivars of pepper and two different growth conditions. The results proved that PAW outperformed water-soaked seeds in terms of both growth conditions and cultivars. PAW treatment improved all growth parameters measured as well as fruit quality parameters compared with water-soaked treatment. In this research, PAW treatment influenced the values of measured parameters of pepper more than greenhouse conditions in regards to open fields. Therefore, it is concluded that PAW treatment is the adequate treatment for pepper seeds in terms of improving production and yield simultaneously.

PAW can be utilized when it is desired to ensure the yield of plants if they are exposed to extreme temperatures and relative humidity. Cultivation in the greenhouse increased the pepper yield compared to the open field, which would be a benefit for producers.

This study suggests that PAW is a useful technique for growing pepper plants in greenhouses and open fields; however, in future studies, more variants of PAW and cultivars should be tested to find out the best characteristics for different ranges of cultivars. It is also recommended that PAW studies should be conducted during more growing seasons in different climatic conditions for evaluation of their performance and effectiveness over time.

Author Contributions: Conceptualization, B.J.-P.; methodology, B.J.-P., K.M. and T.B.-L.; investigation, B.J.-P., R.B., N.R.F., T.B.-L., S.A. and M.J.; resources, B.J.-P.; data processing, B.J.-P. and R.B.; writing-original draft preparation, B.J.-P.; writing-review and editing, K.M. All authors have read and agreed to the published version of the manuscript.

Funding: This work was supported by the project Adaptation of vegetables to new agrometeorological conditions in Slavonia (AVACS), KK.05.1.1.02.0004. The project was financed by the European Union from European Regional Development Fund.

Institutional Review Board Statement: Not applicable.

Informed Consent Statement: Not applicable.

Data Availability Statement: Not applicable.

Acknowledgments: Authors express their sincere gratitude to Katica Šimunović from the University of Slavonski Brod for her advice in the statistical analysis of experimental data. Moreover, the authors express their sincere gratitude to Slobodan Milošević and Mario Rakić from the Institute of Physics for PAW preparation.

Conflicts of Interest: The authors declare no conflict of interest.

References

1. Carmen Pinero, M.; Perez-Jimenez, M.; Lopez-Marin, J.; del Amor, F.M. Fruit Quality of Sweet Pepper as Affected by Foliar Ca Applications to Mitigate the Supply of Saline Water under a Climate Change Scenario. *J. Sci. Food Agric.* **2018**, *98*, 1071–1078. [CrossRef]
2. Stagnari, F.; Campanelli, G.; Galieni, A.; Platani, C.; Bertone, A.; Ficcadenti, N. Adaptive Responses to Nitrogen and Light Supplies of a Local Varieties of Sweet Pepper from the Abruzzo Region, Southern Italy. *Agronomy-Basel* **2021**, *11*, 1343. [CrossRef]
3. Mahmoud, A.M.A.; El-Eslamboly, A.A.S.A. Production and Evaluation of High Yielding Sweet Pepper Hybrids under Greenhouse Conditions. *Am. Eurasian J. Agric. Environ. Sci.* **2015**, *15*, 573–580.
4. Hou, Y.; Li, A.; Li, Y.; Jin, D.; Tian, Y.; Zhang, D.; Wu, D.; Zhang, L.; Lei, W. Analysis of Microclimate Characteristics in Solar Greenhouses under Natural Ventilation. *Build. Simul.* **2021**, *14*, 1811–1821. [CrossRef]
5. Singh, B.; Biwalkar, N.; Chhina, R.S. Response of Sweet Pepper (*Capsicum annuum*) under Varying Fertigation and Irrigation Applications Grown in Naturally Ventilated Greenhouse. *J. Krishi Vigyan.* **2020**, *8*, 236–241. [CrossRef]
6. Ge, J.; Zhao, L.; Gong, X.; Lai, Z.; Traore, S.; Li, Y.; Long, H.; Zhang, L. Combined Effects of Ventilation and Irrigation on Temperature, Humidity, Tomato Yield, and Quality in the Greenhouse. *Hortscience* **2021**, *56*, 1080–1088. [CrossRef]
7. Kwack, Y.; An, S.; Kim, S.K. Development of Growth Model for Grafted Hot Pepper Seedlings as Affected by Air Temperature and Light Intensity. *Sustainability* **2021**, *13*, 5895. [CrossRef]
8. Chowdhury, M.; Kiraga, S.; Islam, M.N.; Ali, M.; Reza, M.N.; Lee, W.-H.; Chung, S.-O. Effects of Temperature, Relative Humidity, and Carbon Dioxide Concentration on Growth and Glucosinolate Content of Kale Grown in a Plant Factory. *Foods* **2021**, *10*, 1524. [CrossRef]
9. der Ploeg, A.; Heuvelink, E. The Influence of Temperature on Growth and Development of Chrysanthemum Cultivars: A Review. *J. Hortic. Sci. Biotechnol.* **2006**, *81*, 174–182. [CrossRef]
10. Angmo, P.; Phuntsog, N.; Namgail, D.; Chaurasia, O.P.; Stobdan, T. Effect of Shading and High Temperature Amplitude in Greenhouse on Growth, Photosynthesis, Yield and Phenolic Contents of Tomato (*Lycopersicum esculentum* Mill.). *Physiol. Mol. Biol. Plants* **2021**, *27*, 1539–1546. [CrossRef]
11. Kang, M.H.; Jeon, S.S.; Shin, S.M.; Veerana, M.; Ji, S.-H.; Uhm, H.-S.; Choi, E.-H.; Shin, J.H.; Park, G. Dynamics of Nitric Oxide Level in Liquids Treated with Microwave Plasma-Generated Gas and Their Effects on Spinach Development. *Sci. Rep.* **2019**, *9*, 1011. [CrossRef] [PubMed]

12. Maher, A.; Kamel, E.; Enrico, F.; Atif, I.; Abdelkader, M. An Intelligent System for the Climate Control and Energy Savings in Agricultural Greenhouses. *Energy Effic.* **2016**, *9*, 1241–1255. [CrossRef]
13. Thomaier, S.; Specht, K.; Henckel, D.; Dierich, A.; Siebert, R.; Freisinger, U.B.; Sawicka, M. Farming in and on Urban Buildings: Present Practice and Specific Novelties of Zero-Acreage Farming (ZFarming). *Renew. Agric. Food Syst.* **2015**, *30*, 43–54. [CrossRef]
14. Mussa, A.; Shinichi, K. Effect of Planting Space and Shoot Pruning on the Occurrence of Thrips, Fruit Yield and Quality Traits of Sweet Pepper (*Capsicum annum* L.) under Greenhouse Conditions. *J. Entomol. Zool. Stud.* **2019**, *7*, 787–792.
15. Pandey, P.; Irulappan, V.; Bagavathiannan, M.V.; Senthil-Kumar, M. Impact of Combined Abiotic and Biotic Stresses on Plant Growth and Avenues for Crop Improvement by Exploiting Physio-Morphological Traits. *Front. Plant Sci.* **2017**, *8*, 537. [CrossRef] [PubMed]
16. Susmita, C.; Kumar, S.P.J.; Chintagunta, A.D.; Lichtfouse, E.; Naik, B.; Ramya, P.; Kumari, K.; Kumar, S. Non-Thermal Plasmas for Disease Control and Abiotic Stress Management in Plants. *Environ. Chem. Lett.* **2022**, *20*, 2135–2164. [CrossRef]
17. Staric, P.; Vogel-Mikus, K.; Mozetic, M.; Junkar, I. Effects of Nonthermal Plasma on Morphology, Genetics and Physiology of Seeds: A Review. *Plants* **2020**, *9*, 1736. [CrossRef]
18. Puac, N.; Gherardi, M.; Shiratani, M. Plasma Agriculture: A Rapidly Emerging Field. *Plasma Process. Polym.* **2018**, *15*, 1700174. [CrossRef]
19. Puac, N.; Skoro, N.; Spasic, K.; Zivkovic, S.; Milutinovic, M.; Malovic, G.; Petrovic, Z.L. Activity of Catalase Enzyme in Paulownia Tomentosa Seeds during the Process of Germination after Treatments with Low Pressure Plasma and Plasma Activated Water. *Plasma Process. Polym.* **2018**, *15*, 1700082. [CrossRef]
20. Attri, P.; Ishikawa, K.; Okumura, T.; Koga, K.; Shiratani, M. Plasma Agriculture from Laboratory to Farm: A Review. *Processes* **2020**, *8*, 1002. [CrossRef]
21. Ranieri, P.; Sponsel, N.; Kizer, J.; Rojas-Pierce, M.; Hernandez, R.; Gatiboni, L.; Grunden, A.; Stapelmann, K. Plasma Agriculture: Review from the Perspective of the Plant and Its Ecosystem. *Plasma Process. Polym.* **2021**, *18*, 2000162. [CrossRef]
22. Kocira, S.; Perez-Piza, M.C.; Bohata, A.; Bartos, P.; Szparaga, A. Cold Plasma as a Potential Activator of Plant Biostimulants. *Sustainability* **2022**, *14*, 495. [CrossRef]
23. Graves, D.B.; Bakken, L.B.; Jensen, M.B.; Ingels, R. Plasma Activated Organic Fertilizer. *Plasma Chem. Plasma Process.* **2019**, *39*, 1–19. [CrossRef]
24. Adhikari, B.; Adhikari, M.; Ghimire, B.; Park, G.; Choi, E.H. Cold Atmospheric Plasma-Activated Water Irrigation Induces Defense Hormone and Gene Expression in Tomato Seedlings. *Sci. Rep.* **2019**, *9*, 16080. [CrossRef] [PubMed]
25. Bourke, P.; Ziuzina, D.; Boehm, D.; Cullen, P.J.; Keener, K. The Potential of Cold Plasma for Safe and Sustainable Food Production. *Trends Biotechnol.* **2018**, *36*, 615–626. [CrossRef]
26. Ingels, R.; Graves, D.B. Improving the Efficiency of Organic Fertilizer and Nitrogen Use via Air Plasma and Distributed Renew-Able Energy. *Plasma Med.* **2015**, *5*, 257–270. [CrossRef]
27. Kucerova, K.; Henselova, M.; Slovakova, L.; Bacovcinova, M.; Hensel, K. Effect of Plasma Activated Water, Hydrogen Peroxide, and Nitrates on Lettuce Growth and Its Physiological Parameters. *Appl. Sci.* **2021**, *11*, 1985. [CrossRef]
28. Safari, N.; Iranbakhsh, A.; Oraghi Ardebili, Z. Non-Thermal Plasma Modified Growth and Differentiation Process of *Capsicum annuum* PP805 Godiva in in Vitro Conditions. *Plasma Sci. Technol.* **2017**, *19*, 055501. [CrossRef]
29. Nalwa, C.; Thakur, A.K.; Vikram, A.; Rane, R.; Vaid, A. Effect of Cold Plasma Treatment and Priming in Bell Pepper (*Capsicum annuum* L.). *Int. J. Bio-Resour. Stress Manag.* **2017**, *8*, 535–538. [CrossRef]
30. Iranbakhsh, A.; Ardebili, Z.O.; Ardebili, N.O.; Ghoranneviss, M.; Safari, N. Cold Plasma Relieved Toxicity Signs of Nano Zinc Oxide in *Capsicum annuum* Cayenne via Modifying Growth, Differentiation, and Physiology. *Acta Physiol. Plant.* **2018**, *40*, 154. [CrossRef]
31. Stepanova, V.; Slavicek, P.; Kelar, J.; Prasil, J.; Smekal, M.; Stupavska, M.; Jurmanova, J.; Cernak, M. Atmospheric Pressure Plasma Treatment of Agricultural Seeds of Cucumber (*Cucumis sativus* L.) and Pepper (*Capsicum annuum* L.) with Effect on Reduction of Diseases and Germination Improvement. *Plasma Process. Polym.* **2018**, *15*, 1700076. [CrossRef]
32. Thisawech, M.; Saritnum, O.; Sarapirom, S.; Prakrajang, K.; Phakham, W. Effects of Plasma Technique and Gamma Irradiation on Seed Germination and Seedling Growth of Chili Pepper. *Chiang Mai J. Sci.* **2020**, *47*, 73–82.
33. Kasih, T.P.; Purwondho, R.; Danil, D.; Radjagukguk, R.; Bagaskara, A. Germination Enhancement of Green Bell Pepper (*Capsicum annuum* L.) by Using Non Thermal Argon Plasma. In *IOP Conference Series: Earth and Environmental Science*; IOP Publishing: Bristol, UK, 2020; p. 12131.
34. Sivachandiran, L.; Khacef, A. Enhanced Seed Germination and Plant Growth by Atmospheric Pressure Cold Air Plasma: Combined Effect of Seed and Water Treatment. *RSC Adv.* **2017**, *7*, 1822–1832. [CrossRef]
35. Shapira, Y.; Bormashenko, E.; Drori, E. Pre-Germination Plasma Treatment of Seeds Does Not Alter Cotyledon DNA Structure, nor Phenotype and Phenology of Tomato and Pepper Plants. *Biochem. Biophys. Res. Commun.* **2019**, *519*, 512–517. [CrossRef] [PubMed]
36. Kutasi, K.; Popovic, D.; Krstulovic, N.; Milosevic, S. Tuning the Composition of Plasma-Activated Water by a Surface-Wave Microwave Discharge and a KHz Plasma Jet. *Plasma Sources Sci. Technol.* **2019**, *28*, 095010. [CrossRef]
37. Gierczik, K.; Vukusic, T.; Kovacs, L.; Szekely, A.; Szalai, G.; Milosevic, S.; Kocsy, G.; Kutasi, K.; Galiba, G. Plasma-Activated Water to Improve the Stress Tolerance of Barley. *Plasma Process. Polym.* **2020**, *17*, 1900123. [CrossRef]
38. Kutasi, K.; Krstulovic, N.; Jurov, A.; Salamon, K.; Popovic, D.; Milosevic, S. Controlling: The Composition of Plasma-Activated Water by Cu Ions. *Plasma Sources Sci. Technol.* **2021**, *30*, 045015. [CrossRef]
39. Gruda, N.; Savvas, D.; Colla, G.; Rouphael, Y. Impacts of Genetic Material and Current Technologies on Product Quality of Selected Greenhouse Vegetables—A Review. *Eur. J. Hortic. Sci.* **2018**, *83*, 319–328. [CrossRef]

40. Gungor, F.; Yildirim, E. Effect of different growing media on quality, growth and yield of pepper (*Capsicum annuum* L.) under greenhouse conditions. *Pakistan J. Bot.* **2013**, *45*, 1605–1608.
41. Ropokis, A.; Ntatsi, G.; Kittas, C.; Katsoulas, N.; Savvas, D. Effects of Temperature and Grafting on Yield, Nutrient Uptake, and Water Use Efficiency of a Hydroponic Sweet Pepper Crop. *Agronomy-Basel* **2019**, *9*, 110. [CrossRef]
42. Bakker, J.C. The Effects of Air Humidity on Growth and Fruit Production of Sweet Pepper (*Capsicum annuum* L.). *J. Hortic. Sci.* **1989**, *64*, 41–46. [CrossRef]
43. Khaitov, B.; Yun, H.J.; Lee, Y.; Ruziev, F.; Le, T.H.; Umurzokov, M.; Bo, A.B.; Cho, K.M.; Park, K.W. Impact of Organic Manure on Growth, Nutrient Content and Yield of Chilli Pepper under Various Temperature Environments. *Int. J. Environ. Res. Public Health* **2019**, *16*, 3031. [CrossRef] [PubMed]
44. Rathore, V.; Tiwari, B.S.; Nema, S.K. Treatment of Pea Seeds with Plasma Activated Water to Enhance Germination, Plant Growth, and Plant Composition. *Plasma Chem. Plasma Process.* **2022**, *42*, 109–129. [CrossRef]
45. Rashid, M.; Rashid, M.M.; Reza, M.A.; Talukder, M.R. Combined Effects of Air Plasma Seed Treatment and Foliar Application of Plasma Activated Water on Enhanced Paddy Plant Growth and Yield. *Plasma Chem. Plasma Process.* **2021**, *41*, 1081–1099. [CrossRef]
46. Holubova, L.; Kyzek, S.; Durovcova, I.; Fabova, J.; Horvathova, E.; Sevcovicova, A.; Galova, E. Non-Thermal Plasma–A New Green Priming Agent for Plants? *Int. J. Mol. Sci.* **2020**, *21*, 9466. [CrossRef] [PubMed]
47. Kucerova, K.; Henselova, M.; Slovakova, L.; Hensel, K. Effects of Plasma Activated Water on Wheat: Germination, Growth Parameters, Photosynthetic Pigments, Soluble Protein Content, and Antioxidant Enzymes Activity. *Plasma Process. Polym.* **2019**, *16*, 1800131. [CrossRef]

sustainability MDPI

Article

Melatonin as a Foliar Application and Adaptation in Lentil (*Lens culinaris* Medik.) Crops under Drought Stress

Sidra Yasmeen [1], Abdul Wahab [2], Muhammad Hamzah Saleem [3,*], Baber Ali [4], Kamal Ahmad Qureshi [5,*] and Mariusz Jaremko [6]

1 Department of Botany, Government College University, Faisalabad 38000, Pakistan
2 Shanghai Center for Plant Stress Biology, CAS Center for Excellence in Molecular Plant Sciences, Chinese Academy of Sciences, Shanghai 200032, China
3 College of Plant Science and Technology, Huazhong Agricultural University, Wuhan 430070, China
4 Department of Plant Sciences, Quaid-i-Azam University, Islamabad 45320, Pakistan
5 Department of Pharmaceutics, Unaizah College of Pharmacy, Qassim University, Unaizah 51911, Saudi Arabia
6 Smart-Health Initiative (SHI) and Red Sea Research Center (RSRC), Division of Biological and Environmental Sciences and Engineering (BESE), King Abdullah University of Science and Technology (KAUST), Thuwal 23955, Saudi Arabia
* Correspondence: saleemhamza312@webmail.hzau.edu.cn (M.H.S.); ka.qurishe@qu.edu.sa (K.A.Q.)

Abstract: Here, we grow two different varieties of lentil (lentil-2009 and lentil-93) under different drought levels and with different applications of melatonin. Increasing the levels of soil water deficit significantly decreased numerous morphological and biochemical characteristics, including shoot length, total chlorophyll content, and transpiration rate, in both varieties of lentil. Contrastingly, drought stress increased the concentrations of malondialdehyde (MDA) and hydrogen peroxide (H_2O_2), and electrolyte leakage, an indicator of oxidative damage to membrane-bound organelles. The activities of enzymatic antioxidants and osmolytes were initially increased up to a drought level of 80% water field capacity (WFC) but gradually decreased with higher levels of drought stress (60% WFC) in the soil. At the same time, the results also showed that the lentil-2009 is more tolerant to drought stress than lentil-93. The negative impact of drought stress can be overcome by the application of melatonin. Melatonin increased plant growth and biomass, photosynthetic pigments, gas exchange characteristics, and enhanced the activities of various enzymatic and non-enzymatic antioxidants and proline content by decreasing oxidative stress. We conclude that foliar application of melatonin offers new possibilities for promoting lentil drought tolerance.

Keywords: melatonin; drought stress; legume family; oxidative stress; proline

check for updates

Citation: Yasmeen, S.; Wahab, A.; Saleem, M.H.; Ali, B.; Qureshi, K.A.; Jaremko, M. Melatonin as a Foliar Application and Adaptation in Lentil (*Lens culinaris* Medik.) Crops under Drought Stress. *Sustainability* **2022**, *14*, 16345. https://doi.org/10.3390/su142416345

Academic Editors: Maroun El Moujabber and Daniel El Chami

Received: 3 July 2022
Accepted: 5 September 2022
Published: 7 December 2022

1. Introduction

Environmental variations due to abiotic stresses, such as drought, heat, cold, and salinity, adversely affect and limit agricultural productivity in developing countries, including Pakistan [1,2]. About 33% of the world's agricultural land is facing water imbalance and promoting drought vulnerability, which may drastically decrease the growth and yield of plants [3–5]. Abiotic stresses, such as drought, can lead to alterations in plant growth and composition and a decrease in growth-related attributes, affecting photosynthetic machinery, which ultimately causes a reduction in the dry biomass of the plant as it is unable to accumulate essential nutrients from the soil [6–8]. In addition, plants are typically exposed to a myriad of biotic and abiotic stresses, including feeding from wild animals and insects, weed infestation, mechanical injury, diseases, low soil fertility, drought, salinity, and others, that can diminish the plant photosynthetic area, and thus the attained total plant biomass or grain yield [9,10]. Water deficiency-induced impairment in photosynthesis is attributed to damage of thylakoid membranes in chloroplasts because the lipid contents of cell membranes are susceptible to the reactive oxygen species (ROS) produced as a consequence of drought [11]. Stress conditions can disturb the dynamic equilibrium of ROS production, which promotes oxidative stress, membrane lipid peroxidation, and disrupts the structure and function of the cell membrane system [12–14]. The rate of photosynthesis decreases in many fodder grasses under drought stress, for example in *Festuca pratensis*, *Lolium perenne*, *Dactylis glomerata*, *Phleum pretense*, and *Arrhenatherum elatius* [15]. Activation of plant stress defensive mechanisms is important for survival.

Melatonin (N-acetyl-5-methoxytryptamine) is a ubiquitous bio-stimulating molecule, whose potential roles in plant growth, development, and stress responses have been progressively investigated in recent studies [16,17]. In plants, melatonin is involved in refining physiological processes such as photosynthesis, senescence, and reproduction [18,19]. Under stress conditions, melatonin mainly functions as a promoter of plant tolerance, and reduces oxidative damages by enhancing the antioxidant defense capacity of organelles, maintaining redox homeostasis [20,21]. It has been reported that exogenous applications of melatonin enhanced plant tolerance by providing protection against abiotic and biotic stresses such as drought stress [22–25]. Lentils (*Lens culinaris* Medik.) are a major cool seasonal food crop in India and the second most important winter-season legume after chickpea (*Cicer arietinum* L.) [26]. Lentils require low temperatures during vegetative growth, while at maturity, warm temperatures are required. The 'optimum' temperature for its best growth has been reported to be 18–30 °C [27]. Of the abiotic stresses experienced by lentils worldwide, drought and heat stress are considered the most important [28]. The susceptibility of lentils to hot and semiarid regions is supported by many studies [29–31]. The objective of the present study is to demonstrate the effect of drought stress, drought tolerance mechanisms, and management measures using melatonin application, for the alleviation of drought stress in lentil varieties. For this purpose, we designed a pot experiment using two varieties of lentil (lentil-2009 and lentil-93) to study: (i) the role of exogenous application of melatonin on growth and biomass, (ii) oxidative stress and antioxidant responses, and (iii) lentil sugar and osmolyte content in the drought-stressed environment. The results suggest that melatonin application may improve plant yield under drought-stressed conditions.

2. Materials and Methods

2.1. Seed Collection and Experimental Setup

Fresh and mature lentil seeds (*Lens culinaris* Medik.) named lentil-2009 and lentil-93 were collected from the Ayub Agriculture Research Institute (AARI) in Faisalabad, Pakistan. Both varieties were surface-sterilized with 0.1% bleaching powder for 10–20 min and washed gently with deionized water and sown in plastic pots (25×35 cm^2). The experiment was conducted at the Department of Botany, Government College University, Faisalabad, Pakistan (coordinates: 31.4162° N, 73.0699° E; elevation m a.s.l.: 186). The seedlings that emerged were thinned to maintain four almost uniform size seedlings per pot, and three pots were used for each treatment. After 21 days of seed germination, drought stress treatments, including control (100% water field capacity (WFC)) and drought-stressed (80% and 60% WFC), were initiated. Water levels were checked and maintained twice a week by weighing and adjusting the moisture level of the pots. After 30 days of drought stress, two levels of melatonin—control (no spray) and 3 mM—were applied as a foliar spray to stressed and non-stressed plants. Each pot was filled with 0.5 kg of sandy loam soil and five seeds were sown per pot. After one week of sowing, germination started. All plants in the glass house territory received natural light, with a day/night temperature of 35/40 °C and day/night humidity of 60/70%. The experiment was arranged in a completely randomized design (CRD) with three replications of each treatment.

2.2. Morphological Traits and Data Collection

After four weeks of foliar-applied melatonin, plant samples were collected from each replicate, and root and shoot fresh weight were measured separately using an electrical balance after harvest. Analyses of different biological parameters were performed at

Government College University, Pakistan. Shoot length was defined as the length of the plant from the surface growth medium line of the pot to the tip of the uppermost shoot, and root length was also measured. Shoot fresh weight was measured with the help of a digital weighing balance and root fresh weight was also measured. After that, plant samples were oven-dried for 1 h at 105 °C, then 65 °C for 72 h until the weight became uniform, and dry biomass was recorded. Roots were washed with distilled water and dipped in 20 mM of Na_2EDTA for 15–20 min, washed thrice with distilled water, and finally with deionized water, and then oven-dried for further analysis. The leaf in each treatment was picked at a rapid growth stage during 09:00–10:30 a.m. The sampled leaves were washed with distilled water, immediately placed in liquid nitrogen, and stored in a freezer at a low temperature (−80 °C) for further analysis.

2.3. Determination of Photosynthetic Pigments

Leaves were collected for determination of their chlorophyll and carotenoid contents. For chlorophylls, 0.1 g of fresh leaf sample was extracted with 8 mL of 95% acetone for 24 h at 4 °C in the dark. The absorbance was measured by a spectrophotometer (UV-2550; Shimadzu, Kyoto, Japan) at 646.6, 663.6, and 450 nm. Chlorophyll content was calculated by the standard method of Arnon [32].

Gas exchange parameters were also measured during the same days. Net photosynthesis (P*n*), leaf stomatal conductance (G*s*), transpiration rate (T*s*), and intercellular carbon dioxide concentration (C*i*) were measured from three different plants in each treatment group. Measurements were conducted between 11:30 and 13:30 on days with a clear sky. Rates of leaf P*n*, G*s*, T*s*, and C*i* were measured with a LI-COR gas exchange system (LI6400; LI-COR Biosciences, Lincoln, NE, USA) with a red–blue LED light source on the leaf chamber. In the LI-COR cuvette, CO_2 concentration was set as 380 mmol mol^{-1} and LED light intensity was set at 1000 mmol m^{-2} s^{-1}, which is the average saturation intensity for photosynthesis in lentils [33].

2.4. Determination of Oxidative Stress Indicators

The degree of lipid peroxidation was evaluated by measuring malondialdehyde (MDA) content. Briefly, 0.1 g of frozen leaves were ground at 4 °C in a mortar with 25 mL of 50 mM phosphate buffer solution (pH 7.8) containing 1% polyethene pyrrole. The homogenate was centrifuged at 10,000× *g* at 4 °C for 15 min. The mixtures were heated at 100 °C for 15–30 min and then quickly cooled in an ice bath. The absorbance of the supernatant was recorded by using a spectrophotometer (xMark™ Microplate Absorbance Spectrophotometer; Bio-Rad, Hercules, CA, USA) at wavelengths of 532, 600, and 450 nm. Lipid peroxidation was expressed as l mol g^{-1} by using the formula: 6.45 (A532–A600)–0.56 A450. Lipid peroxidation was measured by using a method previously published by Heath and Packer [34].

To estimate the H_2O_2 content of plant tissues (root and leaf), 3 mL of sample extract was mixed with 1 mL of 0.1% titanium sulfate in 20% (*v*/*v*) H_2SO_4 and centrifuged at 6000× *g* for 15 min. The yellow color intensity was evaluated at 410 nm. The H_2O_2 level was computed by an extinction coefficient of 0.28 $mmol^{-1}$ cm^{-1}. The contents of H_2O_2 were measured using the method presented by Jana and Choudhuri [35].

Stress-induced electrolyte leakage (EL) of the uppermost stretched leaves was determined by using the methodology of Dionisio-Sese and Tobita [36]. The leaves were cut into minor slices (5 mm length) and placed in test tubes containing 8 mL of distilled water. These tubes were incubated and transferred into a water bath for 2 h prior to measuring the initial electrical conductivity (EC_1). The samples were autoclaved at 121 °C for 20 min, and then cooled down to 25 °C before measuring the final electrical conductivity (EC_2). Electrolyte leakage was calculated by the following formula:

$$EL = (EC_1/EC_2) \times 100$$

2.5. Determination of Antioxidant Enzyme Activities

To evaluate enzyme activities, fresh leaves (0.5 g) were homogenized in liquid nitrogen and 5 mL of 50 mmol sodium phosphate buffer (pH 7.0), including 0.5 mmol ethylenediaminetetraacetic acid (EDTA) and 0.15 mol NaCl. The homogenate was centrifuged at 12,000× *g* for 10 min at 4 °C, and the supernatant was used for the measurement of superoxidase dismutase (SOD) and peroxidase (POD) activities. SOD activity was assayed in a 3 mL reaction mixture containing 50 mM sodium phosphate buffer (pH 7), 56 mM nitro blue tetrazolium, 1.17 mM riboflavin, 10 mM methionine, and 100 μL enzyme extract. Finally, the sample was measured with a spectrophotometer (xMark™ Microplate Absorbance Spectrophotometer; Bio-Rad, Hercules, CA, USA). Enzyme activity was measured using the method of Chen and Pan [37] and expressed as U g^{-1} FW.

POD activity in the leaves was estimated using the method of Sakharov and Ardila [38] by using guaiacol as the substrate. A reaction mixture (3 mL) containing 0.05 mL of enzyme extract, 2.75 mL of 50 mM phosphate buffer (pH 7.0), 0.1 mL of 1% H_2O_2, and 0.1 mL of 4% guaiacol solution was prepared. Increases in the absorbance at 470 nm because of guaiacol oxidation were recorded for 2 min.

Catalase (CAT) activity was analyzed according to Aebi [39]. The assay mixture (3.0 mL) was comprised of 100 μL enzyme extract, 100 μL H_2O_2 (300 mM) and 2.8 mL 50 mM phosphate buffer, with 2 mM ETDA (pH 7.0). CAT activity was measured from the decline in absorbance at 240 nm as a result of H_2O_2 loss (ε = 39.4 mM^{-1} cm^{-1}).

Ascorbate peroxidase (APX) activity was measured according to Nakano and Asada [40]. The mixture containing 100 μL enzyme extract, 100 μL ascorbate (7.5 mM), 100 μL H_2O_2 (300 mM), and 2.7 mL of 25 mM potassium phosphate buffer with 2 mM EDTA (pH 7.0) was used for measuring APX activity. The oxidation pattern of ascorbate was estimated from the variations in wavelength at 290 nm (ε = 2.8 mM^{-1} cm^{-1}).

2.6. Determination of Non-Enzymatic Antioxidant and Proline

Plant ethanol extracts were prepared for the determination of non-enzymatic antioxidants and some key osmolytes. For this purpose, 50 mg of plant dry material was homogenized with 10 mL of ethanol (80%) and filtered through Whatman No. 41 filter paper. The residue was re-extracted with ethanol and the two extracts were pooled together to a final volume of 20 mL. The determination of phenolics [41], ascorbic acid [42], and total sugars [43] was measured in the extracts.

Fresh leaf material (0.1 g) was mixed thoroughly in 5 mL of aqueous sulphosalicylic acid (3%). The mixture was centrifuged at 10,000× *g* for 15 min and a 1 mL aliquot was poured into a test tube having 1 mL of acidic ninhydrin and 1 mL of glacial acetic acid. The reaction mixture was first heated at 100 °C for 10 min and then cooled in an ice bath. The reaction mixture was extracted with 4 mL of toluene and test tubes were vortexed for 20 s and cooled. Thereafter, the light absorbance at 520 nm was measured by using a UV-VIS spectrophotometer (Hitachi U-2910, Tokyo, Japan). Free proline content was determined on the basis of a standard curve at 520 nm absorbance and expressed as μmol (g FW)$^{-1}$ [44].

2.7. Statistical Analysis

Statistical analysis was performed with analysis of variance (ANOVA) by using the statistical program Co-Stat version 6.2 (Cohorts Software, 2003, Monterey, CA, USA). All the data obtained were tested by one-way ANOVA. Thus, the differences between treatments were determined by using ANOVA, and the least significant difference test ($p < 0.05$) was used for multiple comparisons between treatment means. Logarithmic or inverse transformations were performed for data normalization, where necessary, prior to analysis. Pearson's correlation analysis was performed to quantify relationships between various analyzed variables. Graphs were drawn in Origin-Pro 2017 (Systat Software Inc., San Jose, CA, USA). RStudio was used to calculate Pearson's correlation.

3. Results

3.1. Impact of Melatonin Application on Plant Growth and Photosynthesis in Lentil Varieties Grown under Drought Conditions

We measured various growth and photosynthetic parameters in both varieties of lentils grown under the varying levels of drought (100%, 80%, and 60% WFC), both with and without the application of melatonin. Morphological traits are presented in Figure 1 and data regarding photosynthetic pigments and gas exchange attributes are presented in Figure 2. Decreasing soil water levels significantly decreased shoot length, root length, shoot fresh weight, shoot dry weight, root fresh weight, root dry weight, chlorophyll a, chlorophyll b, total chlorophyll, carotenoid, net photosynthesis, stomatal conductance, and transpiration rate in lentil varieties, compared to 100% WFC. Under the same levels of drought in the soil, lentil-2009 showed better growth and development compared to the lentil-93. Melatonin also increased plant growth and biomass and photosynthetic pigments even in the plants grown in the drought-stressed environment. The application of melatonin increased shoot length, root length, shoot fresh weight, shoot dry weight, root fresh weight, root dry weight, chlorophyll a, chlorophyll b, total chlorophyll, carotenoid, net photosynthesis, stomatal conductance, and transpiration rate. However, drought stress did not affect the intercellular CO_2 in both varieties of lentil and the application of melatonin also did not have any significant effect on the levels of intercellular CO_2.

3.2. Impact of Melatonin Application on Oxidative Stress and Antioxidant Enzymes in Lentil Varieties Grown under Drought Conditions

We measured various markers of oxidative stress in both lentil varieties grown in drought conditions, including malondialdehyde, hydrogen peroxide, and electrolyte leakage (EL) (Figure 3A–C). We also measured antioxidant capacity in the form of super oxidase dismutase (SOD), peroxidase (POD), catalase (CAT), and ascorbate peroxidase (APX) (Figure 3D–G, respectively). Increasing levels of drought stress in the soil increased the concentrations of MDA, EL, and H_2O_2 in the tissues of both lentil varieties. The maximum increase was observed by the lentil-sensitive variety, i.e., lentil-2009, compared to the control. The activities of various antioxidant enzymes (SOD, POD, CAT, and APX) initially increased up to a water-deficit level of 80% WFC, but then decreased significantly at the highest level of drought in the soil. In addition, the activities of enzymatic antioxidants were higher in the drought-tolerant varieties, i.e., lentil-2009, compared to the drought-sensitive variety, i.e., lentil-93. However, we also noticed that the application of melatonin decreased the concentrations of MDA, EL, and H_2O_2 in both varieties of lentil, compared to those which did not have melatonin applied. Similarly, increasing levels of melatonin significantly increased the activities of SOD, POD, CAT, and APX in the leaves of both lentil varieties, compared to those which were not treated with melatonin (Figure 3).

3.3. Impact of Melatonin Application on Osmolytes and Proline of Enzymatic Antioxidants in Lentil Varieties Grown under Drought Conditions

We also measured the contents of phenolics, ascorbic acid, soluble sugar, and proline from both varieties of lentil grown under varying levels of drought stress, i.e., 100%, 80%, and 60% WFC, with or without the application of melatonin. Decreasing levels of water (80% and 60% WFC) in the soil significantly induced ($p < 0.05$) the phenolic acid, ascorbic acid, soluble sugar, and proline content of both varieties of lentil, compared to control plants (Figure 4). Phenolics and ascorbic acid contents first increased up to a drought level of 80% WFC but gradually decreased with more drought in the soil (60% WFC). The application of melatonin also increased the content of phenolics, ascorbic acid, soluble sugar, and proline in both varieties of lentil compared to plants grown without the application melatonin.

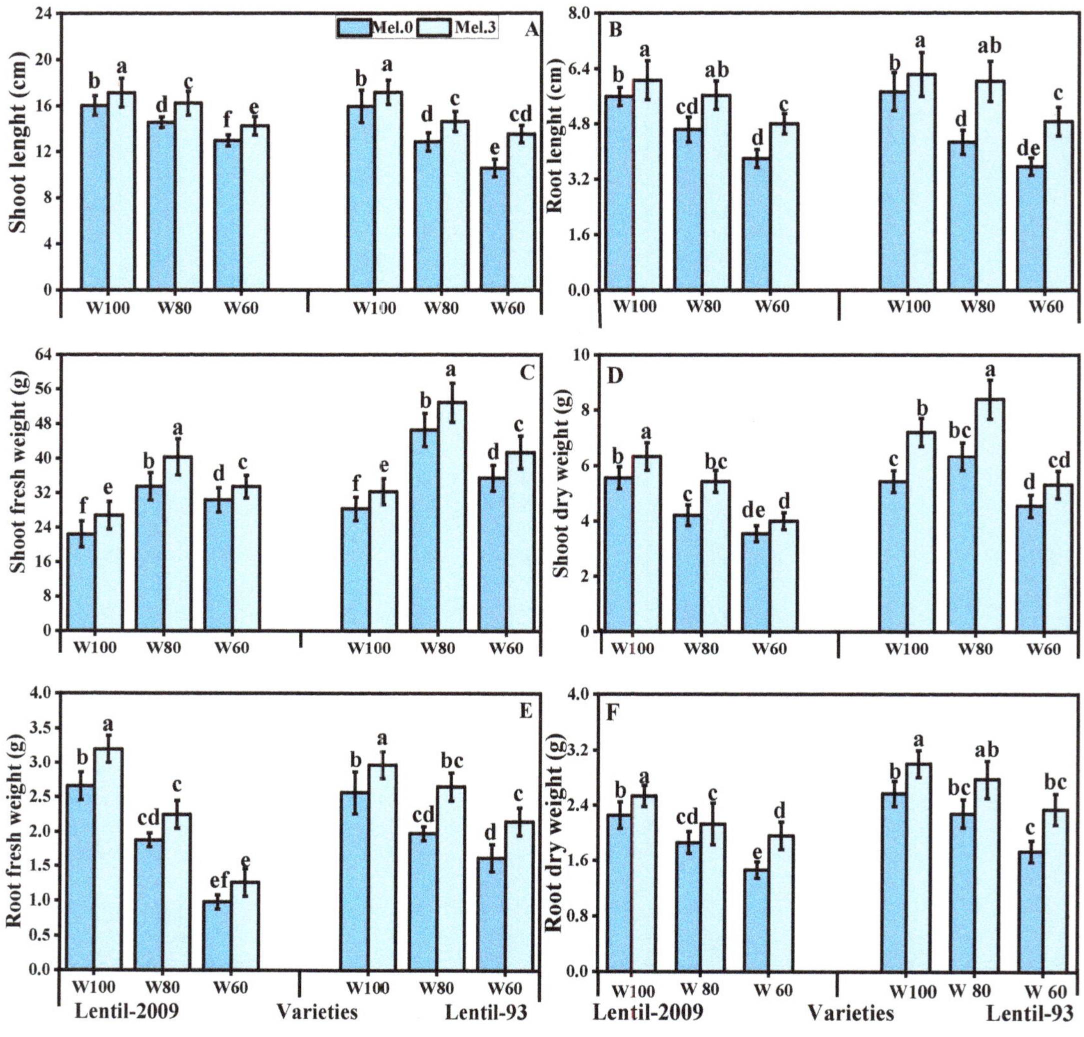

Figure 1. The impact of melatonin application on different morphological traits. The graphs show shoot length (**A**), root length (**B**), shoot fresh weight (**C**), shoot dry weight (**D**), root fresh weight (**E**), and root dry weight (**F**) under various water-deficit conditions (W100 (100% WFC), W80 (80% WFC), and W60 (60% WFC)) in both varieties of lentil (lentil-2009 and lentil-93), either with or without melatonin (0 or 3 mM). Means sharing similar letter(s) within a column for each parameter do not differ significantly at $p < 0.05$. Data in the figures are means of four repeats (n = 4) of just one harvest of lentil varieties ± standard deviation (SD). Different lowercase letters on the error bars indicate significant difference between the treatments.

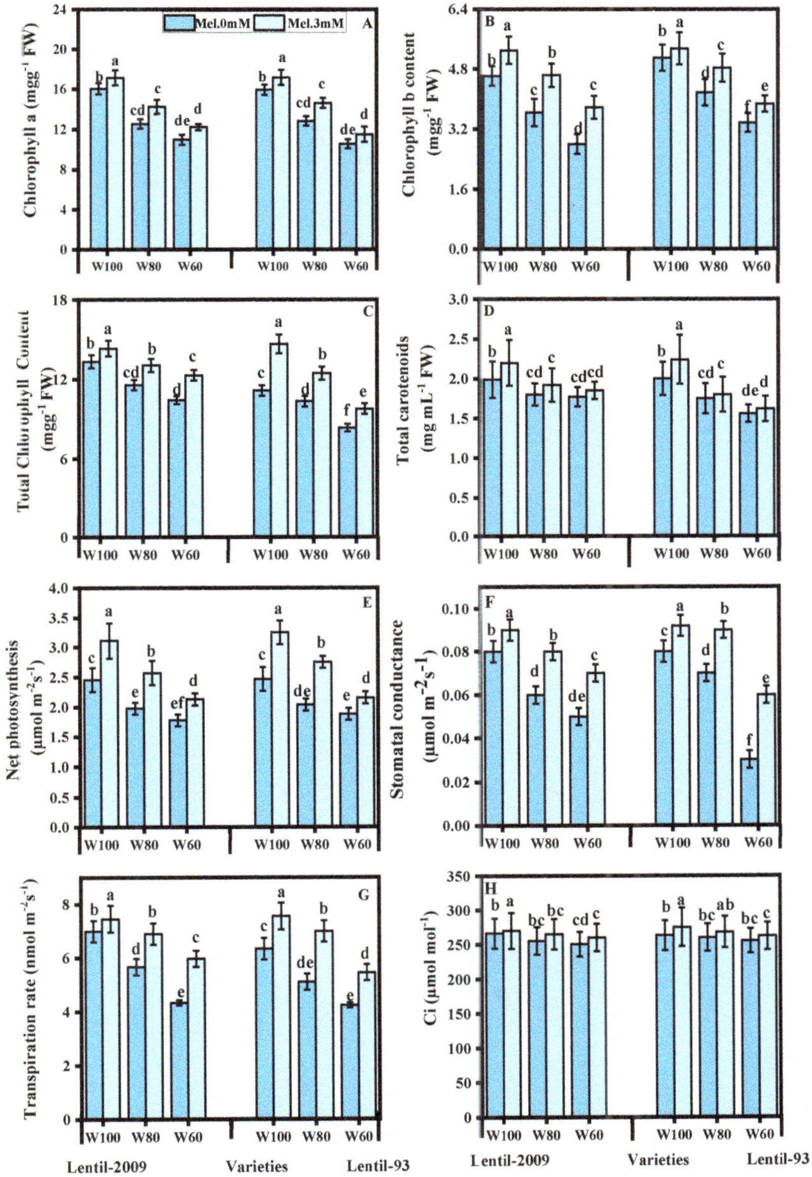

Figure 2. Impact of melatonin application on different photosynthetic pigments and gas exchange attributes. The graphs show chlorophyll "a" (**A**), chlorophyll "b" (**B**), total chlorophyll content (**C**), total carotenoid content (**D**), net photosynthesis (**E**), stomatal conductance (**F**), transpiration rate (**G**), and intercellular CO_2 (**H**) under various water-deficit conditions (W100 (100% WFC), W80 (80% WFC), and W60 (60% WFC)) in both varieties of lentil (lentil-2009 and lentil-93). Means sharing similar letter(s) within a column for each parameter do not differ significantly at $p < 0.05$. Data in the figures are means of four repeats (n = 4) of just one harvest of lentil varieties ± standard deviation (SD). Different lowercase letters on the error bars indicate significant difference between the treatments.

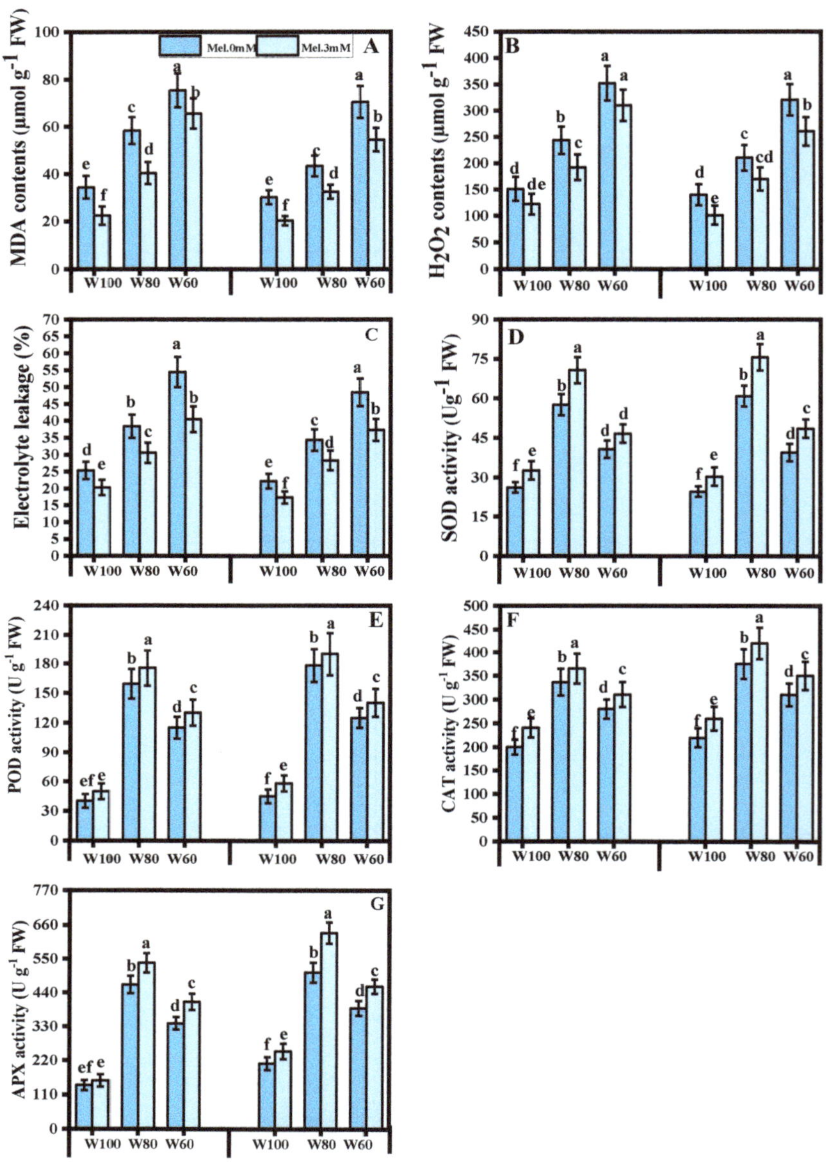

Figure 3. Impact of melatonin application on different markers of oxidative stress. The graphs show malondialdehyde content (**A**), hydrogen peroxide content (**B**), electrolyte leakage (**C**), and enzymatic antioxidants superoxidase dismutase (SOD) (**D**), peroxidase (POD) (**E**), catalase (CAT) (**F**), and ascorbate peroxidase (APX) (**G**) under various water conditions (W100 (100% WFC), W80 (80% WFC), and W60 (60% WFC)) in both varieties of lentil. Means sharing similar letter(s) within a column for each parameter do not differ significantly at $p < 0.05$. Data in the figures are means of four repeats (n = 4) of just one harvest of lentil varieties ± standard deviation (SD). Different lowercase letters on the error bars indicate significant difference between the treatments.

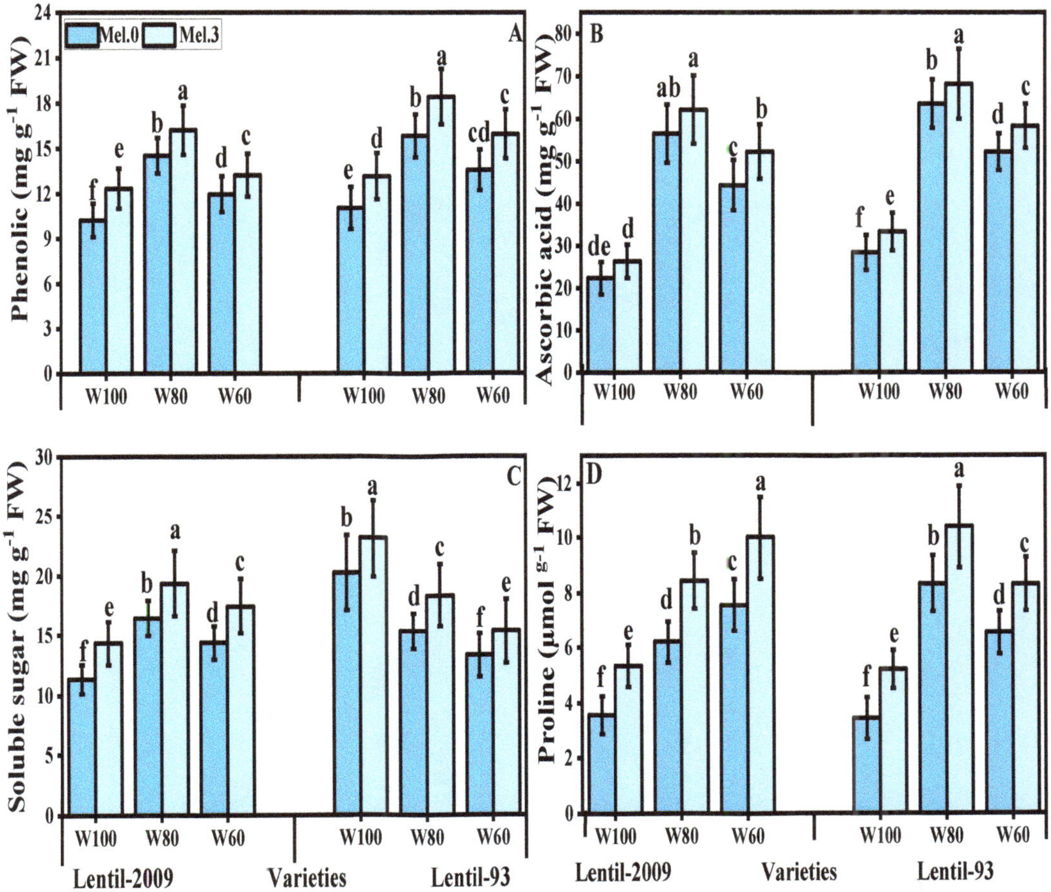

Figure 4. Impact of melatonin application on phenolics content (**A**), ascorbic acid (**B**), soluble sugar (**C**), and proline (**D**) under various water conditions (W100 (100% WFC), W80 (80% WFC), and W60 (60% WFC)) in both varieties of lentil. Means sharing similar letter(s) within a column for each parameter do not differ significantly at $p < 0.05$. Data in the figures are means of four repeats (n = 4) of just one harvest of lentil varieties ± standard deviation (SD). Different lowercase letters on the error bars indicate significant difference between the treatments.

3.4. Relationship between Various Growth Parameters of Lentil

A Pearson's correlation graph depicts the relationship between various growth and physiological parameters in lentils (lentil-2009) (Figure 5). Malondialdehyde was positively correlated with electrolyte leakage, hydrogen peroxide, catalase, ascorbate peroxide, superoxide dismutase, peroxidase, phenolics, ascorbic acid, sugar, and proline. Malondialdehyde was negatively correlated with shoot length, root length, shoot fresh weight, shoot dry weight, root fresh weight, root dry weight, chlorophyll a, chlorophyll b, total chlorophyll, carotenoid, net photosynthesis, stomatal conductance, and transpiration rate. Similarly, electrolyte leakage was positively correlated with malondialdehyde, hydrogen peroxide, catalase, ascorbate peroxide, superoxide dismutase, peroxidase, phenolics, ascorbic acid, sugar, and proline, while negatively correlated with shoot length, root length, shoot fresh weight, shoot dry weight, root fresh weight, root dry weight, chlorophyll a, chlorophyll b, total chlorophyll, carotenoid, net photosynthesis, stomatal conductance, and transpiration rate. This relationship showed a close connection between various attributes of lentil.

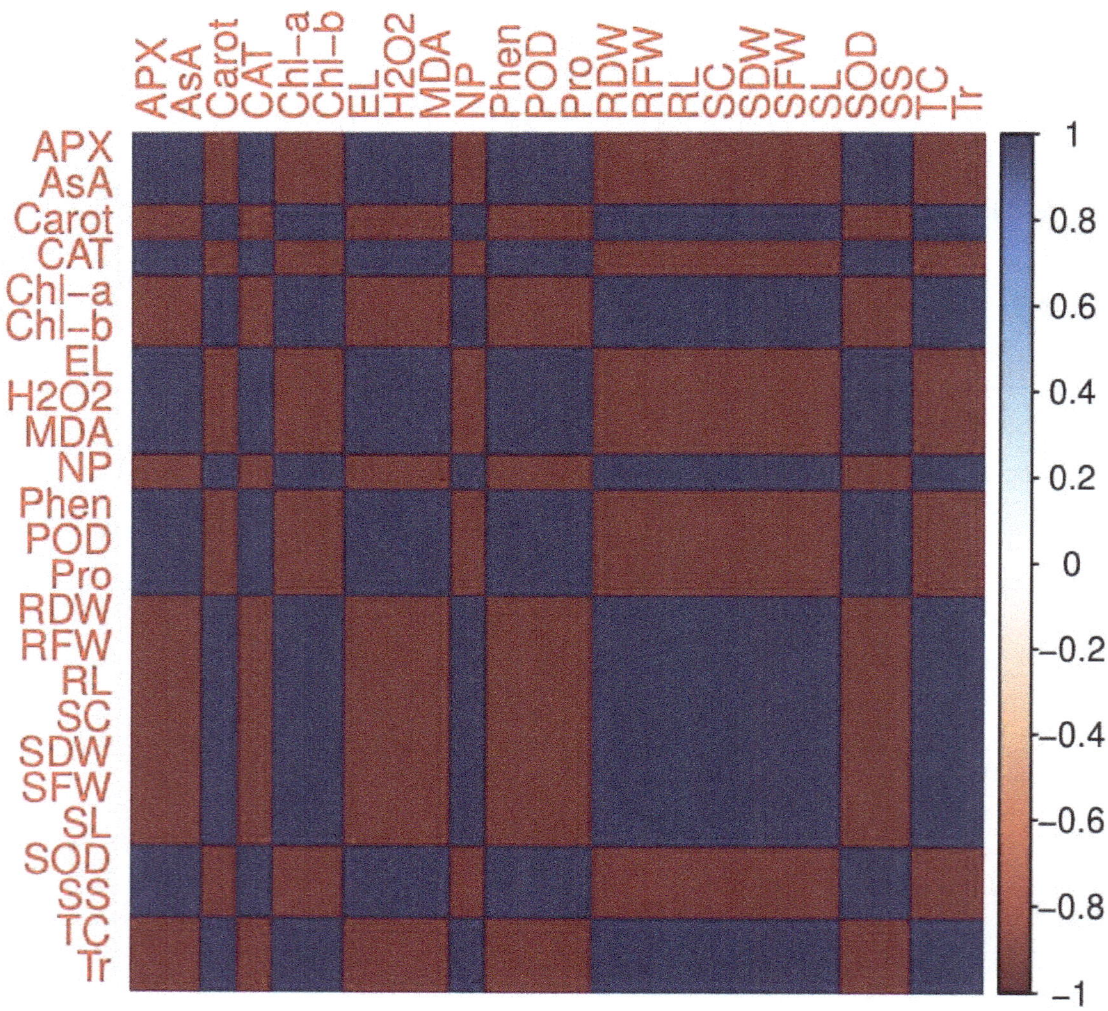

Figure 5. Correlation between various lentil growth parameters under the different water-deficit conditions with or without the application of melatonin. Different abbreviations used are as follows: (MDA) malondialdehyde contents, (EL) electrolyte leakage percentage, (H_2O_2) hydrogen peroxide content, (CAT) catalase activity, (APX) ascorbate peroxide activity, (SOD) superoxide dismutase activity, (POD) peroxidase activity, (Phe) phenolic content, (AsA) ascorbic acid content, (SS) sugar content, (Pro) proline contents, (SL) shoot length, (RL) root length, (SFW) shoot fresh weight, (SDW) shoot dry weight, (RFW) root fresh weight, (RDW) root dry weight, (Chl-a) chlorophyll a content, (Chl-b) chlorophyll b content, (TC) total chlorophyll content, (Carot) carotenoid content, (NP) net photosynthesis, (SC) stomatal conductance, and (Tr) transpiration rate.

4. Discussion

Drought stress adversely affects morphological aspects of plants, such as early germination, plant height, relative root length, root diameter, the total biomass of leaves and roots, the number of leaves, and branch number [11]. We assessed the influence of melatonin on growth, osmolyte accumulation, and enzymatic and non-enzymatic antioxidants in lentil varieties under drought stress conditions. Drought-induced reductions might be due to photosynthesis, respiration, cell extension, and enzymatic activities [6,8] because drought-stressed plants had a diminished number of leaves, and the development of new leaves, stems, and leaflets, and leaf area were reduced compared to those in the control plants. This might be attributed to the impact of water stress on the physiological cycles in plants, such as photosynthesis, leaf zone extension, nucleic acid metabolism, protein synthesis, and the partitioning of assimilates [30,45,46]. In addition to this water deficiency, decreased photosynthesis restricts the mechanism of cell development and cell enlargement closed stomata [47], eventually reducing the yield [27,48].

Water-deficient environments are generally known to initiate oxidative stress in plants by the production of extra reactive oxygen species (ROS) [49–51] and antioxidative enzymes that play a protective role in reducing metal toxicity by scavenging ROS [52,53]. Excessive reactive oxygen species (ROS) production causes oxidative stress, as reported for many crops under drought stress treatment, and is likely to be initiated by molecular oxygen excitation (O_2) to generate singlet oxygen or by electron transfer to O_2 and genesis of free radicals, i.e., O^{2-} and OH^- [11,27]. Plant responses to oxidative stress also depend upon plant species and cultivars, and ROS are removed in plants by a variety of antioxidant enzymes, such as SOD, POD, CAT, and APX [54,55]. The increase in the activities of antioxidant enzymes is concomitant with the generation of extra ROS. It has also been reported that an increase in the activities of various antioxidant enzymes under environmental stress conditions is due to a reduction in the glutathione level [56,57]. Plants produce a variety of antioxidants (ascorbic acid, glutathione, proline, carotenoids, phenolic acids, and flavonoids) that improve tolerance against drought stress [58,59]. Phenolics are potent antioxidants against drought-induced oxidative damage and efficiently scavenge ROS [60]. Proline accumulation in plant tissue/organs is a response to drought stress, which might be associated with signal transduction and prevents membrane distortion [61]. Studies related to our results found that drought increased oxidative stress by increasing MDA, EL, and H_2O_2 in wheat [46], cucumber [62], and canola [63].

Many efforts have been made to mitigate the hazardous impacts of drought stress by using various plant growth regulators, such as salicylic acid [64], polyamine [65], abscisic acid [66], glycine betaine [67], and melatonin [18,23,24]. Melatonin is an amphiphilic biological (indolamine) hormone found throughout the animal and plant kingdoms. Melanin is produced by the shikimate pathway in chloroplasts from tryptophan [68,69]. Melatonin plays a vital part in plant growth and development by regulating plant physiology and root regeneration [17], antioxidant activity [16], photosynthesis [22], senescence of leaves [70], and immunological enhancement [71]. Melatonin might also boost the antioxidative capacity to fortify a variety of plant species from various abiotic stresses, especially drought stress [23,70], by altering the expression of salt tolerance genes, upregulating antioxidant enzymes, and directly scavenging ROS. Previously, melatonin has been reported to enhance resistance against drought stress in various crops, including barley [72], soybean [73], and tomato [74]. Many possibilities could be suggested to explain how melatonin can help plants to alleviate the adverse effects of various environmental conditions. One of the most important defensive mechanisms in this respect is the protection of the photosynthetic apparatus via improving the scavenging efficiency of reactive oxygen species (ROS) and reducing the stress-induced oxidative damages [71,75]. Under drought stress, melatonin joins in the readjustment of the cell osmotic potential and accumulation of osmolytes such as proline and soluble sugars [76]. Moreover, melatonin can maintain the water status of water-stressed plants through regulating stomatal movement [20] and modulating a broad spectrum of anatomical aspects, i.e., preserving the integrity of cell membranes [74] and increasing the cuticle and/or wax accumulation [77]. In addition, it has been confirmed that during the exposure to stress, melatonin has a close linkage in plant signal transduction and can trigger cascades of reprogramming primary metabolites, transcriptomes, and proteomes [16].

5. Conclusions

In this study, we investigated the influence of melatonin on the growth of lentil varieties in well-watered (100% WFC) and water-depleted conditions (80% and 60% WFC). Drought conditions had a harsh impact on plant growth, and as well as influencing photosynthetic measurements, they induced oxidative stress, antioxidant enzymes, and osmoprotectants. The application of melatonin is useful in alleviating oxidative stress by accelerating the activities of antioxidants and increasing the content of soluble sugars and other enzymatic and non-enzymatic antioxidants, even in drought conditions. Hence, we suggest that the application of melatonin offers new opportunities by promoting greater drought tolerance and enhancing the capacity to adapt to future environmental challenges.

Author Contributions: Conceptualization, S.Y.; formal analysis, A.W. and M.J.; funding acquisition, B.A., K.A.Q. and M.J.; investigation, S.Y.; methodology, S.Y.; project administration, M.J.; resources, M.H.S. and M.J.; software, A.W., B.A. and K.A.Q.; validation, M.H.S. and B.A.; visualization, K.A.Q.; M.H.S., S.Y. and A.W.; writing—review and editing, A.W., M.H.S., B.A., K.A.Q. and M.J. All authors have read and agreed to the published version of the manuscript.

Funding: The APC was funded by the King Abdullah University of Science and Technology (KAUST), Thuwal, Saudi Arabia.

Institutional Review Board Statement: Not applicable.

Informed Consent Statement: Not applicable.

Data Availability Statement: Not applicable.

Conflicts of Interest: The authors declare no conflict of interest.

References

1. Nawaz, M.; Wang, X.; Saleem, M.H.; Khan, M.H.U.; Afzal, J.; Fiaz, S.; Ali, S.; Ishaq, H.; Khan, A.H.; Rehman, N.; et al. Deciphering Plantago ovata Forsk Leaf Extract Mediated Distinct Germination, Growth and Physio-Biochemical Improvements under Water Stress in Maize (*Zea mays L.*) at Early Growth Stage. *Agronomy* **2021**, *11*, 1404. [CrossRef]
2. Ali, B.; Hafeez, A.; Ahmad, S.; Javed, M.A.; Sumaira Afridi, M.S.; Dawoud, T.M.; Almaary, K.S.; Muresan, C.C.; Marc, R.A.; Alkhalifah, D.H.M.; et al. Bacillus thuringiensis PM25 ameliorates oxidative damage of salinity stress in maize via regulating growth, leaf pigments, antioxidant defense system, and stress responsive gene expression. *Front. Plant Sci.* **2022**, *13*, 921668. [CrossRef] [PubMed]
3. Siddiqui, H.; Ahmed, K.B.M.; Sami, F.; Hayat, S. Silicon nanoparticles and plants: Current knowledge and future perspectives. In *Sustainable Agriculture Reviews 41*; Springer: Cham, Switzerland, 2020; pp. 129–142.
4. Afridi, M.S.; Javed, M.A.; Ali, S.; De Medeiros, F.H.V.; Ali, B.; Salam, A.; Sumaira; Marc, R.A.; Alkhalifah, D.H.M.; Selim, S.; et al. New opportunities in plant microbiome engineering for increasing agricultural sustainability under stressful conditions. *Front. Plant Sci.* **2022**, *13*, 899464. [CrossRef] [PubMed]
5. Ali, B.; Hafeez, A.; Javed, M.A.; Ahmad, S.; Afridi, M.S.; Sumaira Nadeem, M.; Khan, A.U.R.; Malik, A.; Ullah, A.; Alwahibi, M.S.; et al. Bacterial-mediated salt tolerance in maize: Insights into plant growth promotion, antioxidant defense system, oxidative stress, and surfactant production. *Front. Plant Sci.* **2022**, *13*, 978291. [CrossRef]
6. Parveen, A.; Liu, W.; Hussain, S.; Asghar, J.; Perveen, S.; Xiong, Y. Silicon priming regulates morpho-physiological growth and oxidative metabolism in maize under drought stress. *Plants* **2019**, *8*, 431. [CrossRef]
7. Akram, N.A.; Shahbaz, M.; Ashraf, M. Relationship of photosynthetic capacity and proline accumulation with the growth of differently adapted populations of two potential grasses (*Cynodon dactylon* (L.) Pers. and *Cenchrus ciliaris* L.) to drought stress. *Pak. J. Bot* **2007**, *39*, 777–786.
8. Amna Ali, B.; Azeem, M.A.; Qayyum, A.; Mustafa, G.; Ahmad, M.A.; Javed, M.T.; Chaudhary, H.J. Bio-Fabricated Silver Nanoparticles: A Sustainable Approach for Augmentation of Plant Growth and Pathogen Control. In *Sustainable Agriculture Reviews 53*; Springer: Cham, Switzerland, 2021; pp. 345–371.
9. Bauddh, K.; Singh, R.P. Growth, tolerance efficiency and phytoremediation potential of *Ricinus communis* (L.) and *Brassica juncea* (L.) in salinity and drought affected cadmium contaminated soil. *Ecotoxicol. Environ. Saf.* **2012**, *85*, 13–22. [CrossRef]
10. Solanki, M.K.; Solanki, A.C.; Rai, S.; Srivastava, S.; Kashyap, B.K.; Divvela, P.K.; Kumar, S.; Yandigeri, M.S.; Kashyap, P.L.; Shrivastava, A.K.; et al. Functional interplay between antagonistic bacteria and Rhizoctonia solani in the tomato plant rhizosphere. *Front. Microbiol.* **2022**, *13*, 990850. [CrossRef]

11. Wahab, A.; Abdi, G.; Saleem, M.H.; Ali, B.; Ullah, S.; Shah, W.; Mumtaz, S.; Yasin, G.; Muresan, C.C.; Marc, R.A. Plants' Physio-Biochemical and Phyto-Hormonal Responses to Alleviate the Adverse Effects of Drought Stress: A Comprehensive Review. *Plants* **2022**, *11*, 1620. [CrossRef]
12. Saleem, M.H.; Kamran, M.; Zhou, Y.; Parveen, A.; Rehman, M.; Ahmar, S.; Malik, Z.; Mustafa, A.; Anjum, R.M.A.; Wang, B.; et al. Appraising growth, oxidative stress and copper phytoextraction potential of flax (*Linum usitatissimum* L.) grown in soil differentially spiked with copper. *J. Environ. Manag.* **2020**, *257*, 109994. [CrossRef]
13. Saleem, M.H.; Ali, S.; Rehman, M.; Rana, M.S.; Rizwan, M.; Kamran, M.; Imran, M.; Riaz, M.; Soliman, M.H.; Elkelish, A.; et al. Influence of phosphorus on copper phytoextraction via modulating cellular organelles in two jute (*Corchorus capsularis* L.) varieties grown in a copper mining soil of Hubei Province, China. *Chemosphere* **2020**, *248*, 126032. [CrossRef] [PubMed]
14. Zainab, N.; Amna; Khan, A.A.; Azeem, M.A.; Ali, B.; Wang, T.; Shi, F.; Alghanem, S.M.; Hussain Munis, M.F.; Hashem, M.; et al. PGPR-Mediated Plant Growth Attributes and Metal Extraction Ability of Sesbania sesban L. in Industrially Contaminated Soils. *Agronomy* **2021**, *11*, 1820. [CrossRef]
15. Ghafar, M.A.; Akram, N.A.; Saleem, M.H.; Wang, J.; Wijaya, L.; Alyemeni, M.N. Ecotypic Morphological and Physio-Biochemical Responses of Two Differentially Adapted Forage Grasses, *Cenchrus ciliaris* L. and Cyperus arenarius Retz. to Drought Stress. *Sustainability* **2021**, *13*, 8069. [CrossRef]
16. Mehmood, S.; Khatoon, Z.; Amna Ahmad, I.; Muneer, M.A.; Kamran, M.A.; Ali, J.; Ali, B.; Chaudhary, H.J.; Munis, M.F. Bacillus sp. PM31 harboring various plant growth-promoting activities regulates Fusarium dry rot and wilt tolerance in potato. *Arch. Agron. Soil Sci.* **2021**, *2021*, 1971654.
17. Chen, L.; Liu, L.; Lu, B.; Ma, T.; Jiang, D.; Li, J.; Zhang, K.; Sun, H.; Zhang, Y.; Bai, Z.; et al. Exogenous melatonin promotes seed germination and osmotic regulation under salt stress in cotton (*Gossypium hirsutum* L.). *PLoS ONE* **2020**, *15*, e0228241. [CrossRef] [PubMed]
18. Sharma, A.; Wang, J.; Xu, D.; Tao, S.; Chong, S.; Yan, D.; Li, Z.; Yuan, H.; Zheng, B. Melatonin regulates the functional components of photosynthesis, antioxidant system, gene expression, and metabolic pathways to induce drought resistance in grafted Carya cathayensis plants. *Sci. Total Environ.* **2020**, *713*, 136675. [CrossRef] [PubMed]
19. Dawood, M.G.; El-Awadi, M.E. Alleviation of salinity stress on Vicia faba L. plants via seed priming with melatonin. *Acta Biológica Colomb.* **2015**, *20*, 223–235. [CrossRef]
20. Bhat, J.A.; Faizan, M.; Bhat, M.A.; Huang, F.; Yu, D.; Ahmad, A.; Bajguz, A.; Ahmad, P. Defense interplay of the zinc-oxide nanoparticles and melatonin in alleviating the arsenic stress in soybean (*Glycine max* L.). *Chemosphere* **2022**, *288*, 132471. [CrossRef]
21. Ali, M.; Kamran, M.; Abbasi, G.H.; Saleem, M.H.; Ahmad, S.; Parveen, A.; Malik, Z.; Afzal, S.; Ahmar, S.; Dawar, K.M.; et al. Melatonin-Induced Salinity Tolerance by Ameliorating Osmotic and Oxidative Stress in the Seedlings of Two Tomato (*Solanum lycopersicum* L.) Cultivars. *J. Plant Growth Regul.* **2020**, *40*, 2236–2248. [CrossRef]
22. Mohamed, I.A.; Shalby, N.; MA El-Badri, A.; Saleem, M.H.; Khan, M.N.; Nawaz, M.A.; Qin, M.; Agami, R.A.; Kuai, J.; Wang, B. Stomata and Xylem Vessels Traits Improved by Melatonin Application Contribute to Enhancing Salt Tolerance and Fatty Acid Composition of *Brassica napus* L. Plants. *Agronomy* **2020**, *10*, 1186. [CrossRef]
23. Khan, M.N.; Zhang, J.; Luo, T. Liu, J.; Rizwan, M.; Fahad, S.; Xu, Z.; Hu, L. Seed priming with melatonin coping drought stress in rapeseed by regulating reactive oxygen species detoxification: Antioxidant defense system, osmotic adjustment, stomatal traits and chloroplast ultrastructure perseveration. *Ind. Crops Prod.* **2019**, *140*, 111597. [CrossRef]
24. Wang, P.; Sun, X.; Li, C.; Wei, Z.; Liang, D.; Ma, F. Long-term exogenous application of melatonin delays drought-induced leaf senescence in apple. *J. Pineal Res.* **2013**, *54*, 292–302. [CrossRef]
25. Bai, Y.; Xiao, S.; Zhang, Z.; Zhang, Y.; Sun, H.; Zhang, K.; Wang, X.; Bai, Z.; Li, C.; Liu, L. Melatonin improves the germination rate of cotton seeds under drought stress by opening pores in the seed coat. *PeerJ* **2020**, *8*, e9450. [CrossRef]
26. Al-Darkazli, M.A.; Al-Saedy, A.J.; Al-Saadi, H.A. The Influence of Interaction between the Phosphorus Fertilizer and Gibberellin on Elements Content of Lentil Crop (Lens culinaris Medic.). *Al-Nahrain J. Sci.* **2011**, *14*, 115–120.
27. Sehgal, A.; Sita, K.; Kumar, J.; Kumar, S.; Singh, S.; Siddique, K.H.; Nayyar, H. Effects of drought, heat and their interaction on the growth, yield and photosynthetic function of lentil (Lens culinaris Medikus) genotypes varying in heat and drought sensitivity. *Front. Plant Sci.* **2017**, *8*, 1776. [CrossRef]
28. Islam, F.; Yasmeen, T.; Ali, Q.; Mubin, M.; Ali, S.; Arif, M.S.; Hussain, S.; Riaz, M.; Abbas, F. Copper-resistant bacteria reduces oxidative stress and uptake of copper in lentil plants: Potential for bacterial bioremediation. *Environ. Sci. Pollut. Res.* **2016**, *23*, 220–233. [CrossRef] [PubMed]
29. Farooq, M.; Romdhane, L.; Al Sulti, M.K.; Rehman, A.; Al-Busaidi, W.M.; Lee, D.J. Morphological, physiological and biochemical aspects of osmopriming-induced drought tolerance in lentil. *J. Agron. Crop Sci.* **2020**, *206*, 176–186. [CrossRef]
30. El Haddad, N.; Choukri, H.; Ghanem, M.E.; Smouni, A.; Mentag, R.; Rajendran, K.; Hejjaoui, K.; Maalouf, F.; Kumar, S. High-temperature and drought stress effects on growth, yield and nutritional quality with transpiration response to vapor pressure deficit in lentil. *Plants* **2021**, *11*, 95. [CrossRef] [PubMed]
31. Sehgal, A.; Sita, K.; Bhandari, K.; Kumar, S.; Kumar, J.; Vara Prasad, P.; Siddique, K.H.; Nayyar, H. Influence of drought and heat stress, applied independently or in combination during seed development, on qualitative and quantitative aspects of seeds of lentil (Lens culinaris Medikus) genotypes, differing in drought sensitivity. *Plant Cell Environ.* **2019**, *42*, 198–211. [CrossRef]
32. Arnon, D.I. Copper enzymes in isolated chloroplasts. Polyphenoloxidase in *Beta vulgaris*. *Plant Physiol.* **1949**, *24*, 1–15. [CrossRef]
33. Austin, R.B. Prospects for genetically increasing the photosynthetic capacity of crops. *Plant Biol.* **1990**, *10*, 395–409.

34. Heath, R.L.; Packer, L. Photoperoxidation in isolated chloroplasts: I. Kinetics and stoichiometry of fatty acid peroxidation. *Arch. Biochem. Biophys.* **1968**, *125*, 189–198. [CrossRef] [PubMed]
35. Jana, S.; Choudhuri, M.A. Glycolate metabolism of three submersed aquatic angiosperms: Effect of heavy metals. *Aquat. Bot.* **1981**, *11*, 67–77. [CrossRef]
36. Dionisio-Sese, M.L.; Tobita, S. Antioxidant responses of rice seedlings to salinity stress. *Plant Sci.* **1998**, *135*, 1–9. [CrossRef]
37. Chen, C.-N.; Pan, S.-M. Assay of superoxide dismutase activity by combining electrophoresis and densitometry. *Bot. Bull. Acad. Sin.* **1996**, *37*, 107–111.
38. Sakharov, I.Y.; Ardila, G.B. Variations of peroxidase activity in cocoa (*Theobroma cacao* L.) beans during their ripening, fermentation and drying. *Food Chem.* **1999**, *65*, 51–54. [CrossRef]
39. Aebi, H. Catalase in vitro. In *Methods in Enzymology*; Elsevier: Amsterdam, The Netherlands, 1984; Volume 105, pp. 121–126.
40. Nakano, Y.; Asada, K. Hydrogen peroxide is scavenged by ascorbate-specific peroxidase in spinach chloroplasts. *Plant Cell Physiol.* **1981**, *22*, 867–880.
41. Bray, H.; Thorpe, W. Analysis of phenolic compounds of interest in metabolism. *Methods Biochem. Anal.* **1954**, *1*, 27–52.
42. Azuma, K.; Nakayama, M.; Koshioka, M.; Ippoushi, K.; Yamaguchi, Y.; Kohata, K.; Yamauchi, Y.; Ito, H.; Higashio, H. Phenolic antioxidants from the leaves of *Corchorus olitorius* L. *J. Agric. Food Chem.* **1999**, *47*, 3963–3966. [CrossRef]
43. Dubois, M.; Gilles, K.A.; Hamilton, J.K.; Rebers, P.t.; Smith, F. Colorimetric method for determination of sugars and related substances. *Anal. Chem.* **1956**, *28*, 350–356. [CrossRef]
44. Bates, L.S.; Waldren, R.P.; Teare, I. Rapid determination of free proline for water-stress studies. *Plant Soil* **1973**, *39*, 205–207. [CrossRef]
45. Sarker, U.; Oba, S. Drought stress effects on growth, ROS markers, compatible solutes, phenolics, flavonoids, and antioxidant activity in Amaranthus tricolor. *Appl. Biochem. Biotechnol.* **2018**, *186*, 999–1016. [CrossRef] [PubMed]
46. Khan, I.; Awan, S.A.; Ikram, R.; Rizwan, M.; Akhtar, N.; Yasmin, H.; Sayyed, R.Z.; Ali, S.; Ilyas, N. Effects of 24-epibrassinolide on plant growth, antioxidants defense system, and endogenous hormones in two wheat varieties under drought stress. *Physiol. Plant.* **2020**, *172*, 696–706. [CrossRef] [PubMed]
47. Talbi, S.; Rojas, J.A.; Sahrawy, M.; Rodríguez-Serrano, M.; Cárdenas, K.E.; Debouba, M.; Sandalio, L.M. Effect of drought on growth, photosynthesis and total antioxidant capacity of the saharan plant Oudeneya africana. *Environ. Exp. Bot.* **2020**, *176*, 104099. [CrossRef]
48. Mumtaz, S.; Hameed, M.; Ahmad, F.; Ahmad, M.S.A.; Ahmad, I.; Ashraf, M.; Saleem, M.H. Structural and Functional Determinants of Physiological Pliability in Kyllinga brevifolia Rottb. for Survival in Hyper-Saline Saltmarshes. *Water Air Soil Pollut.* **2021**, *232*, 424. [CrossRef]
49. Hussain, I.; Saleem, M.H.; Mumtaz, S.; Rasheed, R.; Ashraf, M.A.; Maqsood, F.; Rehman, M.; Yasmin, H.; Ahmed, S.; Ishtiaq, M.; et al. Choline Chloride Mediates Chromium Tolerance in Spinach (*Spinacia oleracea* L.) by Restricting its Uptake in Relation to Morpho-physio-biochemical Attributes. *J. Plant Growth Regul.* **2021**, *41*, 1594–1614. [CrossRef]
50. Mumtaz, S.; Saleem, M.H.; Hameed, M.; Batool, F.; Parveen, A.; Amjad, S.F.; Mahmood, A.; Arfan, M.; Ahmed, S.; Yasmin, H.; et al. Anatomical adaptations and ionic homeostasis in aquatic halophyte *Cyperus laevigatus* L. under high salinities. *Saudi J. Biol. Sci.* **2021**, *28*, 2655–2666. [CrossRef]
51. Imran, M.; Hussain, S.; Rana, M.S.; Saleem, M.H.; Rasul, F.; Ali, K.H.; Potcho, M.P.; Pan, S.; Duan, M.; Tang, X. Molybdenum improves 2-acetyl-1-pyrroline, grain quality traits and yield attributes in fragrant rice through efficient nitrogen assimilation under cadmium toxicity. *Ecotoxicol. Environ. Saf.* **2021**, *211*, 111911. [CrossRef]
52. Farooq, T.H.; Rafay, M.; Basit, H.; Shakoor, A.; Shabbir, R.; Riaz, M.U.; Ali, B.; Kumar, U.; Qureshi, K.A.; Jaremko, M. Morpho-physiological growth performance and phytoremediation capabilities of selected xerophyte grass species toward Cr and Pb stress. *Front. Plant Sci.* **2022**, *13*, 997120. [CrossRef]
53. Hussain, M.I.; Lyra, D.-A.; Farooq, M.; Nikoloudakis, N.; Khalid, N. Salt and drought stresses in safflower: A review. *Agron. Sustain. Dev.* **2016**, *36*, 4. [CrossRef]
54. Ahlem, A.; Lobna, M.; Mohamed, C. Ecophysiological responses of different ploidy levels (tetraploid and hexaploid), of cenchrus ciliaris to water deficiency conditions. *Pak. J. Bot.* **2021**, *53*, 1997–2002. [CrossRef] [PubMed]
55. Faryal, S.; Ullah, R.; Khan, M.N.; Ali, B.; Hafeez, A.; Jaremko, M.; Qureshi, K.A. Thiourea-Capped Nanoapatites Amplify Osmotic Stress Tolerance in Zea mays L. by Conserving Photosynthetic Pigments, Osmolytes Biosynthesis and Antioxidant Biosystems. *Molecules* **2022**, *27*, 5744. [CrossRef] [PubMed]
56. Ijaz, W.; Kanwal, S.; Tahir, M.H.N.; Razzaq, H. Gene Action Of Yield Related Characters Under Normal And Drought Stress Conditions In *Brassica Napus* L. *Pak. J. Bot.* **2021**, *53*, 1979–1985. [CrossRef] [PubMed]
57. Naz, S.; Perveen, S. Response of wheat (triticum aestivum l. Var. Galaxy-2013) to pre-sowing seed treatment with thiourea under drought stress. *Pak. J. Bot.* **2021**, *53*, 1209–1217. [CrossRef]
58. Bibi, S.; Ullah, S.; Hafeez, A.; Khan, M.N.; Javed, M.A.; Ali, B.; Din, I.U.; Bangash, S.A.K.; Wahab, S.; Wahid, N.; et al. Exogenous Ca/Mg quotient reduces the inhibitory effects of PEG induced osmotic stress on *Avena sativa* L. *Braz. J. Biol.* **2024**, *84*, e264642. [CrossRef]
59. Ma, J.; Saleem, M.H.; Yasin, G.; Mumtaz, S.; Qureshi, F.F.; Ali, B.; Ercisli, S.; Alhag, S.K.; Ahmed, A.E.; Vodnar, D.C.; et al. Individual and combinatorial effects of SNP and NaHS on morpho-physio-biochemical attributes and phytoextraction of chromium through Cr-stressed spinach (Spinacia oleracea L.). *Front. Plant Sci.* **2022**, *13*, 973740. [CrossRef]

60. Valivand, M.; Amooaghaie, R. Sodium hydrosulfide modulates membrane integrity, cation homeostasis, and accumulation of phenolics and osmolytes in Zucchini under nickel stress. *J. Plant Growth Regul.* **2021**, *40*, 313–328. [CrossRef]
61. Ma, J.; Saleem, M.H.; Ali, B.; Rasheed, R.; Ashraf, M.A.; Aziz, H.; Ercisli, S.; Riaz, S.; Elsharkawy, M.M.; Hussain, I.; et al. Impact of foliar application of syringic acid on tomato (Solanum lycopersicum L.) under heavy metal stress-insights into nutrient uptake, redox homeostasis, oxidative stress, and antioxidant defense. *Front. Plant Sci.* **2022**, *13*, 950120. [CrossRef]
62. Wang, Y.; Wu, Y.; Duan, C.; Chen, P.; Li, Q.; Dai, S.; Sun, L.; Ji, K.; Sun, Y.; Xu, W.; et al. The expression profiling of the CsPYL, CsPP2C and CsSnRK2 gene families during fruit development and drought stress in cucumber. *J. Plant Physiol.* **2012**, *169*, 1874–1882. [CrossRef]
63. Saleem, K.; Asghar, M.A.; Saleem, M.H.; Raza, A.; Kocsy, G.; Iqbal, N.; Ali, B.; Albeshr, M.F.; Bhat, E.A. Chrysotile-Asbestos-Induced Damage in Panicum virgatum and Phleum pretense Species and Its Alleviation by Organic-Soil Amendment. *Sustainability* **2022**, *14*, 10824. [CrossRef]
64. Ali, B.; Wang, X.; Saleem, M.H.; Azeem, M.A.; Afridi, M.S.; Nadeem, M.; Ghazal, M.; Batool, T.; Qayyum, A.; Alatawi, A.; et al. Bacillus mycoides PM35 Reinforces Photosynthetic Efficiency, Antioxidant Defense, Expression of Stress-Responsive Genes, and Ameliorates the Effects of Salinity Stress in Maize. *Life* **2022**, *12*, 219. [CrossRef]
65. Ali, B.; Wang, X.; Saleem, M.H.; Sumaira; Hafeez, A.; Afridi, M.S.; Khan, S.; Zaib-Un-Nisa; Ullah, I.; Amaral Júnior, A.T.; et al. PGPR-Mediated Salt Tolerance in Maize by Modulating Plant Physiology, Antioxidant Defense, Compatible Solutes Accumulation and Bio-Surfactant Producing Genes. *Plants* **2022**, *11*, 345. [CrossRef] [PubMed]
66. Kong, H.; Zhang, Z.; Qin, J.; Akram, N.A. Interactive Effects Of Abscisic Acid (Aba) And Drought Stress On The Physiological Responses Of Winter Wheat (Triticum Aestivum L.). *Pak. J. Bot.* **2021**, *53*, 1545–1551. [CrossRef]
67. Shemi, R.; Wang, R.; Gheith, E.-S.M.; Hussain, H.A.; Hussain, S.; Irfan, M.; Cholidah, L.; Zhang, K.; Zhang, S.; Wang, L. Effects of salicylic acid, zinc and glycine betaine on morpho-physiological growth and yield of maize under drought stress. *Sci. Rep.* **2021**, *11*, 3195. [CrossRef] [PubMed]
68. Yu, R.; Zuo, T.; Diao, P.; Fu, J.; Fan, Y.; Wang, Y.; Zhao, Q.; Ma, X.; Lu, W.; Li, A. Melatonin enhances seed germination and seedling growth of Medicago sativa under salinity via a putative melatonin receptor MsPMTR1. *Front. Plant Sci.* **2021**, *12*, 702875. [CrossRef] [PubMed]
69. Cen, H.; Wang, T.; Liu, H.; Tian, D.; Zhang, Y. Melatonin application improves salt tolerance of alfalfa (Medicago sativa L.) by enhancing antioxidant capacity. *Plants* **2020**, *9*, 220. [CrossRef] [PubMed]
70. Ma, J.; Saleem, M.H.; Alsafran, M.; Jabri, H.A.; Mehwish; Rizwan, M.; Nawaz, M.; Ali, S.; Usman, K. Response of cauliflower (Brassica oleracea L.) to nitric oxide application under cadmium stress. *Ecotoxicol. Environ. Saf.* **2022**, *243*, 113969. [CrossRef]
71. Jahan, M.S.; Guo, S.; Baloch, A.R.; Sun, J.; Shu, S.; Wang, Y.; Ahammed, G.J.; Kabir, K.; Roy, R. Melatonin alleviates nickel phytotoxicity by improving photosynthesis, secondary metabolism and oxidative stress tolerance in tomato seedlings. *Ecotoxicol. Environ. Saf.* **2020**, *197*, 110593. [CrossRef]
72. Ahmad, S.; Kamran, M.; Ding, R.; Meng, X.; Wang, H.; Ahmad, I.; Fahad, S.; Han, Q. Exogenous melatonin confers drought stress by promoting plant growth, photosynthetic capacity and antioxidant defense system of maize seedlings. *PeerJ* **2019**, *7*, e7793. [CrossRef]
73. Imran, M.; Latif Khan, A.; Shahzad, R.; Aaqil Khan, M.; Bilal, S.; Khan, A.; Kang, S.-M.; Lee, I.-J. Exogenous melatonin induces drought stress tolerance by promoting plant growth and antioxidant defence system of soybean plants. *AoB Plants* **2021**, *13*, plab026. [CrossRef]
74. Ibrahim, M.F.; Elbar, O.H.A.; Farag, R.; Hikal, M.; El-Kelish, A.; El-Yazied, A.A.; Alkahtani, J.; El-Gawad, H.G.A. Melatonin counteracts drought induced oxidative damage and stimulates growth, productivity and fruit quality properties of tomato plants. *Plants* **2020**, *9*, 1276. [CrossRef] [PubMed]
75. Castañares, J.L.; Bouzo, C.A. Effect of exogenous melatonin on seed germination and seedling growth in melon (*Cucumis melo* L.) under salt stress. *Hortic. Plant J.* **2019**, *5*, 79–87. [CrossRef]
76. Ahammed, G.J.; Wu, M.; Wang, Y.; Yan, Y.; Mao, Q.; Ren, J.; Ma, R.; Liu, A.; Chen, S. Melatonin alleviates iron stress by improving iron homeostasis, antioxidant defense and secondary metabolism in cucumber. *Sci. Hortic.* **2020**, *265*, 109205. [CrossRef]
77. Liang, B.; Ma, C.; Zhang, Z.; Wei, Z.; Gao, T.; Zhao, Q.; Ma, F.; Li, C. Long-term exogenous application of melatonin improves nutrient uptake fluxes in apple plants under moderate drought stress. *Environ. Exp. Bot.* **2018**, *155*, 650–661. [CrossRef]

sustainability

MDPI

Article

Influence of Seed Treated by Plasma Activated Water on the Growth of *Lactuca sativa* L.

Nataša Romanjek Fajdetić [1], Teuta Benković-Lačić [1], Krunoslav Mirosavljević [1,*], Slavica Antunović [1], Robert Benković [1], Mario Rakić [2], Slobodan Milošević [2] and Božica Japundžić-Palenkić [1]

1 Biotechnical Department, University of Slavonski Brod, 35000 Slavonski Brod, Croatia
2 Institute of Physics, 10000 Zagreb, Croatia
* Correspondence: kmirosavljevic@unisb.hr

Abstract: The aim of this work was to determine if PAW (Plasma Activated Water) seed treatment and growing conditions could have positive effects on lettuce seedlings and growth. The paper presents the results of a pot experiment on lettuce (*Lactuca sativa* L.) cultivation in greenhouse and field conditions after seed treatment with PAW. The experiment was conducted in two consecutive seasons in 2021 and 2022 and the following growth parameters were measured: head mass, rosette height, rosette width, number of leaves, root mass and root length. As a result of the study, it was found that lettuces grown in the greenhouse from PAW treated seeds had higher results in the first measurement for both cultivars (mass 32.26%, diameter 19.01%, number of leaves 13.49% and height 24.01%), while there were no statistically significant effects on the root system. The lowest results were obtained in untreated and field-grown plants. In addition, plant dry matter was measured and it was found that plants grown from PAW treated seeds had a higher percentage of dry matter (11.51% in 2021, and 11.58% in 2022). It was also found that cultivation in greenhouse resulted in a better quality of plants than the cultivation in the open field.

Keywords: growing conditions; lettuce; PAW; seed treatment

check for updates

Citation: Romanjek Fajdetić, N.; Benković-Lačić, T.; Mirosavljević, K.; Antunović, S.; Benković, R.; Rakić, M.; Milošević, S.; Japundžić-Palenkić, B. Influence of Seed Treated by Plasma Activated Water on the Growth of *Lactuca sativa* L.. *Sustainability* **2022**, *14*, 16237. https://doi.org/10.3390/su142316237

Academic Editors: Daniel El Chami and Maroun El Moujabber

Received: 9 November 2022
Accepted: 1 December 2022
Published: 5 December 2022

1. Introduction

Lettuce (*Lactuca sativa* L.) is widely grown vegetable and one of the most important vegetable crops [1]. It is one of the most demanded lettuce crops in both fresh and ready-to-eat markets around the world [2]. The goal of every grower is to achieve higher and better quality yields. At the same time, the elimination of artificial fertilizers, chemical growth promotors and chemical protectants is highly desirable.

A Highly reactive environment with high-energy electrons, radicals, and various gaseous reactive species, including OH, H_2O_2, NO, HNO is generated by various types of atmospheric pressure air discharges generated by cold plasma. PAW (Plasma Activated Water) occurs when plasma comes into contact with water and when these reactive species dissolve in water, creating reactive oxygen and nitrogen species (RONS) [3]. There are many ways in which PAW affects plants: protein synthesis, gene expression, enzyme activity, DNA methylation, DNA demethylation, DNA damage, seed coat morphological and chemical changes, plant hormone balance, germination and seedling growth [4].

The experiments using PAW began in early 21st century. The use of PAW during vegetation also improves resistance to pathogens as well. Plasma treatment of water creates an acidic environment that leads to conductivity and the formation of nitrogen species and reactive oxygen, which change the redox potential. PAW contains nitrites, nitrates, and ammonia which could be nutrients as well as signaling molecules that can promote seed germination and seedling growth [5,6]. These molecules may also help to break dormancy and promote germination. Nitric oxide (NO), one of the reactive nitrogen species produced by plasma, is known to regulate resistance to abiotic and biotic stresses, and plant development [7,8]. According to Thirdumas et al., soaking seeds with PAW not only has an antibacterial effect but also promotes seed germination and plant growth [9]. Nowadays, there are more and more problems in breeding related to climate changes. As a result of global warming there has been a worldwide increase in temperature since 1970 as well as changes in the precipitation regime, leading to several serious consequences for agriculture [10]. Crops are more frequently exposed to stresses of both abiotic and biotic origin, including exposure to unpredictable and extreme climatic events, changes in plant physiology, growing season and phytosanitary hazard, and increased losses of up to 30% and 50% in global agricultural production in the current scenario of rapidly evolving climate change [11]. Predicting the impact of climate change on crop yields is difficult and depends on the crop type and geographic region [12,13] and the type of abiotic stress [14]. Improving crop resilience to stress is critical to maintaining agricultural productivity [15]. Heat waves will occur more frequently, last longer, have higher temperature peaks and longer duration, and begin earlier in the summer than historically [16,17]. Multiple abiotic stress may occur sequentially or simultaneously, generating an important challenge to crop production that requires the analysis of stress combinations in the field [18].

High climate variability and extreme temperature events were experienced in the twentieth century, since the frequency of summer heat waves and spring frost has increased significantly. For plant seedlings, transplanting is an environmentally stressful period as they must adapt to outdoor conditions and quickly establish themselves in the field [14]. Stress to plants caused by adverse environmental conditions can result in lower yields. To meet the requirements of plants, the environment can be artificially modified in such cases, for example in greenhouses [19]. One of the most important components for the development of agriculture has become the greenhouse industry [20]. In a controlled environment in agriculture, the use of greenhouse can help maintain temperature, humidity, and photosynthetically active radiation [21]. To reduce the influence of climate fluctuation, cultivation in protected areas such as greenhouses is often used [22,23]. In addition to climate conditions, seed quality has a significant influence on a good harvest. To achieve high production and productivity rates, the use of high quality seeds is essential [24].

The hypothesis was that seed treatment with PAW and growing conditions (greenhouse and field) will have an effect on the growth and development of two lettuce cultivars and an effect on dry matter. The experiment was conducted with lettuce plants and the parameters evaluated were the visual appearance and growth parameters of the plants (head mass, head height, root mass, root length, number of leaves and rosette diameter) and dry matter. The aim of this work was to study influence of treatment of two lettuce cultivars with PAW on the morphological characteristics of seedlings and lettuce in greenhouse and field conditions.

2. Materials and Methods

The study was conducted with a three-factorial experiment; two varieties of lettuce seeds, PAW treatment for seeds and cultivation under greenhouse and field conditions in four replicates.

2.1. Preparing of Plasma Activated Water

Active species in plasma activated water (PAW) are generated in a plasma reactor based on a single electrode atmospheric pressure plasma jet. The atmospheric pressure plasma jet consists of a quartz tube with outer and inner diameters of 1.5 and 1 mm, respectively, and a copper wire with a diameter of 100 μm that is inserted into the capillary and serves as an electrode. The electrode is powered with a sinusoidal voltage waveform of 28 kHz with a maximum voltage of 12 kV (PVM500-2500 Plasma Power Generator, Information Unlimited), which was selected based on our previous optimization work [25], and the delivered power is typically about 15 W. The discharge is generated in nitrogen gas (99.996% purity), supplied to the capillary at a 500 sccm flow rate. The 215 mL sample in the Berzelius beaker is brought into contact with the plasma jet by placing the liquid surface at a distance of 5 mm from the capillary opening, which insures the comparable production of H_2O_2 and NO_2^- in the PAW. We use a commercial purified pharmaceutical grade water of (Pharmacopoeia Europaea, Ph. Eur. 9) with a pH of 6.5 and a conductivity of 0.98 μS/cm. The treatment time with the nitrogen plasma jet was 40 min. The plasma-activated samples are analyzed immediately after treatment, and several times during their storage. The concentration of NO_2^-, NO_3^- and H_2O_2, as well as the pH of the samples are measured with QUANTOFIX test strips (nitrate/nitrite 500, 10–500 mg L^{-1} NO_3^-, 1–80 mg L^{-1} NO_2^-, nitrate/nitrite 100, 5–100 mg L^{-1} NO_3^-, 0.5–50 mg L^{-1} NO_2^-, peroxide H_2O_2 25, 0.5–25 mg L^{-1}, peroxide H_2O_2 100, 0.5–25 mg L^{-1}; pH-Fix 2.0–9.0). The strips are evaluated using the QUANTOFIX Relax unit optical reader (from Macherey-Nagel, GmbH, Düren, Germany), which allows quantitative analysis with high accuracy. The Nitrate/nitrite strips were calibrated with $NaNO_2$ and NH_4NO_3 solutions of known concentrations, and the calibration was checked with UV–VIS absorption spectroscopy measurements. The use of strips allows the ageing of samples to be monitored with a time resolution of 1 min without significant consumption of the sample (i.e., the sample volume can be kept quasi-constant during ageing). The measurement error was set to less than 10%. The ageing dynamics of PAW depend on the pH, which was controlled by adding metal ion concentration to the water [26,27]. The Mg ions were added by inserting a solid piece of magnesium (5 g) immediately during the plasma treatment and left in the liquid for one hour after the treatment. The conductivity of PAW was measured after treatment and during ageing using the conductivity probe (Metrohm 914 pH/DO/Conductometer Pt 1000/B/O). After treatment with the plasma jet, the conductivity of the sample increases to about 27 μS/cm. Each treatment was monitored by optical emission spectroscopy (OES). The spectra show the presence of NO, N_2*. In addition, the concentrations of peroxide, nitrate/nitrite and pH were measured at 10 min intervals during treatment to ensure consistency of repeated treatment. For the March 2021 experiment, three PAW treatments were performed in volumes of 215 mL, mixed in the 2 L bottle and separated into 200 mL containers for storage and transport. From each treatment 15 mL of PAW was stored for ageing measurements over the next two weeks. PAW was characterized by the following physico-chemical elements, which are listed in Table 1.

Table 1. Physical-chemical characteristics of PAW.

Concentration of (in mg L^{-1})	H_2O_2	NO_2^-	NO_3^-	pH
40 min treatment (t_1, t_2, t_3)	4.1	3.3	11.2	6.1
	3.1	2.8	5.5	6.1
	0	1.4	5.4	5.8

2.2. *Setting up the Experiment*

The experiment was established at 9 March 2021 at the location Slobodnica (45°9′58″ N, 17°57′8″ E, an elevation of 87 m) in eastern Croatia in the trial filed and greenhouse of the University of Slavonski Brod. A three-factor trial was set up; seed varieties (V1, V2), growing conditions (greenhouse (G) and field (F)), and untreated and PAW treated seeds. Two seed varieties of crystal lettuce (Maradone (Lot: N87925, Vilmorin, France) and Minestrone (Lot: N78760, Vilmorin, France)) were obtained from commercial lettuce company. The PAW was prepared at the Institute of Physics in Zagreb and delivered to Slavonski Brod the day before lettuce sowing. A replicate of 4 × 50 seeds of each variety was soaked in PAW for one hour and then sown in modular trays (Pöpellmann TEKU (BP3153/60) 53 cm × 31 cm × 5.6 cm) with 60 sowing sites (volume of each site 76 mL) filled with cultivation medium Potgrond P (Klasmann-Deilmann). The other replicates of 4 × 50 seeds of each variety were soaked in untreated water and planted at the same time as a control batch. Greenhouse conditions were optimized to maximize seedling growth, with a pre-germination temperature of 15 °C and a post-germination temperature (10–12 °C during the day) and (6–10 °C during the night). After the seedlings reached the required growth stage (3–4 leaves) they were hardened off and planted in (Pöpellmann TEKU VTG 9)

9 cm × 6.8 cm (0.27 L) pots filled with ECO (Klasmann-Deilmann) batch cultivation media. The pots were tested under different growing conditions. One hundred pots of Maradone lettuce seeds treated with PAW were under greenhouse conditions and the other hundred pots were under field conditions. The control group was also grown under greenhouse and field conditions. The same was done with the Minestrone variety. The greenhouse was not specially heated or cooled, but was opened during the day and closed at night, as needed. Temperatures were measured daily and varied from 10 °C at night to 23 °C during the day. A total of 400 plants were grown in each experiment and 128 of them were randomly selected for measurements. The following biometric growth indicators were measured; rosette height, number of leaves, head mass, head diameter, root mass and root length. Plant mass was determined using an analytical balance with an accuracy of 0.01 g (PL3002, Mettler-Toledo International Inc., Greifensee, Switzerland Measurements were taken weekly on the following dates; 26 April 2021, 5 May 2021, 10 May 2021, 17 May 2021 and in the year 2022 at the following dates; 22 April 2022, 3 May 2022, 10 May 2022 and 17 May 2022 until lettuce reached commercial maturity. Each time biometric growth indicators of four plants of Maradona and Minestrone cultivars (per treatment and control) were measured in the greenhouse and in the field.

2.3. *Dry Matter*

Lettuce heads were measured four times and each time measured plants were dried at 60 °C for 3–4 days to the constant dry weight. The total dry weight of the plants was measured using an analytical balance an accuracy of 0.01 g (PL3002, Mettler-Toledo International Inc., Greifensee, Switzerland). Each time four plants of Maradona and Minestrone cultivars (per treatment and control) were measured in the greenhouse and in the field and the average dry weight of each plant was calculated by dividing the total dry weight by the number of plants. This was repeated 4 times and averaged again.

2.4. *Statistical Analysis*

The collected research data were statistically processed using the RStudio computer program, while the statistically significant differences (LSD) were calculated with the MS Excel program, 2019. According to Fisher's test on significance of variance analysis least significant differences (LSD) was calculated for $p < 0.05$ by comparing the means.

3. Results and Discussion

In the experiment, the influence of treating the seeds of two lettuce varieties with PAW on the development of the plants in greenhouse and in the field was studied. Each measured parameter (head mass, head diameter, root mass and length, number of leaves and rosette height) is presented in a separate table for 2021 and 2022. The measurement results are presented separately for the above-ground and below-ground parts of the plant, i.e., the root system.

3.1. *Measurement of the Above-Ground Parts*

The results of the measurement of the mass of the lettuce heads are shown in Table 2 for the year 2021 and in Table 3 for the year 2022.

In 2021 and 2022 (Tables 2 and 3), statistical significance was found between treatments C (Control) and PAW in the first measurement. In the other measurements there was no statistical difference between the treatments. In the second, third and fourth measurements, a statistically significant influence of growing conditions was found, and it was found that growing in a greenhouse gave a statistically better result.

The second parameter measured was the diameter of the lettuce head. The results are shown in Table 4 for the year 2021 and in Table 5 for the year 2022.

Table 2. The mass of the lettuce heads in the year 2021 (in g).

Sampling Date		Measurement 1st 26 April 2021		Measurement 2nd 3 May 2021		Measurement 3rd 10 May 2021		Measurement 4th 17 May 2021	
Variety	Growing Conditions	C	PAW	C	PAW	C	PAW	C	PAW
V1	Greenhouse	8.17Bb	13.65aB	36.3A	42.62A	96.96A	98.11B	197.29B	211.48B
V1	Field			25.77B	28.61B	74.63B	92.11B	152.34C	159.47C
V2	Greenhouse	12.52bA	17.61aA	35.63A	44.8A	95.65A	118.29A	234.76A	243.01A
V2	Field			18.39B	21.12B	80.91B	88.6B	195.02B	215.91B
F_V		* ($p < 0.05$, F = 15.52)		ns ($p < 0.05$, F = 1.15)		ns ($p < 0.05$, F = 0.74)		* ($p < 0.05$, F = 19.03)	
F_{GC}				* ($p < 0.05$, F = 27.57)		* ($p < 0.05$, F = 8.53)		* ($p < 0.05$, F = 18.05)	
F_{TS}		* ($p < 0.05$, F = 7.03)		ns ($p < 0.05$, F = 2.85)		ns ($p < 0.05$, F = 3.78)		ns ($p < 0.05$, F = 1.71)	

V1—variety 1; V2—variety 2; C—control; P—PAW; G—greenhouse; F—field; FV—F test lettuce variety; FGC—F test growing conditions; FBS—F test treatment before sowing; *—statistical significance; ns-no statistical significance; Values between the same line followed by different lowercase letters are statistically significantly different (LSD) at $p < 0.05$; Values within the same columns followed by different capital letters are statistically significantly different (LSD) at $p < 0.05$.

Table 3. The mass of the lettuce heads in the year 2022 (in g).

Sampling Date		Measurement 1st 22 April 2022		Measurement 2nd 3 May 2022		Measurement 3rd 10 May 2022		Measurement 4th 17 May 2022	
Variety	Growing Conditions	C	PAW	C	PAW	C	PAW	C	PAW
V1	Greenhouse	9.54bA	14.32aA	38.58A	48.01A	182.95A	201.18A	252.59B	305.57A
V1	Field			33.56B	43.17A	97.92B	74.36B	277.23B	267.72B
V2	Greenhouse	13.11bB	17.86aB	40.16A	46.08A	213.46A	188.14A	368.64A	283.88A
V2	Field			22.48B	35.52B	117.68B	105.16B	234.14B	227.28B
F_V		* ($p < 0.05$, F = 4.55)		ns ($p < 0.05$, F = 0.01)		ns ($p < 0.05$, F = 3.96)		ns ($p < 0.05$, F = 0.04)	
F_{GC}				* ($p < 0.05$, F = 33.67)		* ($p < 0.05$, F = 130.57)		* ($p < 0.05$, F = 13.49)	
F_{TS}		* ($p < 0.05$, F = 5.12)		ns ($p < 0.05$, F = 0.01)		ns ($p < 0.05$, F = 1.60)		ns ($p < 0.05$, F = 0.75)	

* Explanations: see Table 2.

Table 4. The diameter of the lettuce head in the year 2021 (in cm).

Sampling Date		Measurement 1st 26 April 2021		Measurement 2nd 3 May 2021		Measurement 3rd 10 May 2022		Measurement 4th 17 May 2022	
Variety	Growing Conditions	C	PAW	C	PAW	C	PAW	C	PAW
V1	Greenhouse	10.25B	10.5B	19AB	20A	25.13bA	28aA	28.25A	29.25A
V1	Field			18B	17.75B	22.25B	23.5B	23.5B	24.25B
V2	Greenhouse	12.13bA	15.25aA	19.75A	21.5A	27A	28A	27.25A	29.5A
V2	Field			15.75C	17.25B	22.25bB	25.25bB	26.25A	26.75B
F_V		* ($p < 0.05$, F = 65.64)		ns ($p < 0.05$, F = 0.02)		ns ($p < 0.05$, F = 1.02)		* ($p < 0.05$, F = 2.95)	
F_{GC}				* ($p < 0.05$, F = 11.8)		* ($p < 0.05$, F = 37.1)		* ($p < 0.05$, F = 26.51)	
F_{TS}		* ($p < 0.05$, F = 0.27)		ns ($p < 0.05$, F = 1.43)		* ($p < 0.05$, F = 7.48)		ns ($p < 0.05$, F = 2.95)	

* Explanations: see Table 2.

Table 5. The diameter of the lettuce head in the year 2022 (in cm).

Sampling Date		Measurement 1st 22 April 2022		Measurement 2nd 3 May 2022		Measurement 3rd 10 May 2022		Measurement 4th 17 May 2022	
Variety	Growing Conditions	C	PAW	C	PAW	C	PAW	C	PAW
V1	Greenhouse	9.2bB	14.18aB	20.5A	20.5A	29.75aA	27.75bA	24.25CD	29.25B
V1	Field			15.88B	162.5B	21.5B	22.5B	25.25C	23C
V2	Greenhouse	14.13bA	17.25aA	20A	20.25A	28aA	26.25bA	34A	34.25A
V2	Field			18C	20.75A	23.3aB	19.13bB	28.75B	24.25C
F_V		* ($p < 0.05$, F = 9.12)		ns ($p < 0.05$, F = 2.66)		ns ($p < 0.05$, F = 2.92)		* ($p < 0.05$, F = 19.97)	
F_{GC}				* ($p < 0.05$, F = 8.31)		* ($p < 0.05$, F = 76.04)		* ($p < 0.05$, F = 22.07)	
F_{TS}		* ($p < 0.05$, F = 8.28)		ns ($p < 0.05$, F = 0.88)		* ($p < 0.05$, F = 5.33)		ns ($p < 0.05$, F = 0.12)	

* Explanations: see Table 2.

In the first measurement of head diameter in the years 2021 and 2002 statistical significance was found for the following parameters: cultivar and seed treatment method. In the second, third and fourth measurement, a statistically significant difference was found in relation to growing conditions, and it was found that growing in a greenhouse gave a statistically better result.

The third parameter measured was the number of the leaves and results are shown in Table 6 for the year 2021 and in Table 7 for the year 2022.

Table 6. The number of the leaves in the year 2021.

Sampling Date		Measurement 1st 26 April 2021		Measurement 2nd 3 May 2021		Measurement 3rd 10 May 2021		Measurement 4th 17 May 2021	
Variety	Growing Conditions	C	PAW	C	PAW	C	PAW	C	PAW
V1	Greenhouse	9.50b	10.75a	16.5A	17.5A	23A	24.5A	36.25A	36.25A
V1	Field			15.5A	16.25A	20B	20B	30.25B	31C
V2	Greenhouse	8.5b	9.75a	13.25B	13.5B	20B	21.5C	30.5B	33.75B
V2	Field			11.75B	11.75B	18.25B	19B	26.5C	28.5D
F_V		ns ($p < 0.05$, F = 0.83)		* ($p < 0.05$, F = 33.1)		* ($p < 0.05$, F = 4.97)		* ($p < 0.05$, F = 12.01)	
F_{GC}				ns ($p < 0.05$, F = 4.17)		* ($p < 0.05$, F = 8.97)		* ($p < 0.05$, F = 24.01)	
F_{TS}		* ($p < 0.05$, F = 6.21)		ns ($p < 0.05$, F = 0.55)		ns ($p < 0.05$, F = 0.91)		ns ($p < 0.05$, F = 2.06)	

* Explanations: see Table 2.

Table 7. The number of the leaves in the year 2022.

Sampling Date		Measurement 1st 22 April 2022		Measurement 2nd 5 May 2022		Measurement 3rd 10 May 2022		Measurement 4th 17 May 2022	
Variety	Growing Conditions	C	PAW	C	PAW	C	PAW	C	PAW
V1	Greenhouse	8.75b	10.5a	16A	17A	24.5A	24.25A	27.25B	32.75B
V1	Field			17.75A	15.5AB	23.75A	25.25A	36.25A	36.5A
V2	Greenhouse	9.75b	11.25a	12B	14BC	19.75B	21.5B	29B	29.75BC
V2	Field			11B	11.5C	18.5B	20.75B	28.75B	27.5C
F_V		ns ($p < 0.05$, F = 3.39)		* ($p < 0.05$, F = 12.09)		* ($p < 0.05$, F = 17.44)		* ($p < 0.05$, F = 8.84)	
F_{GC}				ns ($p < 0.05$, F = 0.41)		ns ($p < 0.05$, F = 0.18)		ns ($p < 0.05$, F = 2.95)	
F_{TS}		* ($p < 0.05$, F = 7.89)		ns ($p < 0.05$, F = 0.06)		ns ($p < 0.05$, F = 1.62)		ns ($p < 0.05$, F = 0.77)	

* Explanations: see Table 2.

In 2021 and 2022, in the first measurement, seed treatment had statistically significant effect on the number of leaves. Growing conditions didn't have a significant effect, unlike the previous parameters. The LSD test showed statistical significance of the number of leaves between seed treatment C and PAW, which was the same for both cultivars (1.25 pcs).

The fourth parameter measured was the height of the rosette which is shown in Tables 8 and 9.

Table 8. The height of the rosette in the year 2021 (in cm).

Sampling Date		Measurement 1st 26 April 2021		Measurement 2nd 3 May 2021		Measurement 3rd 10 May 2021		Measurement 4th 17 May 2021	
Variety	Growing Conditions	C	PAW	C	PAW	C	PAW	C	PAW
V1	Greenhouse	6.75bB	9.5aB	11A	10.5A	13.5B	15.25A	16.25A	16.75A
V1	Field			8.75B	9.5A	11.63C	12.25C	13.25D	13.63C
V2	Greenhouse	10.63A	10.88A	10.5A	11.75A	15.25A	15.75A	15.38bB	17aA
V2	Field			8.75B	8B	12.25C	13C	14.25C	14.88B
F_V		* ($p < 0.05$, F = 23.42)		ns ($p < 0.05$, F = 0.87)		* ($p < 0.05$, F = 6.26)		ns ($p < 0.05$, F = 1.24)	
F_{GC}				* ($p < 0.05$, F = 16.0)		* ($p < 0.05$, F = 31.9)		*($p < 0.05$, F = 41.3)	
F_{TS}		* ($p < 0.05$, F = 7.65)		ns ($p < 0.05$, F = 0.12)		ns ($p < 0.05$, F = 2.7)		* ($p < 0.05$, F = 4.58)	

* Explanations: see Table 2.

Table 9. The height of the rosette in the year 2022 (in cm).

Sampling Date		Measurement 1st 22 April 2022		Measurement 2nd 5 May 2022		Measurement 3rd 10 May 2022		Measurement 4th 17 May 2022	
Variety	Growing Conditions	C	PAW	C	PAW	C	PAW	C	PAW
V1	Greenhouse	6.5bB	9.7aB	11.5A	11.38A	14.13A	16.75A	16.63aA	14.13b
V1	Field			10B	9.63B	10.88C	10.13C	14.5aB	13b
V2	Greenhouse	7.38bA	10.88aA	9.63B	11.5A	12B	11.75B	17.13aA	13.75b
V2	Field			9.5B	9.63B	7.75D	8.88D	11.63bC	13.25a
F_V		* ($p < 0.05$, F = 7.27)		ns ($p < 0.05$, F = 0,87)		* ($p < 0.05$, F = 30.82)		ns ($p < 0.05$, F = 0.89)	
F_{GC}				* ($p < 0.05$, F = 4.74)		* ($p < 0.05$, F = 67.34)		*($p < 0.05$, F = 12.19)	
F_{TS}		* ($p < 0.05$, F = 13.85)		ns ($p < 0.05$, F = 0.39)		ns ($p < 0.05$, F = 1.76)		* ($p < 0.05$, F = 4.71)	

* Explanations: see Table 2.

Tables 8 and 9 show that rosette height of was significantly affected by the cultivar and seed treatment in the first measurement. In the other measurements the rosette height was significantly influenced by the growing conditions and it was found that growing in the greenhouse gave a statistically better result.

3.2. Measurement of the Root Parameters

In this part, the results of the measurement of the below-ground part of the plants are presented. The first parameter measured was root mass which is presented in Table 10 for the year 2021 and in Table 11 for the year 2022.

As can be seen, root mass was statistically significantly affected by cultivar in 2021 and 2022. The influence of seed treatment with PAW was absent. With the same cultivar, there was no statistically significant difference between treatment C and PAW.

Table 10. The root mass in the year 2021 (in g).

Sampling Date		Measurement 1st 26 April 2021		Measurement 2nd 3 May 2021		Measurement 3rd 10 May 2021	
Variety	**Growing Conditions**	**C**	**PAW**	**C**	**PAW**	**C**	**PAW**
V1	Greenhouse	8.76A	9.08A	36.31	37.96	31.56	37.30
V1	Field			18.52	20.28	27.29	28.32
V2	Greenhouse	4.02B	3.65B	13.82	23.46	27.34	30.40
V2	Field			12.73	13.73	16.16	25.23
F_V		* ($p < 0.05$, F = 9.26)		ns ($p < 0.05$, F = 3.34)		ns ($p < 0.05$, F = 1.74)	
F_{GC}				ns ($p < 0.05$, F = 2.94)		ns ($p < 0.05$, F = 2.38)	
F_{TS}		ns ($p < 0.05$, F = 0.0)		ns ($p < 0.05$, F = 0.27)		ns ($p < 0.05$, F = 0.97)	

* Explanations: see Table 2.

Table 11. The root mass in the year 2022 (in g).

Sampling Date		Measurement 1st 22 April 2022		Measurement 2nd 3 May 2022		Measurement 3rd 10 May 2022	
Variety	**Growing Conditions**	**C**	**PAW**	**C**	**PAW**	**C**	**PAW**
V1	Greenhouse	8.82A	9.12 A	37.24	26.75	38.45	37.96
V1	Field			18.89	20.54	27.48	28.51
V2	Greenhouse	3.82B	3.47B	17.58	22.56	26.84	29.56
V2	Field			12.82	14.52	17.12	15.16
F_V		* ($p < 0.05$, F = 10.13)		ns ($p < 0.05$, F = 4.01)		ns ($p < 0.05$, F = 1.03)	
F_{GC}				ns ($p < 0.05$, F = 3.25)		ns ($p < 0.05$, F = 3.18)	
F_{TS}		ns ($p < 0.05$, F = 0.2)		ns ($p < 0.05$, F = 0.98)		ns ($p < 0.05$, F = 1.84)	

* Explanations: see Table 2.

The second parameter measured was the root length and the measurement results are shown in Tables 12 and 13.

Table 12. The root length in the year 2021 (in cm).

Sampling Date		Measurement 1st 26 April 2021		Measurement 2nd 3 May 2021		Measurement 3rd 10 May 2021	
Variety	**Growing Conditions**	**C**	**PAW**	**C**	**PAW**	**C**	**PAW**
V1	Greenhouse	19.25B	20.00	26.5A	25.75A	25A	26.75A
V1	Field			15.25B	16.B	16.25bB	24.5aA
V2	Greenhouse	24A	24.00	23.25A	235A	2.45bA	29aA
V2	Field			22A	22.5A	22A	22B
F_V		* ($p < 0.05$, F = 4.85)		ns ($p < 0.05$, F = 0.55)		ns ($p < 0.05$, F = 0.69)	
F_{GC}				* ($p < 0.05$, F = 7.2)		* ($p < 0.05$, F = 1.01)	
F_{TS}		ns ($p < 0.05$, F = 0.04)		ns ($p < 0.05$, F = 0.07)		* ($p < 0.05$, F = 4.48)	

* Explanations: see Table 2.

Table 13. The root length in the year 2022 (in cm).

Sampling Date		Measurement 1st 22 April 2022		Measurement 2nd 3 May 2022		Measurement 3rd 10 May 2022	
Variety	Growing Conditions	C	PAW	C	PAW	C	PAW
V1	Greenhouse	20.51B	21.2	26.7A	27.08A	26.89A	26.75A
V1	Field			15.85B	16.88B	19.15B	23.8A
V2	Greenhouse	24.36A	24.45	23.78A	24.89A	26.98A	29.18B
V2	Field			22.15A	23.01A	22.13AB	22.46A
F_V		* ($p < 0.05$, F = 4.26)		ns ($p < 0.05$, F = 1.63)		ns ($p < 0.05$, F = 3.02)	
F_{GC}				* ($p < 0.05$, F = 8.84)		* ($p < 0.05$, F = 4.75)	
F_{TS}		ns ($p < 0.05$, F = 0.47)		ns ($p < 0.05$, F = 0.12)		ns ($p < 0.05$, F = 2.32)	

* Explanations: see Table 2.

3.3. Dry Matter

The next parameter determined was dry matter (Figure 1 for the year 2021 and Figure 2 for the year 2022).

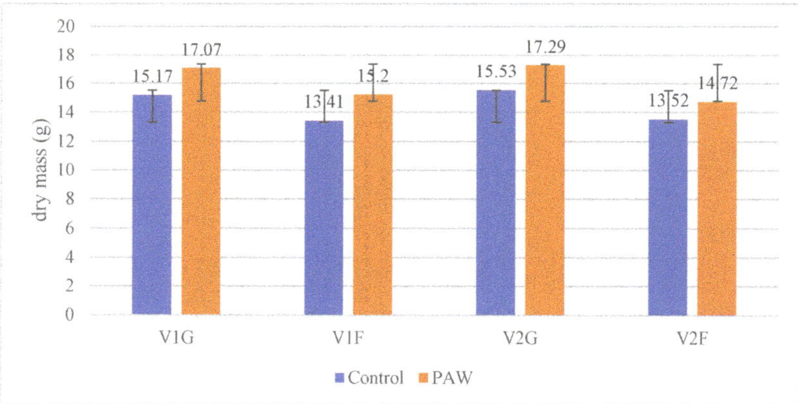

Figure 1. Dry matter content of lettuce head in 2021.

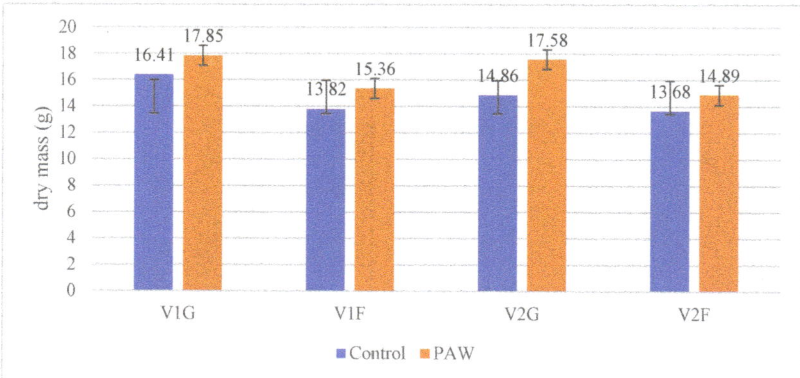

Figure 2. Dry matter content of lettuce head in 2022.

As shown in Figure 1 for 2021, the average dry weight ranged from 13.68 g (lowest value) to 17.85 g (highest value), and no statistically significant influence was found between measurements ($p = 0.2296$).

Figure 2 for the year 2022 shows that the mass of dry matter ranged from 13.82 g as the lowest value to 17.85 g as the highest value and no statistical difference was found ($p = 0.5398$). As can be seen in Figures 1 and 2, the PAW treatment had a positive impact on the dry matter of the measured plants. Figure 1 shows that V1PG had a dry matter 12.52% higher than V1CG and V1PF had a dry matter 13.34% higher than V1CF. V2PG had a dry matter 11.33% higher than V2CG. V2PF had a dry matter 8.87% higher than V2CF. Figure shows that V1PG had a dry matter 8.07% higher than V1CG and V1PF had a dry matter 11.14% higher than V1CF. V2PG had a dry matter 18.3% higher than V2CG. Also, V2PF had a dry matter 8.84% higher than V2CF.

4. Discussion

PAW is used in agriculture to increase protection against biotic and abiotic stress. It is environmentally friendly and can be used in organic farming or as an alternative to chemical biostimulants and artificial fertilizers. In addition, PAW affects plants in various ways; protein synthesis, gene expression, enzyme activity, DNA methylation, DNA demethylation, DNA damage, morphological and chemical changes in seed coat, plant hormonal balance, germination and seedling growth. In the conducted research, we tried to find out the influence of soaking of lettuce seeds in the PAW on further development of plants. According to Holubova et al. [28], there are many studies in which a positive influence of PAW on germination and development of plants was found, and many in which there were no results or they were negative. However, an optimal dose of PAW cannot be generally stated because different plant species respond differently to treatment with PAW. Therefore, we tried to find out how long the seeds must be soaked in PAW to obtain good results, because according to Filatova et al. [29] too short an exposure time has no effect, while too long exposure time can have a detrimental effect. From the obtained results it can be concluded that the seed treatment had an influence on the development of the above-ground part of the plants in the first stage of development, which can be seen from the Tables 2–9. Similar results were obtained by Sarinont et al. [30] who found that the average plant length was higher in plasma-irradiated radish seeds than in the control samples. Dobrin et al. [31] also obtained similar results when they studied wheat. Puač et al. [32] and Sivachandiran et al. [33] also found that the treated plants had better germination, improved enzyme activity and better developed plants. The tables show a statistically significant influence of PAW in the first measurement of the following parameters: mass of the head, number of leaves and height of the rosette, indicating the possibility that PAW treatment improves the growth of early seedling. Comparable results were obtained by Abbaszadeh et al. [34] and Kučerova et al. [35] who reported that PAW treatment on seeds improved the growth of early seedlings. Kučerova et al. [35] found that seeds cultivated at PAW interacted mainly with H_2O_2 at the early growth stages during imbibition and germination, while NO_2^- and NO_3^- were metabolized after the onset of germination and during early plant growth which also confirmed the assumption that the seedlings were still under the influence of PAW treatment in first measurement. From the presented research results, it is also evident that cultivation in a protected area led to significantly better results in seedling growth in terms of lettuce head, which is consistent with the results of Kang et al. [36] whose results indicated that the improvement in plant growth and stress tolerance could be explained by cultivation in greenhouse. The growing conditions in the protected area were not specifically adapted; there was no cooling of the area. The greenhouse was opened in the morning and closed in the evening. Nevertheless, the plants grown in the greenhouse were significantly larger and of better quality than those grown in the field. The increased CO_2 concentration in the air at night and shading during the day probably contributed to this development. Pearson et al. [37] found that an increase in CO_2 concentration resulted in an increase in final head weight of lettuce.

According to Sarinont et al. [30] the growth promoting effects become weaker with time and this could be an explanation for the lack of effect on subsequent development. As for the root part, no positive result was found in the first measurement. Similar results were also obtained by others [30,33,34]. According to Guragin et al. [38] no significant increase in root length was observed in PAW and deionized irrigated radish and pea seedlings. This can be explained by the fact that PAW is rich in nitrogen and not in phosphorus, which is essential for root development. As for dry matter, it was found that the plants under PAW had higher dry matter than the untreated plants. Similar results were obtained by Ling et al. [39] who observed an increase in the shoot and root dry weight (21.95% and 27.51%, respectively) in soybean compared to untreated plants (control). According to Stoleru et al. [40] who found similar results on lettuce this mass could be explained by a stimulation of protein synthesis, induced by PAW. A similar conclusion was reached by Takaki et al. [41] in studying of *Brassica rapa* and Matra [42], who found that plasma treatment increased the dry matter of radish by 9–12%.

The results are also comparable to recent findings by Japundžić-Palenkić et al. [43] but research should certainly be expanded to obtain more detailed information on this topic.

5. Conclusions

According to the research results, it can be concluded that cultivation under greenhouse conditions gave better results for all measured morphological characteristics: head mass, rosette length, number of leaves, rosette diameter, root length and root mass.

Seed treatment with PAW had a positive effect on seedlings growth in the first stage of development only for the above-ground parts (head mass, rosette length, number of leaves, rosette diameter), in the first measurement compared to the control, but in the following measurements this effect was lost. In addition, there was no influence on the below-ground parts (root mass and length).

Lettuce grown from seeds treated with PAW had a higher percentage of dry matter than the control plants.

Author Contributions: Conceptualization, K.M., T.B.-L. and N.R.F.; methodology, N.R.F., B.J.-P., T.B.-L. and S.M.; investigation, N.R.F., B.J.-P., R.B., T.B.-L., S.A., M.R. and S.M.; resources, N.R.F.; data processing R.B.; writing-original draft preparation, N.R.F.; writing-review and editing, K.M. All authors have read and agreed to the published version of the manuscript.

Funding: This work was supported by the project Adaptation of vegetables to new agrometeorological conditions in Slavonia (AVACS), KK.05.1.1.02.0004. The project was financed by European Union from European Regional Development Fund. This work was also supported by COST Action CA19110.

Institutional Review Board Statement: Not applicable.

Informed Consent Statement: Not applicable.

Data Availability Statement: Not applicable.

Acknowledgments: Authors express their sincere gratitude to Katica Šimunović from University of Slavonski Brod for her advices in statistical analysis of experimental data.

Conflicts of Interest: The authors declare no conflict of interest.

References

1. Stoleru, V.; Munteanu NSellito, M.V. *New Approach of Organic Vegetable Systems*; Aracne Publishing House: Rome, Italy, 2014; pp. 24–26.
2. Fallovo, C.; Rouphael, Y.; Rea, E.; Baltistelli, A. Nutrient solution concentration and growing season affect yield and quality of *Lactuca sativa* L. var. acephala in floating raft culture. *J. Sci. Food Agric.* **2009**, *89*, 1682–1689. [CrossRef]
3. Kučerova, K.; Henselova, M.; Slovakova, L.; Bačovinova, M.; Hensel, K. Effect of Plasma Activated Water, Hydrogen Peroxide, and Nitrates on Lettuce Growth and Its Psysiological Parameters. *Appl. Sci.* **2021**, *11*, 1985. [CrossRef]
4. Starič, P.; Vogel-Mikuš, K.; Mozetič, M.; Junkar, I. Effects of Nonthermal Plasma on Morphology, Genetics and Physiology of Seeds: A Review. *Plants* **2020**, *12*, 1736. [CrossRef] [PubMed]

5. Porto, L.C.; Ziuzina, D.; Los, A.; Boehm, D.; Palumbo, F.; Favia, P.; Tiwari, B.; Bourke, P.; Cullen, P. Plasma activated water and airborne ultrasound treatments for enhanced germination and growth of soybean. *Innov. Food Sci. Emerg. Technol.* **2018**, *49*, 13–19. [CrossRef]
6. Kaushik, N.K.; Ghimire, B.; Li, Y.; Adhikari, M.; Veerana, M.; Kaushik, N.; Jha, N.; Adhikari, B.; Lee, S.J.; Masur, K.; et al. Biological and medical applications of plasma-activated media, water and solutions. *Biol. Chem.* **2019**, *400*, 39–62. [CrossRef]
7. Sami, F.; Faizam, M.; Faraz, A.; Siddiqui, H.; Yusuf, M.; Hayat, S. Nitric oxide-mediated integrative alterations in plant metabolism to confer abiotic stress tolerance, NO crosstalk with phytohormones and NO-mediated post translational modifications in modulating diverse plant stress. *Nitric Oxide* **2018**, *73*, 22–38. [CrossRef] [PubMed]
8. Salgado, I.; Martinez, M.C.; Cliviera, H.C.; Frungillo, L.L. Nitric oxide signaling and homeostasis in plants: A focus on nitrate reductase and S-nitrosglutathione reductase in stress-related responses. *Braz. J. Bot.* **2013**, *36*, 89–98. [CrossRef]
9. Thirdumas, R.; Kothakota, A.; Annapure, U.; Siliveru, K.; Blundell, R.; Gatt, R.; Valdramidis, V.P. Plasma activated water (PAW): Chemistry physico-chemical properties, applications in food and agriculture. *Trends Food Sci. Technol.* **2018**, *77*, 21–31.
10. Lotze-Campen, H.; Müller, C.; Popp, A.; Füssel, H. Food security in a changing climate. In *Climate Change, Justice and Sustainability*; Springer: Berlin/Heidelberg, Germany, 2012; pp. 33–43.
11. Sangiorgio, D.; Cellini, A.; Donati, I.; Pastore, C.; Onofrietti, C.; Spinelli, F. Facing Climate Change: Application of Microbial Biostimulants to Mitigate Stress in Horticultural Crops. *Agronomy* **2020**, *10*, 794. [CrossRef]
12. Iizumi, T.; Furuya, J.; Shen, Z.; Kim, W.; Okada, M.; Fujimori, S.; Hasegawa, T.; Nishimori, M. Responses of crop yield growth to global temperature and socioeconomic changes. *Sci. Rep.* **2017**, *7*, 7800. [CrossRef] [PubMed]
13. Zhao, C.; Liu, B.; Piao, S.; Asseng, S. Temperature increase reduces global yields of major crops in four independent estimates. *Proc. Natl. Acad. Sci. USA* **2017**, *114*, 9326–9331. [CrossRef] [PubMed]
14. Beacham, A.M.; Hand, P.; Pink, D.A.C.; Monaghan, J.M. Analysis of Brassica oleracea early stage abiotic stress responses reveal tolerance in multiple crop types and for multiple sources of stress . *J. Sci. Food Agric* **2017**, *97*, 5271–5277. [CrossRef] [PubMed]
15. Zhu, J.K. Abiotic stress signaling and responses in plants. *Pub. Med. Cell* **2016**, *167*, 313–324.
16. Hayhoe, K.; Cayan, D.; Field, C.B.; Frumhoff, P.C.; Maurer, E.P.; Miller, N.L.; Moser, S.C.; Schneider, S.H.; Cahill, K.N.; Clelend, E.E.; et al. Emissions pathways, climate change and impacts on California. *Proc. Natl. Acad. Sci. USA* **2004**, *101*, 12422–12427. [CrossRef]
17. Miller, N.L.; Jin, J.; Hayhoe, K.; Auffhammer, M. Climate change, extreme heat, and electricity demand in California. *J. Appl. Meteorol. Clim.* **2007**, *47*, 1834–1844. [CrossRef]
18. Pandey, P.; Ramegowda, V.; Senthill Kumar, M. Shared and unique responses of plants to multiple individual stresses and stress combinations: Physiological and molecular mechanism. *Front. Plant Sci.* **2015**, *6*, 723. [CrossRef] [PubMed]
19. Abou-Hussain, S.D. Climate change and its impact on the productivity and quality of vegetable crops. *J. Appl. Sci. Res.* **2012**, *8*, 4359–4383.
20. Zhang, J.; Wang, P.; Guo, T. Effect of no tillage system on melon (*Cucumis melo* L.) yield, nutrient uptake and microbial community structures in greenhouse soils. *Folia Hortic.* **2020**, *32*, 265–278. [CrossRef]
21. Pal, B.; Chakraborti, D.; Biswas, P. A genetic algorithm-based hybrid goal programming approach to land allocation problem for optimal cropping plan in agricultural system. In Proceedings of the International Conference on Industrial and Information Systems (ICIIS), Phoenix, AZ, USA, 15–18 December 2009; pp. 181–186.
22. Thomaier, S.; Specht, K.; Henckel, D.; Dierich, A.; Siebert, R.; Freisinger, U.B.; Sawicka, M. Farming in and on urban buildings: Present practice and specific novelties of Zero_Acerage Farming (Z Farming). *Renew. Agric. Food Sys.* **2015**, *30*, 43–54. [CrossRef]
23. Maher, A.; Kamel, E.; Enrico, F.; Atif, I.; Abdelkader, M. An intelligent system for climate control and energy savings in agricultural greenhouses. *Energy Effic.* **2016**, *9*, 1241–1255. [CrossRef]
24. Oliviera, K.R.; Sampaio, F.R.; Siquiera, G.S.; Galvao, I.M.; Bennett, S.J.; Gratao, P.L.; Barbosa, R.M. Physiological quality of soybean seeds grown under different low altitude field environments and storage time. *Plant Soil Environ.* **2021**, *67*, 92–98. [CrossRef]
25. Kutasi, K.; Krstulović, N.; Jurov, A.; Salamon, K.; Popović, D.; Milošević, S. Controlling the composition of plasma-activated water by Cu ions. *Plasma Sources Sci. Technol.* **2021**, *30*, 045015. [CrossRef]
26. Kutasi, K.; Popović, D.; Krstulović, N.; Milošević, S. Tuning the composition of plasma-activated water by a surface-wave microwave discharge and a kHz plasma jet. *Plasma Sources Sci. Technol.* **2019**, *28*, 095010. [CrossRef]
27. Kutasi, K.; Benes, L.; Toth, Z.; Milošević, S. The role of metals in the deposition of long-lived reactive oxygen and nitrogen species into the plasma-activated liquids. *Plasma Processes Polym.* **2022**, in press. [CrossRef]
28. Holubova, L.; Kyzek, S.; Durovcova, I.; Fabova, J.; Horvathova, E.; Ševčovičova, A ; Galova, E. Non-Thermal Plasma—A New Green Priming Agent for plants? *Int. J. Mol. Sci.* **2020**, *21*, 9466. [CrossRef]
29. Filatova, I.; Azharonoki, V.; Kadyrov, M.; Beljavsky, V.; Grozdov, A.; Shik, A.; Antonuk, A. The effect of plasma treatment of seeds of some grain and legumes an their sowing quality and productivity. *Ronanian J. Phys.* **2013**, *56*, 139–143.
30. Sarinont, T.; Anano, T.; Kitazaki, S. Growth enhancement effects of radish sprouts: Atmospheric pressure plasma irradiation vs. heat shock. *J. Phys. Conf. Ser.* **2014**, *518*, 012017. [CrossRef]
31. Dobrin, D.; Magureanu, M.; Mandache, N.B.; Ionita, M.D. The effect of non-thermal plasma treatment on wheat germination and early growth. *Innov. Food Sci. Emerg. Technol.* **2015**, *29*, 255–260. [CrossRef]
32. Puač, N.; Gherardi, M.; Shiratani, M. Plasma agriculture: A rapidly emerging field. *Plasma Process Polym.* **2017**, *15*, 1700174. [CrossRef]

33. Sivachandiran, L.; Khacef, A. Enhanced seed germination and plant growth by atmospheric pressure cold air plasma: Combined effect of seed and water treatment. *RSC Adv.* **2016**, *7*, 1822–1832. [CrossRef]
34. Abbaszadeh, R.; Nia, P.K.; Fattahi, M.; Marzdashti, H.G. The effects of three plasma-activated water generation systems on lettuce seed germination. *Res. Agric. Eng.* **2021**, *67*, 131–137. [CrossRef]
35. Kučerova, K.; Henselova, M.; Slovakova, L.; Hensel, K. Effects of plasma activated water on wheat: Germination, growth parameters, photosynthetic pigments, soluble protein content, and antioxidant enzymes activity. *Plasma Process. Polym.* **2019**, *16*, 1800131. [CrossRef]
36. Kang, M.H.; Jeon, S.S.; Shin, S.M.; Veerana, M.; Ji, S.H.; Uhm, H.S.; Choi, E.H.; Shin, J.H.; Park, G. Dynamics of nitric oxide level in liquids treated with microwave plasma-generated gas and their effects on spinach development. *Sci. Rep.* **2019**, *9*, 1011. [CrossRef]
37. Pearson, N.; Palmer, M.R. Atmospheric carbon dioxide concentrations over the past 60 million years. *Nature* **1997**, *406*, 695–699. [CrossRef]
38. Guragin, R.P.; Baniya, H.B.; Pradhan, S.P.; Pandey, B.P.; Subedi, D.P. Influence of PAW on the germination of radish, fenugreek and pea seeds. *AIP Adv.* **2021**, *11*, 125304. [CrossRef]
39. Ling, L.; Jiangang, L.; Michong, S.; Xin, H.; Hangliang, S.; Yuanhoa, D. Effects of cold plasma treatment on seed germination and seedling growth of soybean. *Sci. Rep.* **2014**, *4*, 5859. [CrossRef]
40. Stoleru, V.; Burlica, R.; Mihalache, G.; Dirlau, D.; Padureanu, S.; Teliban, G.C.; Astanei, D.; Cojocaru, A.; Beniuga, O.; Patras, A. Plant growth promotion effect of plasma activated water on *Lactuca sativa* L. cultivated in two different volumes of substrate. *Sci. Rep.* **2020**, *10*, 20920. [CrossRef]
41. Takaki, K.; Takahata, J.; Watanabe, S.; Satta, N.; Yamada, O.; Fujio, T.; Sasaki, Y. Improvements in plant growth rate using underwater discharge. *J. Phys. Conf. Ser.* **2013**, *418*, 012140. [CrossRef]
42. Matra, K. Non-thermal Plasma for Germination Enhancement of Radish Seeds. *Procedia Comput. Sci.* **2016**, *86*, 132–135. [CrossRef]
43. Japundžić-Palenkić, B.; Benković, R.; Benković-Lačić, T.; Antunović, S.; Japundžić, M.; Romanjek Fajdetić, N.; Mirosavljević, K. Pepper growing modified by plasma activated water and growth conditions. *Sustainability* **2022**, *14*, 15967. [CrossRef]

sustainability MDPI

Article

Physiological Comparison of Wheat and Maize Seedlings Responses to Water Stresses

Agnieszka Ostrowska * and Tomasz Hura

Polish Academy of Sciences, The Franciszek Górski Institute of Plant Physiology, Niezapominajek 21, 30-239 Kraków, Poland; t.hura@ifr-pan.edu.pl
* Correspondence: a.ostrowska@ifr-pan.edu.pl

Abstract: The aim of the study was to investigate specific responses of spring wheat (C_3 photosynthesis) and maize (C_4 photosynthesis) to drought and flooding stress. Analyses of water content, gas exchange intensity, photosynthetic apparatus activity, chlorophyll content, plant height and biological membrane integrity were performed on the 10th day of drought and flooding in both species at the third leaf stage. A specific response of wheat under both drought and flooding conditions involved an increase in ET_o/RC ratio, describing electron transport flux converted into a single reaction center in PSII. Correlations between electrolyte leakage and the probability of electron transport beyond the plastoquinone Q_A, and the amount of energy used for the electron transport were also found. A specific response of maize during flooding was the increase of stomatal conductance. Additionally, a significant correlation between P_N/C_i and relative water content was exhibited. Furthermore, the parameters differentiating the studied species only under stressful conditions were rendered. The application of such parameters can be widely used, e.g., for studying the reaction of a potential cultivars to drought and flooding. Providing such information to potential farmers can help better select cultivars for their environmental conditions.

Keywords: chlorophyll fluorescence; C_3; C_4; drought; flooding; maize; spring wheat

Citation: Ostrowska, A.; Hura, T. Physiological Comparison of Wheat and Maize Seedlings Responses to Water Stresses. *Sustainability* **2022**, *14*, 7932. https://doi.org/10.3390/su14137932

Academic Editors: Daniel El Chami and Maroun El Moujabber

Received: 20 April 2022
Accepted: 25 June 2022
Published: 29 June 2022

1. Introduction

Most forecasts on climate changes predict increasing water shortages in the areas of cereal cultivation as well as weather anomalies such as periodic flooding, larger floods, or whirlwinds [1,2]. This is why the demand for species and cultivars with enhanced tolerance to these environmental stresses is on the rise [3].

Soil drought is a serious risk to plant productivity. A decrease in the quantity and quality of yields is the final result of soil drought, as a consequence of many physiological disturbances, such as increasing the degree of cell membrane damage and chlorophyll degradation or reducing the photochemical activity of PSII by disordering the energy flow in PSII [4]. The occurrence of too heavy a rainfall after a long drought is also not favorable. Too much rain results in excess soil water, limited oxygen and nutrient availability and therefore reduced plant growth. Finally, agricultural lands can be subject to the inundation of water over-spilling from ditches, rivers, and streams during periods of intense rainfall [5,6]. Our previous research shows that plant aging induced by soil drought can be continued and even intensified by too rapid irrigation in the rehydration phase [7]. Water availability is critical for wheat production, and drought is the major cause of yield losses [8,9]. This is a serious problem, because wheat is one of the most important foods and is a major source of energy and nutrition for many people, for example, bread wheat is the staple food for about 35% of the world's population [10–12]. Maize is also of great importance for human and animal nutrition, and has been reported to be very sensitive to drought stress. Although maize is perceived as preferring warm areas, such as the Mediterranean, which would suggest its low water requirements, in fact its cultivation in these areas often has to be supported by additional irrigation. There is a group of current studies that focuses on modeling the actual needs of maize to reduce unnecessary water losses [13,14]. This is another proof that plants are very often subjected to opposing effects, such as lack and excess of water.

Therefore, our research focused on two opposing stresses to investigate the specific responses of spring wheat (C_3 photosynthesis) and maize plants (C_4 photosynthesis) to drought and flooding stress at early stages of growth. As the species differ in their type of photosynthesis, we assumed they would also react in different ways to these two stress factors. To verify this hypothesis, we investigated and compared their gas exchange, photosynthetic apparatus activity, chlorophyll content, growth dynamics, water ratios, and damage to biological membranes.

The studied species differ not only in the type of photosynthesis, but also in height and development of the plant. For this reason, comparing their stress response, without specifying the control groups, is virtually impossible. Therefore, the aim of the study was also to select the parameters differentiating both species under stressful conditions and with insignificant differences under control conditions. The knowledge of such parameters may be useful for comparative studies of species in natural conditions. We suppose that the parameters of the chlorophyll fluorescence may be the most useful. As compared with measuring gas exchange, chlorophyll fluorescence measurement is quick, inexpensive and simple to perform. Its huge advantage is its high sensitivity that allows for detecting early stress-related changes prior to the visible symptoms.

The novelty of these studies will be the selection of parameters for the evaluation of the reactions of morphologically different species with a different type of photosynthesis, based on the same level of these parameters under optimal conditions and showing differences only under water stresses. Then, it will be possible to conclude that these differences do not result from differences between these species, but from their sensitivity to the stress tested.

Currently, the registration of new cultivars in Poland does not require providing information on a detailed response to drought and flooding. The changing climate becoming more unpredictable may suggest that the inclusion of such information in the characteristics of cultivars is highly justified. The potential selection of such quick and reliable indicators of plant conditions after stress would facilitate this process, especially in field conditions.

2. Materials and Methods

2.1. Plants Material, Growth Conditions, Treatments

In the experiment, spring wheat (*Triticum aestivum* L.) cv. 'Nimfa' and maize (*Zea mays* L.) cv. 'Kosmo230' were used as plant material. The wheat cv. 'Nimfa' (Plant Breeding Strzelce Station) is a mid–early cultivar, with perfect baking quality and a large mass of a thousand grain, as compared with other Polish cultivars. The producer describe the Nimfa as a very good facultative spring wheat cultivar (for late autumn sowing). In addition to possible better winter hardiness, it may be combined with a better use of water from winter rainfall and a greater tolerance to flooding. A mid–early cv. 'Kosmo 230' (Małopolska Plant Breeding Station Ltd., Kraków, Poland) was intended for both silage and grain yield. Additionally, it is a high yield cultivar with a good yield structure. The producer describes cv. 'Kosmo230' as repetitively and stably yielding in various environments. At the same time, it was mentioned that it is a cultivar suitable for cultivation on weaker soils, but rich in water, which suggests its greater demand for water.

The plants were grown in a growth chamber in 4 L pots (8 pots per treatment, 12 plants per pot), filled with a mixture of commercial peat soil and sand (1:1, *v/v*). The base of the substrate was soil with highly milled peat deacidified with chalk and supplemented with a starting dose of 0.6 kg m^{-3} of NPK+Mg compound fertilizer. The pH of the substrate was 5.5–6.5, and the salinity was below 1.9 NaCl dm^3. Vegetation was maintained at a 15 h photoperiod, an irradiance of 450 μmol (photon) m^{-2} s^{-1} (provided by high pressure sodium lamps, 400 W; Philips SON-T AGRO, Belgium), a temperature of 23/18 °C (day/night), and 40% air humidity. The plants were irrigated with full-strength Hoagland's nutrient solution once a week. Soil moisture was controlled every day to maintain 70% of field water capacity (FWC) by weighing the pots and adding enough water to achieve weight corresponding to 70% FWC. The weights assigned to specific FWC levels were calculated from a proportion based on a previous determination of the maximum and minimum water capacity. Additionally, soil humidity was measured with a HydroSense Soil Water Content Measurement System (Campbell Scientific Australia, Inc., Garbutt, QLD, Australia) at different points of the pot to check the homogeneity of water content.

Water stress was initiated from the third leaf stage of wheat and maize. For plants exposed to drought stress, water level in the pots was maintained at 30–35% FWC for 10 days. The FWC level was maintained by weight by adding water, as necessary. The flooded plants and water level (pH 7.5, dGH = 273 mg/dm^3) of 0.5 cm above the soil surface was maintained for 10 days. Analyses of water content, gas exchange intensity, photosynthetic apparatus activity, chlorophyll content, plant height, and biological membrane integrity were performed on the tenth day of the stress treatment. For all measurements, one replicate means one plant.

2.2. *Relative Water Content (RWC)*

The measurements involved one 6 cm middle fragment from the second fully developed leaf from the top. RWC was determined according to the Barrs and Weatherley [15] formula:

$$RWC = [(FW - DW)/(TW - DW)] \times 100\% \quad (1)$$

where FW represents fresh weight, DW dry weight and TW turgid weight. First, a leaf fragment was weighed to determine the fresh weight (FW). Then, this fragment was placed in darkness for 24 h in vials containing water to allow complete rehydration. After that time the TW was determined. Next, the leaf fragment was dried at 85 °C for 48 h and then DW was measured. The measurements were taken in 7 replicates.

2.3. *Photochemical Efficiency*

Measurements involved the central part of the first fully expanded leaf and were performed with a Handy PEA (Hansatech Ltd., King's Lynn, UK). They were taken after 30 min of leaf adaptation to darkness. The excitation irradiance was 3000 µmol (quantum) m^{-2} s^{-1} (peak at 650 nm). Changes in fluorescence were registered automatically by the fluorimeter during irradiation between 10 µs and 1 s. During the initial 2 ms, data were collected every 10 µs with a 12-bit resolution. After this period, the frequency of the automatic data set was dropped. The collected data were analyzed with a JIP test, based on the theory of energy flow in PSII [16–19]. The following parameters were calculated: absorbed energy flux (ABS), trapped energy flux (TR_o), electron transport flux (ET_o), and dissipated energy flux (DI_o), as well as the area of the photosynthetic sample (RC/CS_m). The parameters are classified as specific, i.e., expressed as calculated per a reactive center (RC), and phenomenological, i.e., expressed in terms of the excited photosynthetic surface (cross section, CS) of the sample. Finally, we calculated the following yield ratios: the probability (at t = 0) of electron transport beyond the plastoquinone Q_A (ψ_o), the quantum yield of electron transport (at t = 0) (φ_{Eo}), the maximum quantum yield of PSII photochemistry (at t = 0) (F_v/F_m). PI (overall performance index of PSII photochemistry) was also determined. Measurements involved 16 replicates.

2.4. *Gas Exchange*

Gas exchange was measured using an infrared gas analyzer LCpro-SD (ADC BioScientific Ltd., Hoddesdon, UK) automatically controlling the measurement conditions. The following parameters were measured: photosynthetic rate (P_N), transpiration (E), stomatal conductance (g_s), as well as the intercellular concentration of CO_2 (C_i). The instantaneous water use efficiency index (WUE) was determined based on the quotient of photosynthetic rate and transpiration (P_N/E). Apparent carboxylation efficiency (P_N/C_i) was also calculated. The conditions in the measurement chamber were as follows: carbon dioxide concentration 360 μmol mol^{-1} air, air humidity equal to ambient humidity, temperature 23 °C, PAR intensity 600 μmol photons m^{-2} s^{-1}. The measurements were carried out between 10:00 a.m. and 12:30 p.m. and involved the first fully expanded leaf in 7 replicates.

2.5. Chlorophyll Content

Measurements based on the leaf greenness index were performed in the first fully developed leaf from the top with a hand-held chlorophyll meter, SPAD-502 (Konica-Minolta, Inc., Tokyo, Japan). Light absorption at 650 nm (absorbed by Chl) and 940 nm (absorbed by the other structures) was determined. The readings were converted using a microprocessor and presented in contractual units, called SPAD readings (with maximum at 60 units). SPAD readings are directly proportional to Chl content. The measurements were taken in 18 replicates.

2.6. Plant Height Analysis

Plant height was determined by measuring the entire aboveground part (from the ground level to the end of the longest leaf).

2.7. Electrolyte Leakage (EL)

For each cultivar, three 3 cm middle leaf fragments were excised from the first fully developed leaf from the top. The samples were washed twice in deionized water and immersed in 25 cm^3 of deionized water. After 24 h (t_1) of shaking at room temperature, the samples were frozen at −35 °C for 24 h, then heated and shaken again (24 h, room temperature, t_2). EL was calculated as follows:

$$EL = (EL_1/EL_2) \times 100\%, \quad (2)$$

where EL_1 and EL_2 represent the specific electrical conductance at t_1 and t_2, respectively. Measurements of electrical conductance were performed with a microcomputer conductivity meter CC-317 (Elmetron, Zabrze, Poland). They were taken with 6 replicates.

2.8. Statistical Analysis

A two-way analysis (stress/control and species) of variance using the Duncan's test at $p < 0.05$ was performed in order to determine the significance of differences within treatment (drought, waterlogging), measured separately for each parameter. All data were checked for normality before analysis using the Shapiro–Wilk test. All data were analyzed using Statistica 13.1 software (Statsoft, Inc., Tulsa, OK, USA).

3. Results and Discussion

3.1. Plant Height, Relative Water Content (RWC) and Electrolyte Leakage (EL)

Table 1 shows the results on plant height, relative water content (RWC) and electrolyte leakage (EL). Only in maize did both soil drought and flooding significantly decrease plant height (Table 1). Plant growth inhibition is a visible effect of environmental stresses, including drought and flooding [20–22]. A significant positive correlation between net photosynthesis of winter wheat and its height difference was noted by the authors, indicating that the height was affected by photosynthesis. However, the impact of drought on both these parameters depends on the development phase of wheat. Bread wheat research has shown that plant height also has a significant positive correlation with grain yield [23,24]. However, in maize, both stressors resulted in a significant drop in leaf water content (RWC) (Table 1), and we found it more likely that plant growth under both stresses was reduced as a result of changes in photosynthesis and photosynthetic apparatus activity. It has been shown that photosynthesis and photosynthetic apparatus activity are negatively affected by changes in RWC [25].

Table 1. Effects of 10-day drought and root flooding on plant height [cm] ($n = 20$), relative water content (RWC) [%] ($n = 7$) and electrolyte leakage (EL) [%] ($n = 6$) of wheat and maize leaves. Means indicated with the same letters within drought or flooding treatment for each parameter are not significantly different at $p < 0.05$. The grey areas point to differences between species within each parameter under stress conditions with no differences under control conditions.

Parameter	Treatment	Drought		Flooding	
		Wheat	Maize	Wheat	Maize
Plant height	Control	46.9 ± 0.90 [bc]	64.1 ± 1.63 [a]	46.9 ±0.90 [c]	64.1 ± 1.63 [a]
	Stress	43.6 ± 0.85 [c]	49.3 ± 1.66 [b]	44.5 ± 0.85 [c]	59.0 ± 1.78 [b]
RWC	Control	96.0 ± 1.11 [a]	97.0 ± 1.09 [a]	96.0 ± 1.11 [a]	97.0 ± 1.09 [a]
	Stress	84.3 ± 2.09 [b]	88.1 ± 1.41 [b]	90.7 ± 2.08 [b]	90.9 ± 1.15 [b]
EL	Control	6.2 ± 0.28 [c]	8.7 ± 0.45 [bc]	6.2 ± 0.28 [c]	8.7 ± 0.45 [b]
	Stress	9.0 ± 0.38 [b]	11.6 ± 1.68 [a]	6.6 ± 0.54 [bc]	35.4 ± 1.46 [a]

The results on relative water content (RWC) indicated a similar level of leaf hydration in both species experiencing drought and flooding (Table 1). According to Hsiao [26], this level of RWC drop corresponds to mild or moderate water stress. Soil drought-induced decrease in water content in wheat and maize was also reported by other authors [27,28]. Azizi, et al. [29] explained a significantly lower leaf RWC in flooded *Populus euphratica* by limited water uptake by the plant root system. A similar decline in leaf water content during flooding stress was seen in sorghum and soybean [30,31].

Drought stress was also associated with increased damage to the biological membranes (higher EL) observed in both investigated species (Table 1). The same response of wheat and maize to drought was witnessed in other studies [4,32].

The integrity of biological membranes was particularly sensitive to flooding in maize, as manifested by the highest value of EL (35.4 %) (Table 1). Similar disturbances to the biological membrane integrity in flooded maize plants were also observed by other authors [33–35]. Yordanova and Popova [36] proved that the increase in electrolyte leakage under flooding stress was linked to root oxygen deficiency that induced photooxidative damage in maize leaves. Additionally, ROS accumulation induced by flooding was another destructive factor for cell membranes [37].

Significant correlations between electrolyte leakage (EL) and the probability (at t = 0) of electron transport beyond the plastoquinone Q_A (ψ_o) (Figure 1A), and the amount of energy used for the electron transport (ET_o/CS_m) were found (Figure 1B). The results point to the fact that flooding conditions disturb electron transport between PSII and PSI in maize.

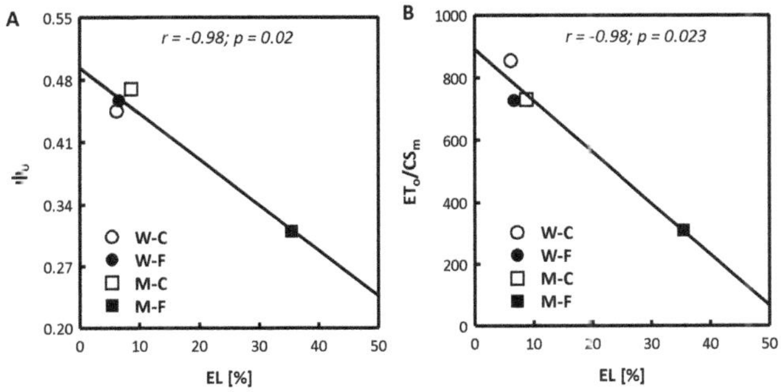

Figure 1. Correlation between electrolyte leakage (EL) and the probability (at t = 0) of electron transport beyond plastoquinone Q_A (ψ_o) (**A**), and the amount of energy used for the electron transport (ET_o/CS_m) (**B**) for wheat (W) and maize (M) under control (C) and flooding (F) conditions. Correlation between measured parameters was tested at a probability of $p < 0.05$.

3.2. *Leaf Gas Exchange and Water Use Efficiency (WUE)*

During drought, both wheat and maize experienced a significant decrease in net photosynthesis (P_N), transpiration intensity (E), and stomatal conductance (gs) (Table 2). No major changes were found in the intercellular concentration of CO_2 (C_i) or water use efficiency (WUE). Drought stress induced alterations in apparent carboxylation efficiency (P_N/C_i) only in wheat (Table 2). However, a significant correlation between P_N/C_i and RWC for both species was found (Figure 2).

Table 2. Effects of 10-day drought and flooding of roots on leaf gas exchange (n = 7) in wheat and maize. P_N—net photosynthesis intensity (μmol(CO_2) m^{-2} s^{-1}), E—transpiration intensity (mmol(H_2O) m^{-2} s^{-1}), g_s—stomatal conductance [mol(H_2O) m^{-2} s^{-1}], C_i—intracellular concentration of CO_2—(μmol(CO_2) mol(air)$^{-1}$), WUE—photosynthetic ratio of water use (μmol(CO_2)mmol^{-1}(H_2O)), P_N/C_i—apparent carboxylation efficiency (mol(air) m^{-2} s^{-1}). Means indicated with the same letters within drought or flooding treatment, for each parameter are not significantly different at $p < 0.05$. The grey areas point to differences between species within each parameter under stress conditions with no differences under control conditions.

Parameters	Treatments	Drought		Flooding	
		Wheat	Maize	Wheat	Maize
P_N	Control	14.25 ± 2.09 ^{ab}	17.85 ± 1.58 ^{a}	14.25 ± 2.09 ^{b}	17.85 ± 1.58 ^{a}
	Stress	10.47 ± 1.79 ^{c}	12.64 ± 1.20 ^{b}	13.46 ± 1.91 ^{b}	16.59 ± 1.85 ^{a}
E	Control	4.95 ± 0.75 ^{b}	5.67 ± 0.83 ^{a}	4.95 ± 0.75 ^{ab}	5.67 ± 0.83 ^{a}
	Stress	3.17 ± 0.43 ^{c}	3.12 ± 0.54 ^{c}	4.34 ± 0.28 ^{b}	4.93 ± 0.40 ^{ab}
g_s	Control	0.046 ± 0.0036 ^{b}	0.064 ± 0.0042 ^{a}	0.046 ± 0.0036 ^{b}	0.064 ± 0.0042 ^{ab}
	Stress	0.025 ± 0.0045 ^{c}	0.036 ± 0.0021 ^{b}	0.039 ± 0.0021^{c}	0.075 ± 0.0062 ^{a}
C_i	Control	371.3 ± 23.12 ^{bc}	468.3 ± 31.43 ^{a}	371.3 ± 23.12 ^{c}	468.3 ± 31.43 ^{a}
	Stress	369.2 ± 19.51 ^{c}	406.3 ± 25.18 ^{ab}	382.2 ± 31.65 ^{c}	411.2 ± 27.78 ^{b}
WUE	Control	2.88 ± 0.47 ^{b}	3.15 ± 0.27 ^{ab}	2.88 ± 0.47 ^{b}	3.15 ± 0.27 ^{ab}
	Stress	3.30 ± 0.39 ^{ab}	4.05 ± 0.62 ^{a}	3.10 ± 0.31 ^{ab}	3.37 ± 0.19^{a}
P_N/C_i	Control	0.038 ± 0.0051 ^{a}	0.039 ± 0.0047 ^{a}	0.038 ± 0.0051 ^{a}	0.039 ± 0.0047 ^{a}
	Stress	0.029 ± 0.0039 ^{b}	0.032 ± 0.0035 ^{ab}	0.036 ± 0.0029 ^{a}	0.041 ± 0.0040 ^{a}

A drought-induced drop in the photosynthetic activity of wheat and maize leaves was also reported in other studies [32,38]. In both species exposed to drought and experiencing reduced photosynthetic activity, the water use efficiency was maintained at the control level, which may be associated with the partial closure of stomata and limited transpiration. It has been shown that WUE may differ depending not only between species, but also on the intensity of the drought [39].

Flooding caused a significant reduction in stomatal conductivity (gs) in wheat and a significant decrease in the intercellular concentration of carbon dioxide (Ci) in maize. The other parameters of gas exchange and WUE were similar in both species (Table 2). Stomatal closure under flooding stress has been already explained by negative hydraulic message and a negative ABA message from oxygen-deficient roots. Both effects result from limited water conductance into the roots and an inhibition of the oxygen-depending ABA biosynthetic pathway. The consequence of the negative hydraulic message was a loss of leaf turgor that induced stomata to close [40,41].

3.3. *Photosynthetic Apparatus Activity and Chlorophyll Content*

Changes in chlorophyll fluorescence parameters indicated a clear reduction in the activity of the photosynthetic apparatus in both species exposed to soil drought (Table 3). Out of 13 chlorophyll fluorescence parameters, 11 changed significantly in response to wheat leaf dehydration (except for ABS/CS_m and ψ_o). In maize, the changes were significant for 12 parameters, except for the quantum yield of electron transport φ_{Eo} (Table 3). Flooding resulted in significant alterations in all chlorophyll fluorescence parameters in maize (Table 3), and 10 parameters in wheat (except for ABS/CS_m, ψ_o and φ_{Eo}). The parameter that differentiated the species under both stresses was ET_o/RC, as it significantly increased in stressed wheat but decreased in drought- and flooding-exposed maize.

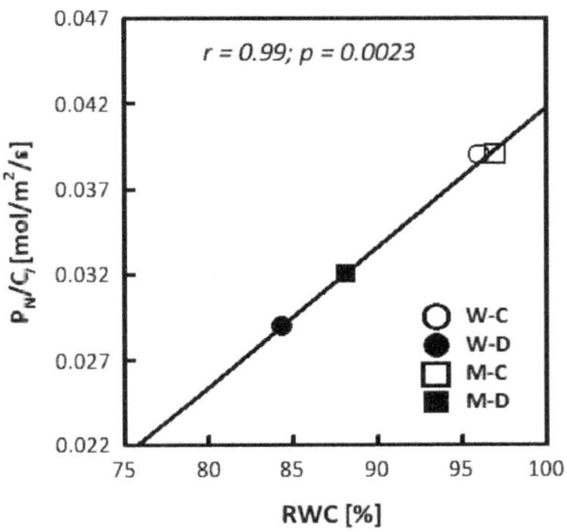

Figure 2. Correlation between apparent carboxylation efficiency (P_N/C_i) and relative water content (RWC) for wheat (W) and maize (M) under control (C) and drought (D) conditions. The correlation between the measured parameters was tested at a probability of $p < 0.05$.

Table 3. Effects of 10-day drought and flooding of wheat and maize roots on chlorophyll fluorescence parameters ($n = 16$) and chlorophyll content ($n = 18$). Means indicated with the same letters within drought or flooding treatment for each parameter are not significantly different at $p < 0.05$. The grey areas point to differences between species within each parameter under stress conditions and with no differences under control conditions.

Parameters	Treatments	Drought		Flooding	
		Wheat	Maize	Wheat	Maize
ABS/CS_m	Control	2409 ± 61.3 [a]	1945 ± 52.0 [b]	2409 ± 61.3 [a]	1945 ± 52.0 [b]
	Stress	2187 ± 116.9 [a]	1481 ± 71.3 [c]	2248 ± 68.5 [a]	1426 ± 69.9 [c]
TR_o/CS_m	Control	1914 ± 52.9 [a]	1553 ± 44.8 [b]	1914 ± 52.9 [a]	1553 ± 44.8 [b]
	Stress	1453 ± 85.4 [b]	959 ± 72.7 [c]	1606 ± 56.9 [b]	940 ± 67.7 [c]
ET_o/CS_m	Control	857 ± 38.6 [a]	731 ± 28.5 [b]	857 ± 38.6 [a]	731 ± 28.5 [b]
	Stress	619 ± 46.6 [c]	304 ± 41.2 [d]	729 ± 29.8 [b]	310 ± 34.2 [c]
DI_o/CS_m	Control	496 ± 15.1 [b]	393 ± 17.9 [c]	496 ± 15.1 [b]	393 ± 17.9 [c]
	Stress	734 ± 49.9 [a]	522 ± 14.1 [b]	642 ± 26.6 [a]	487 ± 30.1 [b]
RC/CS_m	Control	687 ± 27.0 [a]	650 ± 17.4 [a]	687 ± 27.0 [a]	650± 17.4 [a]
	Stress	425 ± 29.9 [b]	383 ± 23.9 [b]	514 ± 25.4 [b]	368 ± 27.7 [c]
ABS/RC	Control	3.54 ± 0.076 [c]	3.00 ± 0.044 [d]	3.54 ± 0.076 [c]	3.00 ± 0.044 [d]
	Stress	5.30 ± 0.230 [a]	3.92 ± 0.083 [b]	4.45 ± 0.141 [a]	3.98 ± 0.128 [b]
TR_o/RC	Control	2.81 ± 0.047 [b]	2.39 ± 0.035 [c]	2.81 ± 0.047 [b]	2.39 ± 0.035 [d]
	Stress	3.47 ± 0.076 [a]	2.48 ± 0.052 [c]	3.16 ± 0.059 [a]	2.56 ± 0.055 [c]

Table 3. *Cont.*

Parameters	Treatments	Drought		Flooding	
		Wheat	Maize	Wheat	Maize
ET_o/RC	Control	1.24 ± 0.020 [b]	1.12 ± 0.022 [b]	1.24 ± 0.020 [b]	1.12 ± 0.022 [b]
	Stress	1.47 ± 0.083 [a]	0.79 ± 0.049 [c]	1.45 ± 0.087 [a]	0.78 ± 0.057 [c]
DI_o/RC	Control	0.73 ± 0.030 [c]	0.61 ± 0.011 [c]	0.73 ± 0.030 [b]	0.61 ± 0.011 [b]
	Stress	1.83 ± 0.158 [a]	1.44 ± 0.105 [b]	1.29 ± 0.088 [a]	1.41 ± 0.143 [a]
F_v/F_m	Control	0.79 ± 0.005 [a]	0.80 ± 0.002 [a]	0.79 ± 0.005 [a]	0.80 ± 0.002 [a]
	Stress	0.66 ± 0.016 [b]	0.64 ± 0.020 [b]	0.71 ± 0.010 [b]	0.63 ± 0.024 [c]
ψ_o	Control	0.45 ± 0.012 [ab]	0.47 ± 0.011 [a]	0.45 ± 0.012 [ab]	0.47 ± 0.011 [a]
	Stress	0.42 ± 0.019 [b]	0.30 ± 0.018 [c]	0.46 ± 0.021 [a]	0.31 ± 0.025 [b]
φ_{Eo}	Control	0.35 ± 0.010 [a]	0.38 ± 0.009 [a]	0.35 ± 0.010 [ab]	0.38 ± 0.019 [a]
	Stress	0.28 ± 0.013 [b]	0.20 ± 0.015 [c]	0.32 ± 0.012 [b]	0.20 ± 0.020 [c]
PI	Control	0.92 ± 0.083 [b]	1.19 ± 0.062 [a]	0.92 ± 0.083 [b]	1.19 ± 0.062 [a]
	Stress	0.31 ± 0.040 [c]	0.24 ± 0.039 [c]	0.50 ± 0.039 [c]	0.27 ± 0.055 [d]
Chl	Control	32.1 ± 0.37 [bc]	38.0 ± 0.79 [a]	32.1 ± 0.37 [c]	38.0 ± 0.79 [a]
	Stress	31.1 ± 0.79 [c]	34.2 ± 0.98 [b]	30.3 ± 0.95 [c]	35.4 ± 0.32 [b]

Chlorophyll fluorescence measurements revealed substantial disturbances in the performance of the photosynthetic apparatus in stressed wheat and maize plants. The parameter is currently commonly used to compare intraspecific responses of the photosynthetic apparatus to environmental stresses and it serves as a quick and reliable tool for the selection of plants with improved tolerance to abiotic stresses [4,42,43].

Although chlorophyll fluorescence is often used to assess the condition of plants under stress, there are no studies using this interspecies comparison tool that were used in our research. Despite the many changes in the response of both species, we chose ET_o/RC, ψ_o, φ_{Eo} as the ones that differentiate the response of wheat and maize to both water stresses. In addition, under the stress of drought, the differentiating parameter was DI_o/RC and under flooding conditions they were F_v/F_m and RC/CS_m. We discarded all those parameters that differed under stress conditions, and also those that differed under control conditions, because they are due to interspecies differences. The ET_o/RC parameter turned out to be the reason why it was possible to observe the specific response of the studied species in the stress condition (increase in wheat and decrease in maize). Environmental stresses, including drought and flooding, limit the photosynthetic metabolism of carbon and reduce the utilization of light-phase products ($NADPH_2$, ATP), which is why some of the absorbed light energy may damage PSII [44,45]. Most changes in chlorophyll fluorescence parameters, especially those expressed per excited surface (CS) of the photosynthetic sample, corroborated those reported in other studies [3,4,46,47].

A specific response involving a significant spike in ET_o/RC was observed in drought- and flooding-exposed wheat, while maize responded to the stresses with a drop in this parameter (Table 3). In drought-stressed leaves, the increase of ET_o/RC may indicate that other sinks for electrons from PSII, such as photorespiration and/or Mehler peroxidase reactions, were sustaining the electron flow [48,49]. Another, more probable explanation for the ET_o/RC spike under soil drought may be not so much increased energy supply to the reaction centers as the reduced number of the active reaction centers (RC/CS_m) [50]. The drop in energy flow through thylakoid membranes accompanied by a reduction in RC/CS_m observed in drought- and flooding-exposed maize was also reported by Golebiowska-Pikania, et al. [51]. PSII reaction centers are inactivated following the degradation of D1 protein [52]. This inhibits the electron transport per active reaction center (ET_o/RC) and limits the probability of electron transport beyond QA (ψ_o). Overall, these effects disturb the transfer of the excitation energy, a parameter that is connected with the increase in non-photochemical dissipation energy (DI_o/RC) and the reduction in the quantum yield of PSII photochemistry (F_v/F_m) (Table 3). Similar changes in the quantum yield of PSII photochemistry and energy dissipation under stress conditions were described in other studies [53–55]. Still, Antonkiewicz and Rapacz [56] claimed that F_v/F_m should not be treated as an indicator of damage to a photosynthetic apparatus but rather as a symptom of its adaptive changes under stress conditions.

A significant drop in Chl content was found only in flooding-exposed maize (Table 3). A decrease in this pigment levels during drought and flooding was probably another factor limiting the intensity of photosynthesis (Table 2) and the activity of the photosynthetic apparatus (Table 3). Apart from non-photochemical dissipation of energy (DI_o/RC), lowering chlorophyll content may be a defense response that allows for limiting light absorption (lower values of ABS/CS_m in drought- and flooding-exposed maize) and diminishing photoinhibition damage to the photosynthetic apparatus [46,57,58]. Plant responses in the form of chlorophyll content drop under drought and flooding was reported in other studies [3,33,47]. However, it has been shown that plants tolerant to flooding do not exhibit a decrease in chlorophyll content or that the decrease is smaller than in susceptible plants [29]. Despite the widespread use of chlorophyll content to assess plant responses to stress, our research indicates that it should not be used for interspecies comparisons, as these differences may be due to factors other than stress factors.

4. Conclusions

Our experimental system involving a 10-day long exposure of wheat and maize plants at three-leaf stage to drought or flooding allowed us to observe some specific responses to these stress factors. A specific response of wheat, both under drought and flooding, was the increase in ET_o/RC reflecting electron transport flux per single reaction center in PSII. Additionally, the parameters differentiating the studied species only under stressful conditions were rendered. During a drought, these were EL, P_N, ET_o/RC, DI_o/RC, ψ_o, φ_{Eo} and during flooding conditions, these were g_s, RC/CS_m, ET_o/RC, F_v/F_m, ψ_o, φ_{Eo}. For most of these parameters, the direction of these changes between the control and stress plants is the same in both species, and the differences are due to the level of these differences. A specific response of maize during flooding was the increase in the level of stomatal conductance which was not observed in the second species.

Further studies with more advanced methodologies and measurements are necessary to prove the potential role of these findings. However, the experimental design presented here and the specific responses of wheat and maize to drought and flooding seem to be promising tools for assessing a large pool of plants for the selection of genotypes with an improved tolerance to both stress factors at early stages of growth. To this end, follow-up large-scale experiments with high number of genotypes, lines, or strains are needed.

Selected indicators of plant responses to drought and flooding in the future may help in a more complete characterization of the registered cultivars.

Author Contributions: Conceptualization, A.O.; Methodology, A.O. and T.H.; Software, T.H.; Validation, A.O. and T.H.; Formal Analysis, A.O. and T.H.; Investigation, A.O. and T.H.; Resources, A.O. and T.H.; Writing—Original Draft Preparation, A.O.; Writing—Review & Editing, A.O. and T.H.; Visualization, T.H.; Supervision, A.O. All authors have read and agreed to the published version of the manuscript.

Funding: This research received no external funding.

Institutional Review Board Statement: Not applicable.

Informed Consent Statement: Not applicable.

Data Availability Statement: Not applicable.

Conflicts of Interest: The authors declare no conflict of interest.

References

1. Borowski, P.F. Nexus between water, energy, food and climate change as challenges facing the modern global, European and Polish economy. *AIMS Geosci.* **2020**, *6*, 397–421. [CrossRef]
2. Eckstein, D.; Künzel, V.; Schäfer, L.; Winges, M. Global climate risk index 2020—Briefing Paper. *Germanwatche. V Bonn.* **2019**. Available online: https://germanwatch.org/sites/default/files/20-2-01e%20Global%20Climate%20Risk%20Index%202020_15.pdf (accessed on 11 March 2022).
3. Hura, T.; Tyrka, M.; Hura, K.; Ostrowska, A.; Dziurka, K. QTLs for cell wall-bound phenolics in relation to the photosynthetic apparatus activity and leaf water status under drought stress at different growth stages of triticale. *Mol. Genet. Genom.* **2017**, *292*, 415–433. [CrossRef]
4. Ostrowska, A.; Biesaga-Koscielniak, J.; Grzesiak, M.T.; Hura, T. Physiological responses of spring wheat to 5-aminolevulinic acid under water stress applied at seedling stage. *Cereal Res. Commun.* **2019**, *47*, 32–41. [CrossRef]
5. Hess, T.; Knox, J.; Holman, I.; Sutcliffe, C. Resilience of Primary Food Production to a Changing Climate: On-Farm Responses to Water-Related Risks. *Water* **2020**, *12*, 2155. [CrossRef]
6. Li, Y.; Guan, K.; Schnitkey, G.D.; DeLucia, E.; Peng, B. Excessive rainfall leads to maize yield loss of a comparable magnitude to extreme drought in the United States. *Glob. Change Biol.* **2019**, *25*, 2325–2337. [CrossRef] [PubMed]
7. Hura, T.; Hura, K.; Ostrowska, A.; Gadzinowska, J.; Fiust, A. Water stress-induced flag leaf senescence may be accelerated by rehydration. *J. Plant Physiol.* **2019**, *236*, 109–116. [CrossRef] [PubMed]
8. Daryanto, S.; Wang, L.; Jacinthe, P.A. Global synthesis of drought effects on cereal, legume, tuber and root crops production: A review. *Agric. Water Manag.* **2017**, *179*, 18–33. [CrossRef]
9. Ding, J.; Huang, Z.; Zhu, M.; Li, C.; Zhu, X.; Guo, W. Does cyclic water stress damage wheat yield more than a single stress? *PLoS ONE* **2018**, *13*, e0195535. [CrossRef]
10. Soares, J.C.; Santos, C.S.; Carvalho, S.M.; Pintado, M.M.; Vasconcelos, M.W. Preserving the nutritional quality of crop plants under a changing climate: Importance and strategies. *Plant Soil* **2019**, *443*, 1–26. [CrossRef]
11. Tadesse, W.; Sanchez-Garcia, M.; Assefa, S.G.; Amri, A.; Bishaw, Z.; Ogbonnaya, F.C.; Baum, M. Genetic gains in wheat breeding and its role in feeding the world. *Crop. Breed. Genet. Genom.* **2019**, *1*, e190005. [CrossRef]
12. Jat, M.L.; Stirling, C.M.; Jat, H.S.; Tetarwal, J.P.; Jat, R.K.; Singh, R.; Lopez-Ridaura, S.; Shirsath, P.B. Soil processes and wheat cropping under emerging climate change scenarios in South Asia. *Adv. Agron.* **2018**, *148*, 111–171. [CrossRef]
13. Djaman, K.; O'Neill, M.; Owen, C.K.; Smeal, D.; Koudahe, K.; West, M.; Allen, S.; Lombard, K.; Irmak, S. Crop evapotranspiration, irrigation water requirement and water productivity of maize from meteorological data under semiarid climate. *Water* **2018**, *10*, 405. [CrossRef]
14. Durodola, O.S.; Mourad, K.A. Modelling maize yield and water requirements under different climate change scenarios. *Climate* **2020**, *8*, 127. [CrossRef]
15. Barrs, H.D.; Weatherley, P.E. A re-examination of the relative turgidity techniques for estimating water deficits in leaves. *Aust. J. Biol. Sci.* **1962**, *15*, 413–428. [CrossRef]
16. Srivastava, A.; Strasser, R.J. Constructive and destructive actions of light on the photosynthetic apparatus. *J. Sci. Ind. Res.* **1997**, *56*, 133–148.
17. Lazar, D. Chlorophyll a fluorescence induction. *Biochim. Biophys. Acta* **1999**, *1412*, 1–28. [CrossRef]
18. Strasser, R.J.; Srivastava, A.; Tsimilli-Michael, M. The fluorescence transient as a tool to characterize and screen photosynthetic samples. In *Probing Photosynthesis: Mechanisms, Regulation and Adaptation*; Yunus, M., Pathre, U., Mohanty, P., Eds.; Taylor and Francis: London, UK, 2000; pp. 445–483.
19. Appenroth, K.J.; Stockel, J.; Srivastava, A.; Strasser, R.J. Multiple effects of chromate on the photosynthetic apparatus of *Spirodelapolyrhiza* as probed by OJIP chlorophyll a fluorescence measurements. *Environ. Pollut.* **2001**, *115*, 49–64. [CrossRef]
20. Khalid, M.F.; Hussain, S.; Ahmad, S.; Ejaz, S.; Zakir, I.; Ali, M.A.; Ahmed, N.; Anjum, M.A. Impacts of abiotic stresses on growth and development of plants. In *Plant Tolerance to Environmental Stress*; CRC Press: Boca Raton, FL, USA, 2019; pp. 1–8.
21. Masoumi, Z.; Haghighi, M.; Jalali, S.A.H. Flooding or drought which one is more offensive on pepper physiology and growth? *Mol. Biol. Rep.* **2021**, *48*, 4233–4245. [CrossRef]
22. Feng, Z.; Ding, C.; Li, W.; Wang, D.; Cui, D. Applications of metabolomics in the research of soybean plant under abiotic stress. *Food Chem.* **2020**, *310*, 125914. [CrossRef]
23. Baye, A.; Berihun, B.; Bantayehu, M.; Derebe, B. Genotypic and phenotypic correlation and path coefficient analysis for yield and yield-related traits in advanced bread wheat (*Triticum aestivum* L.) lines. *Cogent Food Agric.* **2020**, *6*, 1752603. [CrossRef]
24. Zhao, W.; Liu, L.; Shen, Q.; Yang, J.; Han, X.; Tian, F.; Wu, J. Effects of water stress on photosynthesis, yield, and water use efficiency in winter wheat. *Water* **2020**, *12*, 2127. [CrossRef]
25. Mutava, R.N.; Prince, K.S.J.; Syed, N.H.; Song, L.; Babu, V.; Chen, W.; Nguyen, H.T. Understanding abiotic stress tolerance mechanisms in soybean: A comparative evaluation of soybean response to drought and flooding stress. *Plant Physiol. Biochem.* **2015**, *86*, 109–120. [CrossRef] [PubMed]
26. Hsiao, T.C. Plant responses to water stress. *Annu. Rev. Plant Physiol. Plant Mol. Biol.* **1973**, *24*, 519–570. [CrossRef]
27. Karimpour, M. Effect of drought stress on RWC and chlorophyll content on wheat (*Triticum durum* L.) genotypes. *World. Ess. J.* **2019**, *7*, 52–56.

28. Azizi, S.; Tabari, M.; Striker, G.G. Growth, physiology, and leaf ion concentration responses to long-term flooding with fresh or saline water of *Populus euphratica*. *S. Afr. J. Bot.* **2017**, *108*, 229–236. [CrossRef]
29. Garcia-Sanchez, F.; Syvertsen, J.P.; Gimeno, V.; Botia, P.; Perez-Perez, J.G. Responses to flooding and drought stress by two citrus rootstock seedlings with different water-use efficiency. *Physiol. Plant.* **2007**, *130*, 532–542. [CrossRef]
30. Zhang, R.; Zhou, Y.; Yue, Z.; Chen, X.; Cao, X.; Ai, X.; Jiang, B.; Xing, Y. The leaf-air temperature difference reflects the variation in water status and photosynthesis of sorghum under waterlogged conditions. *PLoS ONE* **2005**, *14*, e0219209. [CrossRef]
31. Sathi, K.S.; Masud, A.A.C.; Falguni, M.R.; Ahmed, N.; Rahman, K.; Hasanuzzaman, M. Screening of soybean genotypes for waterlogging stress tolerance and understanding the physiological mechanisms. *Adv. Agric.* **2022**, *2022*, 5544665. [CrossRef]
32. Grzesiak, M.T.; Janowiak, F.; Szczyrek, P.; Kaczanowska, K.; Ostrowska, A.; Rut, G.; Hura, T.; Rzepka, A.; Grzesiak, S. Impact of soil compaction stress combined with drought or waterlogging on physiological and biochemical markers in two maize hybrids. *Acta Physiol. Plant* **2016**, *38*, 109. [CrossRef]
33. Lama, R.; Jaishee, N.; Chakraborty, U.; Roy, A. Responses of seven maize genotypes during flooding stress and identification of cultivars most tolerant to flooding conditions. *Plant Arch.* **2020**, *20*, 3244–3249.
34. Tian, L.; Bi, W.; Liu, X.; Sun, L.; Li, J. Effects of waterlogging stress on the physiological response and grain-filling characteristics of spring maize (*Zea mays* L.) under field conditions. *Acta Physiol. Plant.* **2019**, *41*, 1–14. [CrossRef]
35. Xiong, R.; Sang, L.; Liu, R.; Cheng, R.; Li, P.; Huang, L.; Cao, G. Effects of waterlogging on maize seedling growth during seed germination. In *IOP Conference Series: Earth and Environmental Science*; IOP Publishing: Bristol, UK, 2020; Volume 598, p. 012075.
36. Paradiso, A.; Caretto, S.; Leone, A.; Bove, A.; Nisi, R.; De Gara, L. ROS production and scavenging under anoxia and re-oxygenation in *Arabidopsis* cells: A balance between redox signalling and impairment. *Front. Plant Sci.* **2016**, *7*, 1803. [CrossRef] [PubMed]
37. Perdomo, J.A.; Capo-Bauca, S.; Carmo-Silva, E.; Galmes, J. Rubisco and Rubisco activase play an important role in the biochemical limitations of photosynthesis in rice, wheat, and maize under high temperature and water deficit. *Front. Plant Sci.* **2017**, *8*, 490. [CrossRef]
38. Todorova, D.; Aleksandrov, V.; Anev, S.; Sergiev, I. Photosynthesis alterations in wheat plants induced by herbicide, soil drought or flooding. *Agronomy* **2022**, *12*, 390. [CrossRef]
39. Zhang, J.; Jiang, H.; Song, X.; Jin, J.; Zhang, X. The responses of plant leaf CO_2/H_2O exchange and water use efficiency to drought: A meta-analysis. *Sustainability* **2018**, *10*, 551. [CrossRef]
40. Else, M.A.; Coupland, D.; Dutton, L.; Jackson, M.B. Decreased root hydraulic conductivity reduces leaf water potential, initiates stomatal closure and slows leaf expansion in flooded plants of castor oil (*Ricinus communis*) despite diminished delivery of ABA from the roots to shoots in xylem sap. *Physiol. Plant.* **2001**, *111*, 46–54. [CrossRef]
41. Tong, C.; Hill, C.B.; Zhou, G.; Zhang, X.Q.; Jia, Y.; Li, C. Opportunities for improving waterlogging tolerance in cereal crops—physiological traits and genetic mechanisms. *Plants* **2021**, *10*, 1560. [CrossRef]
42. Bussotti, F.; Gerosa, G.; Digrado, A.; Pollastrini, M. Selection of chlorophyll fluorescence parameters as indicators of photosynthetic efficiency in large scale plant ecological studies. *Ecol. Indic.* **2020**, *108*, 105686. [CrossRef]
43. Kalaji, H.M.; Rastogi, A.; Živčák, M.; Brestic, M.; Daszkowska-Golec, A.; Sitko, K.; Alsharafa, K.Y.; Lotfi, R.; Stypiński, P.; Samborska, I.A.; et al. Prompt chlorophyll fluorescence as a tool for crop phenotyping: An example of barley landraces exposed to various abiotic stress factors. *Photosynthetica* **2018**, *56*, 953–961. [CrossRef]
44. Yin, C.Y.; Berninger, F.; Li, C.Y. Photosynthetic responses of *Populus przewalski* subjected to drought stress. *Photosynthetica* **2006**, *44*, 62–68. [CrossRef]
45. Gadzinowska, J.; Ostrowska, A.; Hura, K.; Dziurka, M.; Pawlowska, B.; Hura, T. Physiological traits determining high adaptation potential of sweet briar (*Rosa rubiginosa* L.) at early stage of growth to dry lands. *Sci. Rep.* **2019**, *9*, 19390. [CrossRef] [PubMed]
46. Hura, T.; Hura, K.; Ostrowska, A.; Dziurka, K. Rapid plant rehydration initiates permanent and adverse changes in the photosynthetic apparatus of triticale. *Plant Soil* **2015**, *397*, 127–145. [CrossRef]
47. Biehler, K.; Fock, H. Evidence for the contribution of the Mehler-peroxidase reaction in dissipating excess electrons in drought-stressed wheat. *Plant Physiol.* **1996**, *112*, 265–272. [CrossRef]
48. Foyer, C.H.; Noctor, G. Oxygen processing in photosynthesis: Regulation and signalling. *New Phytol.* **2000**, *146*, 359–388. [CrossRef]
49. Fghire, R.; Anaya, F.; Ali, O.I.; Benlhabib, O.; Ragab, R.; Wahbi, S. Physiological and photosynthetic response of quinoa to drought stress. *Chil. J. Agric. Res.* **2015**, *75*, 174–183. [CrossRef]
50. Golebiowska-Pikania, G.; Kopec, P.; Surowka, E.; Janowiak, F.; Krzewska, M.; Dubas, E.; Nowicka, A.; Kasprzyk, J.; Ostrowska, A.; Malaga, S.; et al. Changes in protein abundance and activity induced by drought during generative development of winter barley (*Hordeum vulgare* L.). *J. Proteom.* **2017**, *169*, 73–86. [CrossRef]
51. Liu, M.; Li, J.; Niu, J.; Wang, R.; Song, J.; Lv, J.; Zong, X.; Wang, S. Interaction of drought and 5-aminolevulinic acid on growth and drought resistance of *Leymuschinensis* seedlings. *Acta Ecol. Sin.* **2016**, *36*, 180–188. [CrossRef]
52. Demetriou, G.; Neonaki, C.; Navakoudis, E.; Kotzabasis, K. Salt stress impact on the molecular structure and function of the photosynthetic apparatus: The protective role of polyamines. *Biochim. Biophys. Acta* **2007**, *1767*, 272–280. [CrossRef]
53. Kalaji, H.M.; Govindjee; Bosa, K.; Koscielniak, J.; Zuk-Golaszewska, K. Effects of salt stress on photosystem II efficiency and CO_2 assimilation of two Syrian barley landraces. *Environ. Exp. Bot.* **2011**, *73*, 64–72. [CrossRef]

54. Lu, C.; Vonshak, A. Effects of salinity stress on photosystem II function in cyanobacterial *Spirulina platensis* cells. *Physiol. Plant.* **2002**, *114*, 405–413. [CrossRef]
55. Antonkiewicz, J.; Rapacz, M. Assessment of photosynthetic activity of plants grown on stubble sediments and furnaceash. *Zesz. Probl. Post. NaukRol.* **2006**, *509*, 187–196. (In Polish)
56. Kudoh, H.; Sonoike, K. Irreversible damage to photosystem I by chilling in the light: Cause of the degradation of chlorophyll after returning to normal growth temperature. *Planta* **2002**, *215*, 541–548. [CrossRef] [PubMed]
57. Lu, C.M.; Zhang, J.H. Photosynthetic CO_2 assimilation, chlorophyll fluorescence and photoinhibition as affected by nitrogen deficiency in maize plants. *Plant Sci.* **2000**, *151*, 135–143. [CrossRef]
58. Park, S.U.; Lee, C.J.; Kim, S.E.; Lim, Y.H.; Lee, H.U.; Nam, S.S.; Kim, H.S.; Kwak, S.S. Selection of flooding stress tolerant sweetpotato cultivars based on biochemical and phenotypic characterization. *Plant Physiol. Biochem.* **2020**, *155*, 243–251. [CrossRef]

sustainability

MDPI

Article

Biological Approaches Promise Innovative and Sustainable Management of Powdery Mildew in Lebanese Squash

Michel Frem [1,2,*], Franco Nigro [2,3], Serge Medawar [1] and Maroun El Moujabber [4]

1 Lebanese Agricultural Research Institute, Zone El Roumieh, Qleiat, Keserwan, Lebanon; sergemedawar@gmail.com
2 Department of Soil, Plant and Food Sciences, University of Bari Aldo Moro, 70126 Bari, Italy; franco.nigro@uniba.it
3 Center of Research, Training and Experimentation in Agriculture "Basile Caramia", 70010 Bari, Italy
4 Mediterranean Agronomic Institute, CIHEAM Bari, 70010 Bari, Italy; elmoujabber@iamb.it
* Correspondence: mefrem@lari.gov.lb

Abstract: Biological management techniques act as a promising and sustainable alternative to alleviate pathogen-induced losses, improve ecosystem functions, and reinforce the resilience of agricultural systems. Lebanese squash production has been threatened by powdery mildew disease caused by the fungus *Podosphaera xanthii*. Very few studies, even unpublished ones, stress the evaluation of biological control approaches in the Lebanese agriculture sector. Here, we have aimed to evaluate the effect of five safe biological treatments (olive soap, sodium bicarbonate, garlic extract, horsetail, and compost tea) in the management of powdery mildew on Lebanese squash in organic open field conditions. Plants were treated after the first spots of powdery mildew appeared on leaves. We then examined the leaves to evaluate disease incidence and severity, and to compare the ability of the five treatments to reduce powdery mildew disease and incidence, in comparison with the untreated control. Plants treated with sodium bicarbonate and garlic extract were the least affected by powdery mildew regarding disease incidence and severity, while tea compost proved to be the least effective product. Organic management of vegetable crops is extremely important in order to ensure global food security and reduce pesticide applications.

Keywords: alternative control measures; climate resilience; Lebanon; organic agriculture; organic pathogen control; powdery mildew; sustainable agriculture

Citation: Frem, M.; Nigro, F.; Medawar, S.; Moujabber, M.E. Biological Approaches Promise Innovative and Sustainable Management of Powdery Mildew in Lebanese Squash. *Sustainability* **2022**, *14*, 2811. https://doi.org/10.3390/su14052811

Academic Editor: Imre J. Holb

Received: 30 December 2021
Accepted: 16 February 2022
Published: 28 February 2022

1. Introduction

In the 21st century, climate change has been a crucial challenge favoring biological invasion and threatening social and ecological development worldwide [1]. Human activities over the past 50 years have led to extreme and unpredictable variations in weather patterns such as temperature and rainfall levels. The average global temperature is rising by almost 0.8 °C per year, and is expected to increase further to reach an approximate value between 0.9 and 3.5 °C by 2100 [2]. According to the consensus of climate models, the Eastern Mediterranean region, including Lebanon, will warm at a rapid rate, which can raise the global temperature average by 1.3 to 2.7 °C and favor plant diseases [3,4]. In addition, annual precipitation levels will decrease by approximately 10 to 20% by the 2050s [5]. There has also been a dramatic increase in the concentration of greenhouse gases (GHG) in the atmosphere [6], particularly of atmospheric CO_2, which has increased by almost 30% in the last few decades [7]. In general, agriculture is vulnerable because it is greatly influenced by climatic variations [8] and subject to crop pests and diseases [5,9]. However, this sector is itself considered a potential contributor to climate change, because it accounts for 24 to 30% of global GHG emissions [10–12].

Almost 40 to 50% of the global food supply is actually lost through crop pests and diseases. In terms of climate change, extreme temperatures and rainfall variations are significant factors that decisively control the distribution of ectotherm organisms and the survival of pests and pathogens. Precipitation variations can have major effects on crop–pathogen interactions because plant pathogens are favored by warm conditions and flourish under humid conditions [13]. Global warming also enhances the overwintering, development, and dispersal potential of pests and pathogens, which ultimately results in extreme disease severity, causing huge yield losses [14].

Accordingly, the proliferations of disease pressure and pest populations under high temperatures have led to the over-use of fungicides and insecticides in pest management techniques (i.e. higher amounts, doses, and frequencies than those recommended), so that pests and fungi have developed resistance to these agrochemicals. Further, the rise in temperature can also reduce the efficacy of particular pesticides. For instance, the toxic effects of bifenthrin and lambda-cyhalothrin (pyrethroids) and of spinosad (spinosyn) declined as the post-exposure temperature rose in [13].

Powdery mildew is a serious fungal disease worldwide, attacking *Cucurbitaceae* such as melon and zucchini. On the latter, it is caused by *Podosphaera xanthii*, previously known as *Sphaerotheca fuliginea* and *Sphaerotheca fusca* [15]. This disease is a considerable problem in Mediterranean countries, where it causes severe losses of crops grown in open fields and in greenhouses. The disease is generally controlled in commercial cucurbit crops via frequent applications of fungicides; however, heavy fungicide application has led to the development of resistant *Podosphaera xanthii* populations that can no longer be controlled by fungicides [16], and has also raised public concerns about contamination of the ecosystem. Therefore, there is great interest in the development of organic control strategies (using natural extracts or mixtures) that stimulate host plant defense mechanisms with the aim of mitigating and enhancing plant resilience to climate change [17,18] in a sustainable way [19]. For instance, the antimicrobial allicin, extracted from garlic (*Allium sativum*), is extremely active against a broad range of phytopathogenic organisms in vitro and in planta [20]. Successful trials and experiments using garlic sprays containing allicin confirm its effectiveness in treating plant diseases [21]. In addition, the bioactive element silicon (Si) contained in horsetail, (*Equisetum arvense* L.), reduces the toxic effect of water excess on plants that stimulates fungal disease development, and *Equisetum arvense* decoction exhibits highly effective anti-sporulation activity [22]. Moreover, compost tea, as described in various studies, has been efficient in controlling leaf diseases caused by powdery mildew, with a similar effect on conventional fungicides [23]. Furthermore, the bicarbonates widely used in the food industry [24] have been assessed for their ability to control plant pathogens. Sodium, potassium, and ammonium bicarbonates have been found to suppress several fungal diseases, particularly in greenhouse-grown cucumbers [25]. Mixed with any botanical extract, olive soap is usually used as wetting agent to facilitate the dispersion and penetration of each water-based treatment. As such, the latter is more effective and less sensitive to leaching. Therefore, this study investigates the effectiveness, under open-field conditions, of various natural products, including garlic extracts; horsetail extracts; aerated compost tea; olive soap (mixed with these natural products and used alone to elicit its potential fungal effect on the studied biotic stress); and sodium bicarbonate, as promising and sustainable treatments for powdery mildew in Lebanese squash within organic agriculture (OA) systems. According to the Research Institute of Organic Agriculture (www.fibl.com access on 10 November 2021), the average organic area in Lebanon is estimated to be close to 1300 ha for a period of 5 years (2015–2019), while the average organic area of vegetables fruits is around 20 ha over all of the country for the same period. According to the Lebanese Ministry of Agriculture (http://www.agriculture.gov.lb access on 30 December 2021), squash constitute the third vegetable–fruit crop (14.20%), after tomatoes and cucumbers, in terms of cultivated area in 2009 (there is a lack of available, updated and reliable agricultural statistics, which are still needed in the country). This study has practical implications and supports Lebanese farmers for appropriate and efficient organic control strategies of pathogens, and could prompt them to rely on their traditional skills and knowledge to control pests and diseases (i.e., preparation of botanical extracts), while enhancing the resilience and adaptation of OA systems to climate change. Furthermore, the added value of the present research is that it enriches the scientific literature on sustainable agriculture, which is particularly required by Lebanese local growers and the scientific community. Organic control strategies should be further encouraged in this country suffering the effects of climate change as it aims to preserve soil fertility, biodiversity, and ecosystem functions.

2. Material and Methods

2.1. Field Experiment Conditions

The experiment was carried out at the *Kleïaat-Keserwan* (LARI-K) station of the Lebanese Agricultural Research Institute (33.97 N, 35.71 E), where the climate is characterized by cold snowy winters and cool humid summers. The field plot was organic-certified (Supplementary Files S1 and S2), and was protected by *Cypressus* trees, on two boundaries, that serve as a buffer zone (Figures 1 and 2). The experiment took place on around an area of 150 m^2. The weather station at LARI-K recorded the climatic data throughout the 2 months of the experiment. The average minimum and maximum temperatures measured were 17.28 °C and 25.25 °C, respectively. The average minimum and maximum humidity levels were 61.49% and 84.85%, respectively. The recorded solar radiation was, on average, 271.86 W/m^2. Consequently, the correspondent humidity and temperature levels favored the development of powdery mildew [26].

Figure 1. An overview of the experimental field.

Prior to planting, soil testing revealed a sandy clay loam texture without any mineral deficiency, thus, providing optimum conditions for zucchini growth and development. According to the climatic conditions, the irrigation process was scheduled every 2 to 3 days, using perforated pipes every 40 cm. In addition, compost tea was used here as an organic amendment and applied during one of every 2 irrigation water applications. The crops were not treated chemically because LARI-K is intended only for organic farming. The experiment began on 17 May 2018 and continued until 31 July 2018, when spot counting was performed. The climatic data were recorded constantly by the LARI-K weather station.

2.2. Field Experiment Treatments

Organic olive soap (Treatment A): Two hundred grams (200 g) of organic olive soap were finely chopped and dissolved in 2 L of water, then added to 8 L of water at the pulverization stage.

Figure 2. A close-up of the experimental field.

Sodium bicarbonate mixed with organic olive soap (Treatment B): One hundred grams (100 g) of sodium bicarbonate powder (Oxy Med) were weighed and mixed with 8 L of potable water. Then, 2 L of water with 200 g of organic olive soap was added to the mixture to spray directly on plants.

Organic garlic extract mixed with organic olive soap (Treatment C): One kilogram of dry organic garlic (*Allium sativum*) was peeled, and then chopped using an electric mixer. Subsequently, it was added to a container containing 8 L of potable water. The mixture was warmed over a low heat for 5 min, and then cooled to room temperature for about 10 min in order to be filtered in the sprayer. Lastly, 2 L of water with 200 g of organic olive soap was added to the mixture to spray directly on plants.

Natural horsetail plants mixed with organic olive soap (Treatment D): Two hundred and fifty grams (250 g) of dry horsetail (*Equisetum arvense*) from the shores of a lake near LARI-K were washed and cut into small pieces, then added to a container containing 8 L of drinking water. The mixture was warmed over a low heat for almost 30 min, then left to cool for 12 h at room temperature. After this, the mixture was drained using a filter to obtain a concentrated solution of horsetail, and 5 L of this solution was added to 3 L of potable water. Lastly, 2 L of water with 200 g of organic olive soap was added to the mixture to spray directly on plants.

Organic compost tea mixed with organic carob molasses (Treatment E): Fifty liters (50 L) of potable water, 2 kg of organic fertilizer (certified organic chicken manure from southern England), and 250 mL of organic carob molasses were mixed in a container and oxygenated for half an hour using an active oxygen pump. The container remained completely covered for 24 hours before spraying. This fermented liquid contained both soluble nutrients and living microorganisms. Lastly, 1 L of the final solution was poured into 9 L of potable water to form a solution for direct spraying onto plants.

2.3. Experimental Setup

In this experiment, ninety squash plants (*Cucurbita pepo*) of the cv Julienne, a variety with extremely small fruits, were planted in three blocks according to the randomized complete block design (RCBD), where three replications of each treatment applied were performed. The six treatments, as prepared and described above, were divided into 3 blocks (Figure 3). Each treated section comprised five squash plants sprayed for the first time (i.e., when the first white powdery spots appeared) on 3 July 2018, then once a week on 10, 17, and 24 July 2018.

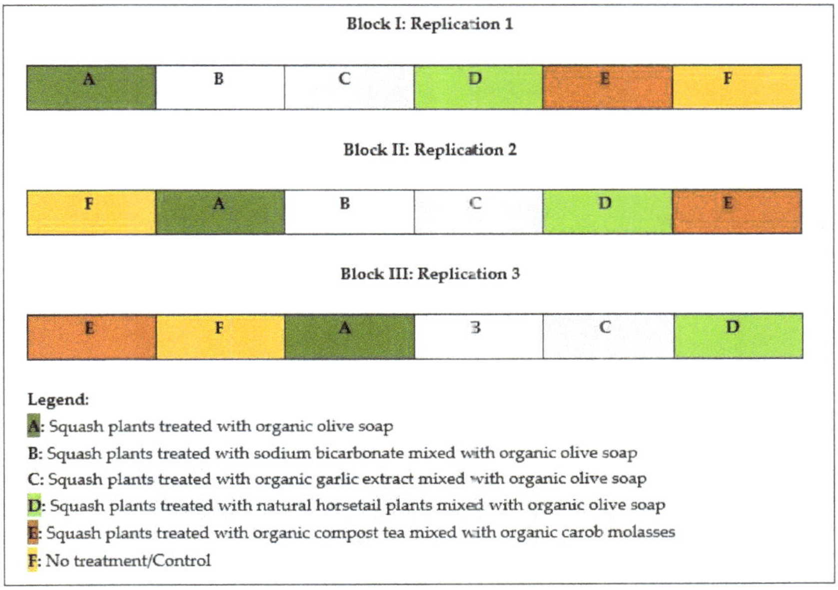

Figure 3. Randomized complete block design.

2.4. Disease Evaluation

Three plants per treatment were randomly selected during the fourth week after the appearance of the first spots on the leaves and were examined to determine the disease incidence and the disease severity. Disease incidence was calculated as the percentage of infected plants, whereas disease severity was assessed by rating the symptoms of the leaves according to an empirical scale covering 8 categories, from 0 (sound leaves) to 7 (more than 75 percent of the leaf area affected by the pathogen). The scale values were used to compute the infection index [27], expressing the weighted average of the disease severity as a percentage of the possible maximum level of disease (100%). Therefore, for each treatment, the disease severity (DS) and its incidence (DI), expressed in %, were calculated [27] according to the equations, as presented in the Supplementary File S3 (see in the Supplementary Materials). Furthermore, a comparison between the efficacy of the treatments implemented (A, B, C, D and E) and the control treatment (F), based on the reduction in severity and incidence of powdery mildew on squash plants, was performed throughout this research study. The reduction percentage of powdery mildew disease severity and incidence as compared to the untreated control were calculated, as reported in the Supplementary File S3 [26].

2.5. Statistical Analysis

The data were subjected to the two-way analysis of variance, and when significant, the means were compared to 5% using Tukey's test. In addition, IBM SPSS statistics software (version 23, 2018) was used for all statistical analyses.

3. Results

3.1. Effects of Treatments on Disease Incidence and Disease Severity

The incidence and severity of the powdery mildew varied between the different treatments, ranging from small powdery white spots on some leaves of the plant, to severe infection of most squash plants. Among all the different organic treatments, the analysis of variance demonstrated a significant statistical difference between the means of the disease incidence, and a very high significance among the means of the disease severity.

Regarding the disease incidence, the post hoc Tukey's Honest Significant Difference (HSD) test divided the population of treatments into three homogenous subsets (Figure 4). Therefore, the plants treated with sodium bicarbonate (Treatment B) were the least affected, whereas the non-treated plants were the most affected by the attack of the powdery mildew, based on disease incidence.

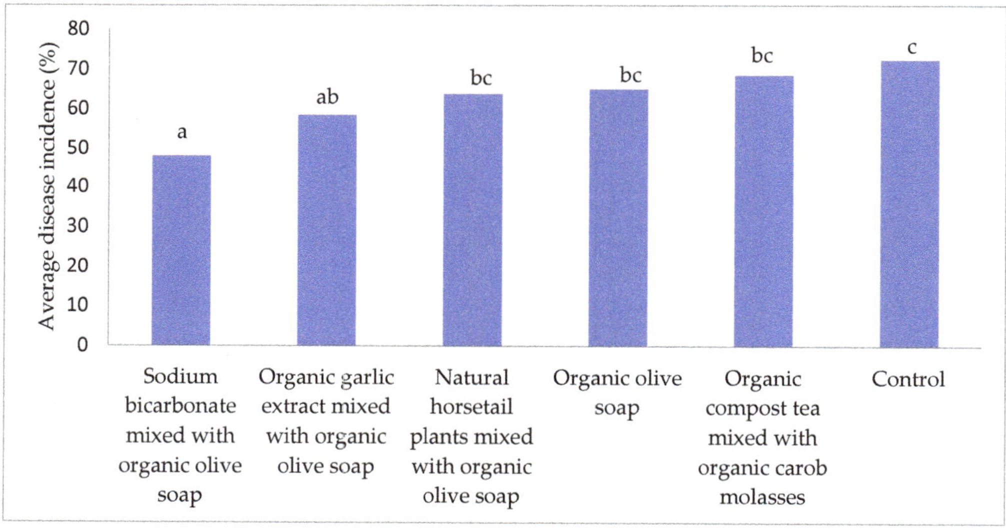

Figure 4. Incidence of powdery mildew on squash plants of the cv Julienne. Values marked with the same letters (a, ab, bc, c) are not statistically different, according to the Tukey's HSD test ($p \leq 0.05$).

Concerning disease severity (Figure 5), the post hoc Tukey's HSD test also showed that the sodium bicarbonate mixed with organic olive soap was able to effectively control the powdery disease on squash. There was no statistical difference between the sodium bicarbonate mixed with organic olive oil or the organic garlic extract mixed with organic olive soap ($p = 0.47$), nor between the others treatments and the control ($p = 0.19$). Thus, the plants treated with sodium bicarbonate (Treatment B) and garlic extract (Treatment C) were the least affected by powdery mildew infection based on disease severity.

3.2. Effects of Treatments on the Reduction in Disease Incidence and Disease Severity

Figures 6 and 7 show the distribution of the various organic treatments according to their effectiveness in terms of the reduction in disease incidence and disease severity, compared to the untreated control, in the management of powdery disease on squash. Among all treatments, sodium bicarbonate mixed with organic olive soap was the most effective treatment in protecting squash plants, with a reduction in powdery mildew incidence and severity. The garlic extract mixed with organic olive soap resulted as the second most effective treatment. Furthermore, the natural horsetail extract mixed with organic olive soap, the organic olive soap, and the organic compost tea mixed with organic carob molasses are classified as the third, fourth, and fifth most effective treatments, respectively.

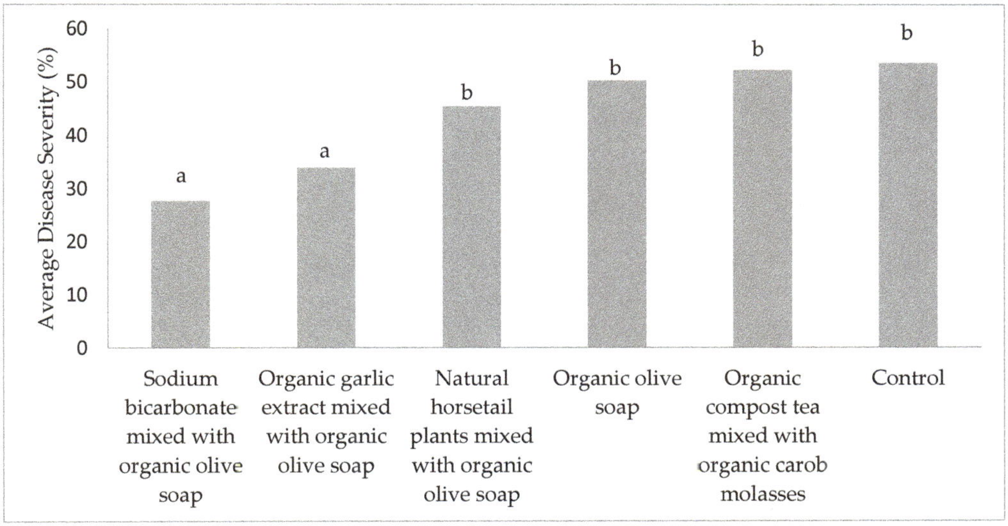

Figure 5. Severity of powdery mildew on squash plants of the cv Julienne. Values marked with the same letters (a, b) are not statistically different, according to the Tukey's HSD test ($p \leq 0.05$).

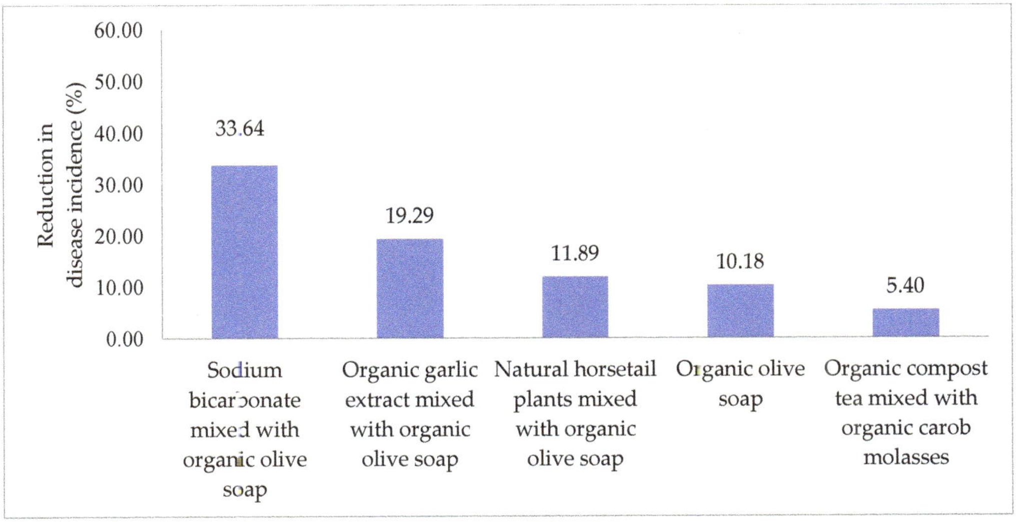

Figure 6. Reduction in powdery mildew incidence on Lebanese squash, cv Julienne, as affected by the different organic treatments.

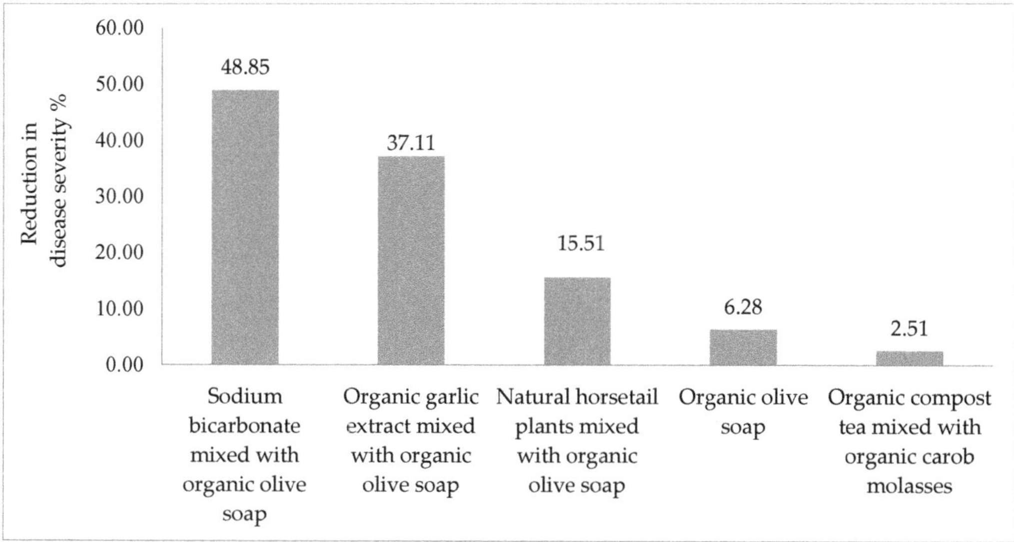

Figure 7. Reduction in powdery mildew disease severity on Lebanese squash, cv Julienne, as affected by the different organic treatments.

4. Discussion

Pests and pathogens thrive and proliferate during long growing seasons in warm climates [28]. Fungal diseases, particularly powdery mildew, are among the most important factors limiting squash yield and causing serious economic losses. Consequently, large farmers tend to rely on extensive fungicide applications to limit this fungal infection and minimize their crop losses; however, this involves emissions of harmful gases that further aggravate climate change, environmental pollution, and human health problems. This challenging issue should prompt scientists to develop modern, sustainable and efficient biological products with low environmental toxicity for use against pests and pathogens, and to encourage their adoption by farmers, in order to enhance food quality and protect natural resources and the ecosystem.

In this experimental study, sodium bicarbonate was found to be a successful and sustainable biological alternative that can protect squash crops from powdery mildew infection. The use of this product in the initial stages of disease, when the first whitish spots appear, damages the cell wall membrane of the fungal spores, causing them to dry out and inducing necrosis [29]. Subsequently, the death of *Sphaerotheca fuliginea (Sf)* spores inhibited the spread of infection, and thus, resulted in low values of disease severity (27.43%) and incidence (48.04%). Our results are in line with other similar studies [30,31]. The first of these [30] applied various polymers and an aqueous solution of sodium or potassium bicarbonate to control powdery mildew on cucurbits. Their results demonstrated that sodium bicarbonate was the most effective treatment and stopped any *Sf* infection in plants. In addition, the second study [31] observed that sodium bicarbonate limited the development of powdery mildew on *Craterostigma pumilum* by 71.42%. Another study evaluated the effect of two different treatments used to control powdery mildew disease in cucurbits: monopotassium phosphate and potassium nitrate. The results demonstrated that additional foliar sprays of sodium bicarbonate (1%) had a considerable effect on suppressing powdery mildew colonies on infected foliage [29,32].

Furthermore, allicin from garlic extract is known for its effective antifungal activity [19], and in this study, the garlic extract actively inhibited powdery mildew spore development and spread, thus limiting disease development. The effectiveness of allicin has been also demonstrated by Seo, who found that garlic extract reduced *Sf* on cucumbers by 74% when studied in vivo [33].

On the other hand, horsetail extract delayed the development of powdery mildew symptoms on plants, but without completely stopping disease development. In fact, *Sf* was able to resist the effect of horsetail silica, which induces the formation of physical and chemical barriers [34]. However, these results differ from those of Bélanger et al. [6], who reported horsetail extract as providing a complete and effective protection for cucumber plants against *Sf* infection.

The powdery mildew disease incidence and severity in the experiment conducted at LARI-K were 50% and 65%, respectively, when the squash plants were treated with organic olive soap. These results do not correspond to those obtained by Muhanna et al. [35]. The latter revealed that potassium-based products induce systemic resistance in squash plants against powdery mildew, and thereby reduce the severity of the disease. This discrepancy can be explained by the fact that organic olive soap must be applied before the plants become infected by powdery mildew in order to act effectively [16], whereas organic olive soap was applied after the onset of the first symptoms during our experimental procedure.

The efficacy of the compost tea was very limited. This limitation matches the results reported by De Bacco [36] when assessing the effect of compost tea on *Podosphaera xanthii* of pumpkin [36]. Certainly, the adoption of biological products to control powdery mildew disease would reduce management costs for farmers. With the exception of sodium bicarbonate (10 US$/Kg), the largest relative reduction in production costs could be obtained by using garlic extract mixed with olive soap (3 US$/Kg). However, this study does not evaluate these products in terms of per-hectare yield, overall cost estimates per cropping system, or farm returns. Ultimately, extension services, scientists, governmental organizations, and public- and private-sector stakeholders should coordinate their efforts in order to reinforce this biological approach to pest and disease management—thus reducing the reliance of agriculture on massive chemical inputs [26]—and also to control the evolving climate-change-induced diseases, as well as pest threats and invasions, sustainably and more efficiently.

5. Conclusions

In this study, the biological treatments contain active substances (i.e., allicin in garlic; silica in horsetail; potassium in olive soap) that can successfully control powdery mildew on squash plants. As sodium bicarbonate is alkaline (pH above 7), it inhibits and creates an unsuitable environment for the growth of the fungal pathogen. The results obtained in this study suggest that sodium bicarbonate can be a safe and effective organic bio-agent for the control of powdery mildew in squash plants.

The organic management of crop diseases, such as powdery mildew in squash, is extremely important in order to ensure global food security, to reduce pesticide applications and GHG emissions, to maintain sustainable production practices, and to minimize the vulnerability of farmers to the negative impacts of climate change [17,37]. It is imperative to recognize the importance of using biological alternatives to control crop pests and diseases without relying solely on chemicals; this is because the biological approach can help to sustain crop production and maintain farmers' livelihoods, mitigating the impacts of global climate change.

Further research is needed in order to comprehensively assess the effectiveness of different concentrations of the different treatments used in this study and their effectiveness "in vitro", and also to give greater recognition to the role of bio-agent compounds for their worldwide authentication and adaptation for use against plant diseases.

Supplementary Materials: The following supporting information can be downloaded at: https://www.mdpi.com/article/10.3390/su14052811/s1, Supplementary Material Files S1: LARI's organic certificate (2017), Supplementary Material Files S2: LARI's organic certificate (2019), Supplementary Material Files S3: Equations used for parameters evaluation.

Author Contributions: Conceptualization, M.F., F.N. and M.E.M.; methodology, M.F., F.N. and M.E.M.; software, M.F. and S.M.; validation, F.N. and M.E.M.; formal analysis, M.F. and F.N.; investigation, M.F. and S.M.; data curation, S.M.; writing—original draft preparation, M.F. and S.M.; writing—review and editing, M.F., F.N. and M.E.M.; visualization, S.M.; supervision, F.N. and M.E.M.; project administration, M.F. All authors have read and agreed to the published version of the manuscript.

Funding: This research received no external funding.

Institutional Review Board Statement: Not applicable.

Informed Consent Statement: Not applicable.

Data Availability Statement: Not applicable.

Acknowledgments: Thanks are due to Nidale El Hachem and Maroun Zgheib for helping with data collection, to Tracy Khalil for sourcing references and to Sarah Jane Christopher for translation.

Conflicts of Interest: The authors declare that there is no conflict of interest.

References

1. Frem, M.; Santeramo, F.G.; Lamonaca, E.; El Moujabber, M.; Choueiri, E.; La Notte, P.; Nigro, F.; Bozzo, F.; Fucilli, V. Landscape restoration due to *Xylella fastidiosa* invasion in Italy: Assessing the hypothetical public's preferences. *NeoBiota* **2021**, *66*, 31–54. [CrossRef]
2. Das, T.; Majumdar, M.H.D.; Devi, R.T.; Rajesh, T. Climate change impacts on plant diseases. *J. Agric.* **2016**, *14*, 200–209. [CrossRef]
3. Abou Kubaa, R.; Choueiri, E.; De Stradis, A.; Jreijiri, F.; Saponari, M.; Cillo, F. Occurrence and distribution of major viruses infecting eggplant in Lebanon and molecular characterization of a local potato virus X isolate. *Agriculture* **2021**, *11*, 126. [CrossRef]
4. Choueiri, E.; Jreijiri, F.; Saponari, M.; Abou Kubaa, R. Tomato spotted wilt virus associated with lettuce dieback in Bekaa valley, Lebanon. *J. Plant Pathol.* **2021**, *103*, 387. [CrossRef]
5. World Bank. *The World Bank Annual Report 2013*; World Bank: Washington, DC, USA, 2013; Available online: https://openknowledge.worldbank.org/handle/10986/16091 (accessed on 15 October 2021).
6. Bélanger, R.; Bowen, P.A.; Ehret, D.L.; Menzies, J.G. Soluble silicon: Its role in crop and disease management of greenhouse crops. *Plant Dis.* **1995**, *79*, 329–336. [CrossRef]
7. Chakraborty, S.; Datta, S. How will plant pathogens adapt to host plant resistance at elevated CO_2 under a changing climate? *New Phytol.* **2013**, *159*, 733–742. [CrossRef] [PubMed]
8. Reilly, J. Climate change and agriculture: Recent findings and issues. *Am. J. Econ.* **1995**, *367*, 118–119. [CrossRef]
9. Gullino, M.L.; Tabone, G.; Gilardi, G.; Garibaldi, A. Effects of elevated atmospheric CO_2 and temperature on the management of powdery mildew of zucchini. *J. Phytopathol.* **2020**, *168*, 405–415. [CrossRef]
10. IPCC. *Climate Change 2001: Synthesis Report. A Contribution of Working Groups I, II, and III to the Third Assessment Report of the Intergovernmental Panel on Climate Change (IPCC)*; Watson, R.T., The Core Writing Team, Eds.; Cambridge University Press: Cambridge, UK; New York, NY, USA, 2001; p. 398.
11. IPCC. *Managing the Risks of Extreme Events and Disasters to Advance Climate Change Adaptation. A Special Report of Working Groups I and II of the Intergovernmental Panel on Climate Change (IPCC)*; Field, C.B., Barros, V., Stocker, T.F., Qin, D., Dokken, D.J., Ebi, K.L., Mastrandrea, M.D., Mach, K.J., Plattner, G.-K., Allen, S.K., et al., Eds.; Cambridge University Press: Cambridge, UK; New York, NY, USA, 2012; p. 582.
12. Gomiero, T.; Paoletti, M.G.; Pimentel, D. Energy and environmental issues in organic and conventional agriculture. *CRC Crit. Rev. Plant Sci.* **2008**, *27*, 239–254. [CrossRef]
13. Heeb, L.; Jenner, E.; Cock, M.J.W. Climate-smart pest management: Building resilience of farms and landscapes to changing pest threats. *J. Pest Sci.* **2019**, *92*, 951–969. [CrossRef]
14. Harvell, C.D.; Mitchell, C.E.; Ward, J.R.; Altizer, S.; Dobson, A.P.; Ostfeld, R.S. Climate warming and disease risks for terrestrial and marine biota. *Science* **2002**, *296*, 2158–2162. [CrossRef]
15. Garibaldi, A.; Gilardi, G.; Gullino, M.L. First report of powdery mildew caused by *Podosphaera xanthii* on *Calendula officinalis* in Italy. *Plant Dis.* **2008**, *92*, 174. [CrossRef] [PubMed]
16. McGrath, M.T.; Shishkoff, N. Evaluation of biocompatible products for managing cucurbit powdery mildew. *J. Crop Prot.* **1999**, *18*, 471–478. [CrossRef]
17. El Chami, D. Towards sustainable organic farming systems. *Sustainability* **2020**, *12*, 9832. [CrossRef]
18. Ntonifor, N.N. Potentials of tropical African spices as sources of reduced-risk pesticides. *J. Entomol.* **2011**, *8*, 16–26. [CrossRef]
19. Leontiev, R.; Hohaus, N.; Jacob, C.; Gruhlke, M.C.H.; Slusarenko, A.J. A comparison of the antibacterial and antifungal activities of thiosulfinate analogues of allicin. *Sci. Rep.* **2018**, *8*, 6763. [CrossRef]
20. Curtis, H.; Noll, U.; Strmann, J.; Slusarenko, A.J. Broad-spectrum activity of the volatile phytoanticipin allicin in extracts of garlic (*Allium sativum* L.) against plant pathogenic bacteria, fungi and Oomycetes. *Physiol. Mol. Plant Pathol.* **2004**, *65*, 7989. [CrossRef]

21. Russell, P.E.; Mussa, A.E.A. The use of garlic (*Allium sativum*) extracts to control foot rot of *Phaseolus vulgaris* caused by Fusarium *Solani f.* sp. *phaseoli*. *Ann. Appl. Biol.* **1977**, *86*, 369–372. [CrossRef]
22. Marchand, P.A. Basic substances under EC 1107/2009 phytochemical regulation: Experience with nonbiocide and food products as biorationals. *J. Plant Prot. Res.* **2016**, *56*, 312–318. [CrossRef]
23. Litterick, A.M.; Harrier, L.; Wallace, P.; Watson, C.A.; Wood, M. The role of uncomposted materials, composts, manures, and compost extracts in reducing pest and disease incidence and severity in sustainable temperate agricultural and horticultural crop production review. *CRC Crit. Rev. Plant Sci.* **2004**, *23*, 453–479. [CrossRef]
24. Lindsay, R.C. Food additives. In *Food Chemistry*; Fennema, O.R., Ed.; Marcel Dekker, Inc.: New York, NY, USA, 1985; Chapter 10.
25. Ziv, O.; Zitter, T.A. Effects of bicarbonates and film-forming polymers on cucurbit foliar diseases. *Plant Dis.* **1992**, *76*, 513–517. [CrossRef]
26. Dik, A.J.; Verhaar, M.A.; Bélanger, R.R. Comparison of three biological control agents against cucumber powdery mildew (*Sphaerotheca fuliginea*) in semi-commercial-scale glasshouse trials. *Eur. J. Plant Pathol.* **1998**, *104*, 413–423. [CrossRef]
27. Surhanee, A.M.A. Efficiency of the extracts of some plants on squash powdery mildew. *Mediterr. J. Soc. Sci.* **2013**, *4*, 39–49. [CrossRef]
28. Frem, M.; Chapman, D.; Fucilli, V.; Choueiri, E.; Moujabber, M.E.; La Notte, P.L.; Nigro, F. *Xylella fastidiosa* invasion of new countries in Europe, the Middle East and North Africa: Ranking the potential exposure scenarios. *NeoBiota* **2020**, *59*, 77–97. [CrossRef]
29. Lahoz, E.; Contillo, R.; Nicoletti, R.; Oliva, A. Efficacy of rue extracts sodium bicarbonate and fungicides at reduced rates for control of powdery mildew on tobacco. *Atti Giornate Fitopatol.* **2000**, *122*, 16–20.
30. Ziv, O.; Hagiladi, A. Controlling powdery mildew in Euonymus with polymer coatings and bicarbonate solutions. *Hortic. Sci.* **1993**, *28*, 124–126. [CrossRef]
31. Tammasorn, N.; Wanasiri, N ; Kuntasup, W.; Cheewangkoon, R.; McGovern, R.J.; ToAnun, C. Controlling powdery mildew disease of *Craterostigma pumilum* Hochst. Ornamental plant. *J. Agric. Technol.* **2017**, *13*, 213–226.
32. Reuveni, M.; Agapov, V.; Reuveni, R. Controlling powdery mildew caused by *Sphaeromeci Juliginea* in cucumber by foliar sprays of phosphate and potassium salts. *J. Crop Prot.* **1996**, *15*, 49–53. [CrossRef]
33. Seo, S.T.; Lee, J.S.; Park, J.H.; Han, K.S.; Jang, H.I. Control of powdery mildew by garlic oil in cucumber and tomato. *Res. Plant Dis.* **2006**, *12*, 51–54. [CrossRef]
34. Pozza, E.A.; Pozza, A.A.A.; Botelho, D.M. Silicon in plant disease control. *Rev. Ceres* **2015**, *62*, 323–331. [CrossRef]
35. Muhanna, N.A.S.; Raghab, S.S.M.; Mohamed, G.M. Mineral salts in controlling powdery mildew of squash. *Egypt. J. Agric. Res.* **2011**, *89*, 809–818. [CrossRef]
36. DeBacco, M. Compost Tea and Milk to Suppress Powdery Mildew (*Podosphaera xanthii*) on Pumpkins and Evaluation of Horticultural Pots Made from Recyclable Fibers under Field Conditions. Master's Thesis, University of Connecticut, Storrs, UK, 2011; p. 101. Available online https://opencommons.uconn.edu/gs_theses/101 (accessed on 15 October 2021).
37. El Chami, D.; Daccache, A.; El Moujabber, M. How can sustainable agriculture increase climate resilience? A systematic review. *Sustainability* **2020**, *12*, 3119. [CrossRef]

sustainability

MDPI

Article

Continuous Pest Surveillance and Monitoring Constitute a Tool for Sustainable Agriculture: Case of *Xylella fastidiosa* in Morocco

Kaoutar El Handi [1,2,3,*], Majida Hafidi [1], Miloud Sabri [2,3,4], Michel Frem [3], Maroun El Moujabber [3], Khaoula Habbadi [2], Najat Haddad [4], Abdellatif Benbouazza [2], Raied Abou Kubaa [5] and El Hassan Achbani [2]

1 Laboratoire de Biotechnologie Végétale et Valorisation des Bio-Ressources, Faculté des Sciences, Université Moulay Ismail, Meknes 11201, Morocco; hafidimaj@yahoo.fr

2 Laboratory of Phyto-Bacteriology and Biocontrol, Plant Protection Unit-National Institute of Agronomic Research INRA, Meknes 50000, Morocco; miloud.sabri@uit.ac.ma (M.S.); khaoula405@gmail.com (K.H.); Dacus05@hotmail.com (A.B.); achbaniofficiel@gmail.com (E.H.A.)

3 CIHEAM Bari, Istituto Agronomico Mediterraneo, Via Ceglie 9, 70010 Bari, Italy; mefrem@lari.gov.lb (M.F.); elmoujabber@iamb.it (M.E.M.)

4 Laboratory of Plant, Animal and Agro-Industry Productions, Faculty of Sciences, University Ibn Toufail, Kenitra 14000, Morocco; najat.haddad@uit.ac.ma

5 Consiglio Nazionale Delle Ricerche, Istituto per la Protezione Sostenibile Delle Piante, Sede Secondaria di Bari, 70126 Bari, Italy; Raied.aboukubaa@ipsp.cnr.it

* Correspondence: kaoutar.elhandi@edu.umi.ac.ma

Abstract: Climate and trade changes are reshaping the cartographic distribution of lethal pervasive pathogens. Among serious emerging challenges is *Xylella fastidiosa* (*Xf*), a xylem-limited phytopathogenic bacterium that produces losses and damages to numerous crops of high economic and agronomic importance. Lately, this grave quarantine pathogen has expended its distribution by arriving to several European countries and infecting both wild and cultivated plants, and no cure has been identified so far. Countries without current outbreaks like Morocco, need to monitor theirs crops frequently because detecting diseases in the early stages may reduce the huge losses caused by *Xf*. For that purpose, inspections were managed in different regions in Morocco from March 2020 to July 2021 to assess the presence of *Xf* in several growing areas of vulnerable economic crops (i.e., almond, citrus and olive). To extend the likelihood of detection, hosts have been inspected and sampled randomly over different environments including symptomatic and asymptomatic plants. Each sample was screened for the existence of *Xf* by using the DAS-ELISA commercial kit, while, further analyses were carried out for doubtful samples, by PCR. Results of both tests did not show any positive sample in the investigated areas. This finding is an update on the *Xf* situation in Morocco and confirms that this country is still a free territory from this bacterium, at least in the monitored regions.

Keywords: DAS-ELISA; diagnosis; PCR; *Xylella fastidiosa*; sustainable agriculture; Morocco

Citation: El Handi, K.; Hafidi, M.; Sabri, M.; Frem, M.; El Moujabber, M.; Habbadi, K.; Haddad, N.; Benbouazza, A.; Abou Kubaa, R.; Achbani, E.H. Continuous Pest Surveillance and Monitoring Constitute a Tool for Sustainable Agriculture: Case of *Xylella fastidiosa* in Morocco. *Sustainability* **2022**, *14*, 1485. https://doi.org/10.3390/su14031485

Academic Editor: Hossein Azadi

Received: 13 December 2021
Accepted: 14 January 2022
Published: 27 January 2022

1. Introduction

The latest appearance and spread of *Xylella fastidiosa* (*Xf*) in the Euro-Mediterranean basin, have brought out some fragilities and criticalities of the quarantine system and the nursery sector. Given this, the early detection and continuous surveillance of *Xf* in Morocco constitute the key factors of sustainable management of this alien species with huge social and economic impacts, and ecological sustainability at the potential entry level of the biological invasion process of the bacterium.

Xf belongs to the top strange phytobacteria amongst the vascular bacteria. It is vectored by sap-feeding insects, infecting an extensive range of plant species [1]. The pathogen, which originates in South America, has hardly been investigated because of its association with catastrophic landscape damages [2] and devastating diseases altering the major league crops like grapevine, olive, citrus, coffee, and stone fruits, besides numerous ornamental and forest species. It is the agent responsible for the famous Pierce's disease (PD), olive quick decline syndrome (OQDS) and citrus variegated chlorosis (CVC) [3]. Symptoms caused by infections with *Xf* are usually the outcome of the systemic occupation of the bacterium that can block the xylem vessels and lead to a gradual inability in water movements [4]. Several recent studies highlighted that an isolate of *Xf* subspecies *pauca* (ST53) has been destroying olive plantations in the Apulia region, among the leading olive production zones of Southern Italy [5]. Furthermore, the bacterium is distinguished by six subspecies which differ in biological and genetic properties [6]. Especially, worldwide movements of infected plants for commercial or panorama planting are most likely the major route that contributed to the propagation and establishment of *Xf*, external of the Americas, where it was known to be restricted until 20 years ago, while the presence of *Xf* was announced in Taiwan, Iran and in several outbreaks newly noticed in European countries [7]. The situation in the European territory raised serious concerns since different *Xf* genotypes have been reported, a broad list of plant species was found vulnerable to the infection, and spittlebug is the predominant and widespread European transmitter species up to now, found in European and Mediterranean countries [1]. The Moroccan climate could hardly be a limiting agent for the establishment of *Xf* and its vectors, notably in the littoral areas of northern Morocco. Climatic data per month on temperature and precipitation during the last 10 years in Morocco according to hikersbay.com/climate/Morocco (accessed on 12 December 2021) are suitable for the multiplication of *Xf*, for the development of the disease, and for the intense activity of the insect vectors. Consequently, they could represent a potential danger for the Moroccan plant patrimony (cultivated, forest, and ornamental plants) [8]. Furthermore, the geographical placement of Morocco, with its close proximation to Spain (13 km), and its commercial exchange with several European countries at the source of *Xf* (Spain, France, Italy) increase the potential risk of entry of *Xf* through infected plant materials/or insect vectors [9]. Morocco has a serious potential risk for the entry of *Xf* into its territory. According to Resourcetrade.earth (accessed on 12 December 2021) [10], Spain, a source country of *Xf*, constitutes the largest exporter of plants for planting (i.e., live plants, bulbs, roots and cut flowers) to Morocco. In the same period, Italy and France have exported 693 and 564 tons, respectively, of these potential host plants of the bacterium to the country. Furthermore, the importation from other European countries in which *Xf* has been intercepted or reported—Central and Southern Asia (Iran), Latin America and the Caribbean as well as Northern American countries, exporters of *Xf*—ranges from 13 to 280 Tons (average period of 2015–2019) as follows: Mexico (280 Tons), United States (275 Tones), Portugal (266 Tons), United Kingdom (242 Tons), Belgium (135 Tons), Germany (79 Tons), Iran (18 Tons), Argentina (17 Tons) and Brazil (13 Tons).

Moreover, the presence of potential insect vectors of *Xf* in Morocco like *Philaenus tesselatus* (the principal spittlebug registered with variable occurrences across the Moroccan country) increase the probability of the entering and establishment of *Xf* in Morocco [11].

Consequently, Morocco has introduced strict emergency measures, regularly updated aiming to prevent the potential introduction and spread of *Xf* into the country, and thus avoiding serious agricultural, environmental and social consequences [12]. In addition, surveys for *Xf* are now mandatory in Morocco, inspections, and diagnostic tests are also compulsory at consignments/place of production for the most susceptible species listed in the EU Decision 2017/2352.

Xf is a tardy-growing (fastidious) bacterium that needs special culture media, but some excellent selective media are available [13]. Additionally, symptom monitoring, isolation and culturing, observation and identification of *Xf* depend on various laboratory tests [14].

The focal point of serological approaches is the special properties of bacterial cell surface. For example, the enzyme linked immunosorbent assay (ELISA) [15] which is regularly employed in the case of *Xf* as a screening test for detection and has a high production capacity because of its easy and simple sample preparation. In fact, kits for serological detection of the bacterium are provided by several companies. ELISA kits from Agritest (Italy) and Loewe (Germany) have been verified for grape, olives, citrus, almond, oak, oleander and other species [16]. In addition, direct tissue blot immunoassay (DTBIA) was newly reported, such as a replacement quick screening test, in order to detect *Xf* in olive samples [17]. Molecular techniques are more efficient compared to serological tests and they comprise conventional PCR [18] and numerous protocols of real-time PCR [19–21], and loop-mediated isothermal amplification (LAMP) [22]. Extraction of the bacterium DNA from plants is accomplished by the CTAB method or by standard commercial kits which can be performed manually or on automated platforms. In pest-free areas and buffer zones, molecular approaches are recommended for *Xf* detection because of their great sensitivity, as stated by the European Food Safety Authority [23].

The earliest detection of *Xf* infections is crucial to the management of this serious phytopathogen, thus, the aim of the present work was to conduct a large survey in order to assess the presence of *Xf* in olive, almond and citrus trees in different commercial groves in Morocco, to update Morocco's situation regarding *Xf*, and to update results obtained by the last monitoring in 2018. Furthermore, sustainable control measures of *Xf* must be done regularly because as it is known, Morocco belongs to the countries ranked at a high-risk level for the entry, establishment and spread of *Xf* [2].

2. Materials and Methods

2.1. Study Areas and Collection of Samples

Xf represents a serious threat to several crops in Morocco, thus, an inspection was managed from March 2020 to July 2021 nationwide, covering olive, almond and citrus planted areas. Overall, 51 commercial groves were visited where the typical symptoms of *Xf* (when presented) were inspected. A total of 1007 plants were randomly sampled and only twigs close to the symptomatic portion were collected (to avoid any false positive reaction being given by the symptomatic portion) as following: (1) 657 olive trees from five regions (Tanger, Béni Mellal, Marrakech, Errachidia, and Meknès), (2) 170 citrus trees collected in two regions (Azilal and Meknès) and (3) 180 almond trees from three regions (Meknès, Haouz, and Gharb,) (Table 1 and Figure 1). Each sample, which included 6–8 cuttings/trees (up to 20 cm/each) was kept in a closed plastic bag, labeled with information (date, location, presence/absence of symptoms, etc). In summer and on high temperature days, samples were kept in a cooling box during transport, and later, all samples were conserved at 4 °C in the laboratory before being analyzed.

Table 1. Crops, locations, number of groves and the collected trees.

Crop	Location	N of Groves	N of Collected Trees
Olive	Tanger	4	141
	Béni Mellal	5	110
	Marrakech	4	120
	Errachidia	5	170
	Meknès	4	116
Citrus	Azilal	5	80
	Meknès	10	90
Almond	Gharb	3	50
	Haouz	5	60
	Meknès	6	70
Total		51	1007

2.2. Sample Preparation

For either serological or molecular detection of *Xf*, 0.5–1 g of the plant tissue (leaf petioles and midribs excised from mature leaves) was recovered from different cuttings, representative of the entire sample.

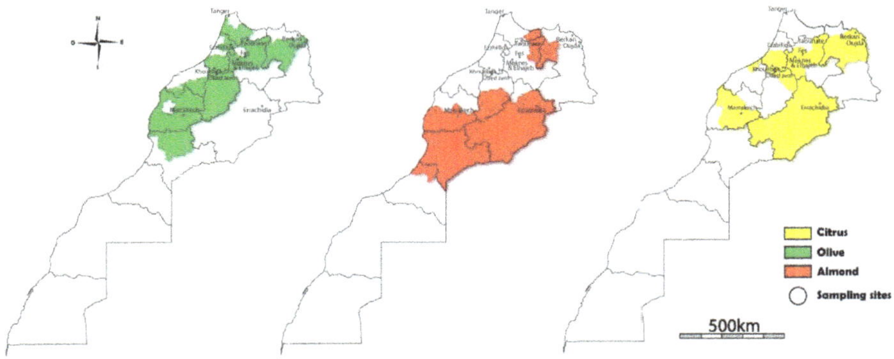

Figure 1. Locations of sampled groves in Morocco during the survey period (March 2020 to July 2021).

Prepared plant tissues were put in extraction bags (BIOREBA, Switzerland) and 5 mL of the extraction buffer was added per bag. Samples were then homogenized using the semi-automatic Homex 6 apparatus (Bioreba, Switzerland) and proceeded according to the extraction protocol developed by Loconsole et al. [24].

2.3. Testing Techniques

2.3.1. Serological Assay

All collected samples were checked by double-antibody sandwich enzyme-linked immunosorbent assay (DAS-ELISA) kit (Agritest, Italy) using a specific antibody and following the steps below: (i) coating the plate; the 96-well ELISA microplates were coated with 200 µL of anti *Xf* IgG diluted 1:200 in coating buffer and then incubated at 37 °C for 4 h. (ii) Antigen incubation and reaction's development; after washing the plates, samples were loaded onto the microplates and kept overnight at 4 °C, then the alkaline- phosphatase-conjugated-anti *Xf* IgG diluted 1:200 was added and plates were then incubated at 37 °C for 4 h prior to adding the substrate (1 mg/mL p-nitrophenyl-phosphate in diethanolamine buffer, pH 9.8). (iii) Absorbance reading; absorbance was calculated 4 times in a total of 3 h using a microplate reader (BioTek ELx 800 UV, Germany) at 405 nm. Positive reactions were determined if after 120 min the absorbance of a sample became 3 times more than the absorbance of the healthy controls [24].

2.3.2. Molecular Assay

- DNA isolation

DNA isolation was carried out using CTAB buffer [25] in a 2 mL micro-centrifuge tube, 1 mL of homogenized extract was put in and the sample was heated at 65 °C for 30 min and then centrifuged at 16,000× *g* for 5 min. In a new 2 mL micro-centrifuge tube, 1 mL of the supernatant was transferred, being careful not to transfer any of the plant tissue debris. 1ml of chloroform-isoamyl-alcohol (24:1) was added, and the sample was thoroughly blended by agitating and then centrifuging at 16,000× *g* for 10 min, then 700 µL of the supernatant was transferred to a 1.5 mL micro-centrifuge tube where 490 µL (approximately 0.7 volume) of cold 2-propanol was adjoined. After combining by upturning twice, the tube was incubated at −20 °C for 20 min. A further 20 min of centrifugation for the samples at 16,000× *g* allowed for the recovery of a pellet that was washed with 1 mL of 70% ethanol followed by extra centrifugation at 16,000× *g* for 10 min. Samples were vacuum dried, and the pellet was resuspended in 100 µL of DNase/ RNase—free water.

- PCR Primers and Cycling Conditions

For some of the samples that returned doubtful results using the ELISA test, the test for the existence of *Xf* in the DNA extracts was carried out by PCR test, using standard primers RST31/33, which are widely used in quarantine programs [26] for the detection of the bacterium. PCR reactions were performed in 20 µL final volume adopting 0.5 µL for either forward and reverse primer, 3 µL of total DNA template and 4 µL of 5× GoTaq polymerase (Promega, Madison, WI, USA).

PCR conditions were: one denaturation step at 95 °C for 5 min followed by 35 cycles of: 30 s denaturation at 94 °C, 30 s for annealing at 55 °C and 40 s for elongation at 72 °C. The reaction was finally extended at 72 °C for 7 min and then amplified bands were observed on 1.2% TAE agarose gel. After electrophoresis, positive and negative samples were inspected [18].

3. Results and Discussion

The early detection of *Xf* infections is crucial to the management of this harmful plant pathogen worldwide [27]. *Xf* has been intensely investigated due its relation to devastating diseases, affecting several major crops such as grapevine, olive, coffee, citrus, and stone fruits, besides several forest and ornamental species [28]. The emergence of *Xf* in novel territory and the ineffective containment of its spread in territories where it previously established highlight the necessity to monitor the progress of large *Xf* outbreaks and to develop exhaustive pest management approaches [8]. Countries without current outbreaks, like Morocco, need to monitor their crops frequently because detecting diseases caused by *Xf* in their early stages may reduce the huge losses caused by pathogen later [29].

All gathered samples were evaluated for the existence/absence of *Xf* by utilizing an ELISA commercial kit (Agritest, Italy). The acquired outcomes did not reveal any positive sample. The ELISA examination was performed correctly. Indeed, the positive control provided with the kit reacted positively, while no color modification was noted with the negative control. Loconsole et al. [24] carried out various laboratory experiments, in which the reactivity and the response of various commercially accessible ELISA kits was paralleled, and revealed that the Loewe kit could detected a greater number of known positive samples with reactions happening in a period of two hours, pursuing manufacturer's instructions and employing the controls delivered with the kit. Therefore, this kit has been used widely to check the presence of *Xf* in several countries worldwide including the previous one in Morocco [30]. Furthermore, by using PCR in the present survey, no amplified DNA was acquired from any of the tested samples, validating the absence of the bacterium in our samples. Although, some positive reactions were expected from some samples, which resulted in doubt in ELISA, only positive controls in each PCR test generated the expected 733-bp amplicons. The employed primers (RST31/33) are broadly recognized for the detection of *Xf* in quarantine programs [31], along with other primers targeting the genomic region 16 S rDNA [31,32], which are more suitable for the proper detection of a bigger number of genetically heterogenous strains of *Xf* [20]. These results are taken as favorable proof, taking into account that *Xf* is absent in Morocco, relative to the surveyed tree crops which is consistent with other negative results obtained from different field surveys, recently carried out on the presence of *Xf* in some other countries such as from Jordan [33] and from Lebanon [5]. Nevertheless, recurrent sizeable surveys in various regions and on various potential host plants are required to hinder its ingress into the country [30]. It should be pointed that the attendance of leaf scorch symptoms that were noticed in many cases during the survey might have myriad origins, biotic or abiotic (salty winds, nutrient toxicity/deficiency, drought, frost damage, fungal pathogens, etc.). Although the risk presented this pathogen in variant hosts (maple, plane, oak … etc.) still needs to be evaluated, thus, plant health service authorities ought to alert presence of these hosts as well submit an avoidable risk.

4. Conclusions

Findings obtained in this investigation clearly indicated that *Xf*—up to this date—was not found in seven investigated areas (Tanger, Béni Mellal, Marrakech, Errachidia, Azilal, Meknès, Haouz, and Gharb) in Morocco, confirming the results obtained from the previous survey carried out during 2018. These results highlight the importance of control measures adopted by phytosaniraty services in Morocco that could, up till now, prevent the introduction and spread of *Xf* in the country. Thus, frequent extended surveys in diverse regions and different host species, and the continious and accurate detection of *Xf* by rapid, sensitive, and reliable laboratory tests are required to avoid any entry of this pathogen into the country. Furthermore, sustainable control measures of *Xf* would include but are not limited to: (i) VSPP (voluntary certification program by the concerned stakeholders, mainly those in the nursery sector); (ii) screening more species and cultivars: promising results of the tolerant/resistant species cultivars as the case of Leccino and FS 17 in Italy); (iii) further research on heat treatment of plant propagation material; and, (iv) continuous surveillance and monitoring on vectors.

Author Contributions: Conceptualization, K.E.H. and E.H.A.; methodology, K.E.H.; software, K.E.H. and M.F.; validation E.H.A. and M.H.; formal analysis, K.E.H.; investigation, E.H.A.; resources, E.H.A.; data curation, K.E.H. and M.S.; writing—original draft preparation, K.E.H. and R.A.K.; writing, review and editing, K.E.H., M.E.M. and N.H.; visualization, K.H. and A.B.; supervision, E.H.A. and M.H. All authors have read and agreed to the published version of the manuscript.

Funding: This research was funded by CURE-*XF*, an EU-funded project, coordinated by CIHEAM Bari (H2020-MSCA-RISE. Reference number: 634353).

Institutional Review Board Statement: Not applicable.

Informed Consent Statement: Informed consent was obtained from all subjects involved in the study.

Data Availability Statement: Not applicable.

Acknowledgments: This research was supported by CURE-*XF*, an EU-funded project, coordinated by CIHEAM Bari (H2020-MSCA-RISE. Reference number: 634353). Authors would like to thank the anonymous reviewers for their valuable comments.

Conflicts of Interest: The authors declare no conflict of interest.

References

1. Rossi, A. Etude de La Prévalence des Vecteurs Potentiels Naturellement Infectés Par *Xylella fastidiosa* subsp. *pauca* dans Les Oliveraies du Sud de L'italie. Master's Thesis, Université de Liège, Liege, Belgium, September 2017.
2. Frem, M.; Chapman, D.; Fucilli, V.; Choueiri, E.; El Moujabber, M.; La Notte, P.; Nigro, F. Xylella fastidiosa invasion of new countries in Europe, the Middle East and North Africa: Ranking the potential exposure scenarios. *NeoBiota* **2020**, *59*, 77–97. [CrossRef]
3. Del Coco, L.; Migoni, D.; Girelli, C.R.; Angilè, F.; Scortichini, M.; Fanizzi, F.P. Soil and Leaf Ionome Heterogeneity in Xylella fastidiosa subsp. *pauca*-Infected, Non-Infected and Treated Olive Groves in Apulia, Italy. *Plants* **2020**, *9*, 760. [CrossRef]
4. Giampetruzzi, A.; Baptista, P.; Morelli, M.; Cameirao, C.; Lino Neto, T.; Costa, D.; D'Attoma, G.; Abou Kubaa, R.; Altamura, G.; Saponari, M.; et al. Differences in the Endophytic Microbiome of Olive Cultivars Infected by *Xylella fastidiosa* across Seasons. *Pathogens* **2020**, *9*, 723. [CrossRef]
5. Abou Kubaa, R.; Giampetruzzi, A.; Altamura, G.; Saponari, M.; Saldarelli, P. Infections of the *Xylella fastidiosa* subsp. *pauca* Strain "De Donno" in Alfalfa (*Medicago sativa*) Elicits an Overactive Immune Response. *Plants* **2019**, *8*, 335. [CrossRef]
6. Marcelletti, S.; Scortichini, M. Genome-wide comparison and taxonomic relatedness of multiple Xylella fastidiosa strains reveal the occurrence of three subspecies and a new Xylella species. *Arch. Microbiol.* **2016**, *198*, 803–812. [CrossRef]
7. Cendoya, M.; Martínez-Minaya, J.; Dalmau, V.; Ferrer, A.; Saponari, M.; Conesa, D.; López-Quílez, A.; Vicent, A. Spatial Bayesian Modeling Applied to the Surveys of Xylella fastidiosa in Alicante (Spain) and Apulia (Italy). *Front. Plant Sci.* **2020**, *11*, 1204. [CrossRef]
8. Frem, M.; Santeramo, F.G.; Lamonaca, E.; El Moujabber, M.; Choueiri, E.; La Notte, P.; Nigro, F.; Bozzo, F.; Fucilli, V. Landscape restoration due to Xylella fastidiosa invasion in Italy: Assessing the hypothetical public's preferences. *NeoBiota* **2021**, *66*, 31–54. [CrossRef]
9. Afechtal, M.; Vicent, A.; Saponari, M.; D'Onghia, A.M. Pest risk analysis on *Xylella fastidiosa* in Morocco. *J. Plant Prot. Res.* **2018**, *58*, 215–219.
10. Available online: https://resourcetrade.earth/data (accessed on 12 December 2021).
11. Haddad, N.; Afechtal, M.; Streito, J.C.; Ouguas, Y.; Benkirane, R.; Lhomme, P.; Smaili, M.C. Occurrence in Morocco of potential vectors of *Xylella fastidiosa* that may contribute to the active spread of the bacteria. *Ann. De La Société Entomol. De Fr.* **2021**, *57*, 359–371. [CrossRef]

12. Cardone, G.; Digiaro, M.; Djelouah, K.; El Bilali, H.; Michel FR, E.M.; Fucilli, V.; Yaseen, T. Potential socio-economic impact of Xylella fastidiosa in the Near East and North Africa (NENA): Risk of introduction and spread, risk perception and socio-economic effects. *New Mediterr. J. Econ. Agric. Environ.* **2021**, *20*, 27.
13. Nunney, L.; Schuenzel, E.L.; Scally, M.; Bromley, R.E.; Stouthamer, R. Large-Scale Intersubspecific Recombination in the Plant-Pathogenic Bacterium Xylella fastidiosa Is Associated with the Host Shift to Mulberry. *Appl. Environ. Microbiol.* **2014**, *80*, 3025–3033. [CrossRef] [PubMed]
14. Loreti, S.; Pucci, N.; Loconsole, G.; Modesti, V.; Potere, O.; Lucchesi, S.; Saponari, M. Organization of ring tests on diagnostic methods among Italian laboratories. *Xylella fastidiosa* & the Olive Quick Decline Syndrome (OQDS). *Options Méditerranéennes* **2017**, *121*, 69.
15. Sherald, J.L.; Lei, J.D. Evaluation of a rapid ELISA test kit for detection of *Xylella fastidiosa* in landscape trees. *Plant Dis.* **1991**, *75*, 200–203. [CrossRef]
16. Marchi, G.; Rizzo, D.; Ranaldi, F.; Ghelardini, L.; Ricciolini, M.; Scarpelli, I.; Surico, G. First detection of *Xylella fastidiosa* subsp. *multiplex* DNA in Tuscany (Italy). *Phytopathol. Mediterr.* **2018**, *57*, 363–364.
17. Djelouah, K.; Frasheri, D.; Valentini, F.; D'Onghia, A.M.; Digiaro, M. Direct tissue blot immunoassay for detection of Xylella fastidiosa in olive trees. *Phytopathol. Mediterr.* **2014**, *53*, 559–564.
18. Minsavage, G.V.; Thompson, C.M.; Hopkins, D.L.; Leite, R.M.V.B.C.; Stall, R.E. Development of a polymerase chain reaction protocol for detection of *Xylella fastidiosa* in plant tissue. *Phytopathology* **1994**, *84*, 456–461. [CrossRef]
19. Francis, M.; Lin, H.; Cabrera-La Rosa, J.; Doddapaneni, H.; Civerolo, E.L. Genome-based PCR primers for specific and sensitive detection and quantification of *Xylella fastidiosa*. *Eur. J. Plant Pathol.* **2006**, *115*, 203–213. [CrossRef]
20. Harper, S.J.; Ward, L.I.; Clover GR, G. Development of LAMP and real-time PCR methods for the rapid detection of *Xylella fastidiosa* for quarantine and field applications. *Phytopathology* **2010**, *100*, 1282–1288. [CrossRef]
21. Li, W.; Xi, B.; Yang, W.; Hawkins, M.; Schubart, U.K. Complex DNA melting profiles of small PCR products revealed using SYBR® Green I. *BioTechniques* **2003**, *35*, 702–706. [CrossRef]
22. Yaseen, T.; Drago, S.; Valentini, F.; Elbeaino, T.; Stampone, G.; Digiaro, M.; D'Onghia, A.M. On-site detection of *Xylella fastidiosa* in host plants and in "spy insects" using the real-time loop-mediated isothermal amplification method. *Phytopathol. Mediterr.* **2015**, *54*, 488–496.
23. European Food Safety Authority (EFSA); Vos, S.; Camilleri, M.; Diakaki, M.; Lázaro, E.; Parnell, S.; Vicent, A. Pest survey card on *Xylella fastidiosa*. *EFSA Supporting Publ.* **2019**, *16*, 1667E.
24. Loconsole, G.; Potere, O.; Boscia, D.; Altamura, G.; Djelouah, K.; Elbeaino, T.; Saponari, M. Detection of *Xylella fastidiosa* in olive trees by molecular and serological methods. *J. Plant Pathol.* **2014**, *96*, 7–14.
25. Rodrigues, C.M.; de Souza, A.A.; Takita, M.A.; Kishi, L.T.; Machado, M.A. RNA-Seq analysis of Citrus reticulata in the early stages of *Xylella fastidiosa* infection reveals auxin-related genes as a defense response. *BMC Genom.* **2013**, *14*, 676. [CrossRef] [PubMed]
26. European and Mediterranean Plant Protection Organization (EPPO). Diagnostic protocols for regulated pests. *Xylella fastidiosa*. *Bull. OEPP/EPPO Bull.* **2004**, *34*, 187–192. [CrossRef]
27. El Handi, K.; Hafidi, M.; Habbadi, K.; El Moujabber, M.; Ouzine, M.; Benbouazza, A.; Achbani, E.H. Assessment of Ionomic, Phenolic and Flavonoid Compounds for a Sustainable Management of *Xylella fastidiosa* in Morocco. *Sustainability* **2021**, *13*, 7818. [CrossRef]
28. Saponari, M.; D'Attoma, G.; Abou Kubaa, R.; Loconsole, G.; Altamura, G.F.; Zicca, S.; Rizzo, D.; Boscia, D. A new variant of Xylella fastidiosa subspecies multiplex detected in different host plants in the recently emerged outbreak in the region of Tuscany, Italy. *Eur. J. Plant Pathol.* **2019**, *154*, 1195–1200. [CrossRef]
29. D'Attoma, G.; Morelli, M.; Saldarelli, P.; Saponari, M.; Giampetruzzi, A.; Boscia, D.; Cobine, P.A. Ionomic differences between susceptible and resistant olive cultivars infected by *Xylella fastidiosa* in the outbreak area of Salento, Italy. *Pathogens* **2019**, *8*, 272. [CrossRef]
30. Afechtal, M.; Ait Friha, A.; Bibi, I. A preliminary survey on the presence of *Xylella fastidiosa* in olive, citrus and grapevine groves in Morocco. *Rev. Maroc. Des. Sci. Agron. Et Vétérinaires* **2017**, *6*, 6–9.
31. EPPO. European and Mediterranean Plant Protection Organization Reporting Service No. 05. Nr. 2016/102. 2017. Available online: https://gd.eppo.int/reporting/article-6070 (accessed on 12 December 2021).
32. Firrao, G.; Bazzi, C. Specific identification of *Xylella fastidiosa* using the polymerase chain reaction. *Phytopathol. Mediterr.* **1994**, *33*, 90–92.
33. AbuObeid, I.; Al-Karablieh, N.; Haddadin, J.; Al Omari, R.; Al-Jabaree, A.M.; Al-Elaumi, L.; Mazahreh, S. Survey on the presence of *Xylella fastidiosa*, the causal agent of olive quick decline syndrome (OQDS) on olives in Jordan. *Arch. Phytopathol. Plant Prot.* **2020**, *53*, 188–197. [CrossRef]

MDPI
St. Alban-Anlage 66
4052 Basel
Switzerland
www.mdpi.com

Sustainability Editorial Office
E-mail: sustainability@mdpi.com
www.mdpi.com/journal/sustainability

www.ingramcontent.com/pod-product-compliance
Lightning Source LLC
Chambersburg PA
CBHW040808250926
51850CB00007BA/1114
9783725804221